2015
开放式专题设计
OPEN DESIGN STUDIO

清华大学建筑 规划 景观设计教学丛书

《开放式专题设计》编委会 编著

中国建筑工业出版社

图书在版编目（CIP）数据

2015 开放式专题设计 / 《开放式专题设计》编委会编著 . -- 北京 : 中国建筑工业出版社 , 2016.10
（清华大学建筑规划景观设计教学丛书）
ISBN 978—7—112—19929—7

I.① 2... II.①开 ... III.①建筑设计—研究 IV.① TU2

中国版本图书馆 CIP 数据核字 (2016) 第 240276 号

责任编辑：陈　桦　王　惠
责任校对：王宇枢　姜小莲
书籍设计：Tofu Design (info@tofu-design.com)

清华大学建筑 规划 景观设计教学丛书
2015 开放式专题设计
《开放式专题设计》编委会 编著

*

中国建筑工业出版社出版、发行（北京西郊百万庄）
各地新华书店、建筑书店经销
北京缤索印刷有限公司印刷

*

开本：787×960 毫米 1/16　印张：20　字数：483 千字
2016 年 10 月第一版　2016 年 10 月第一次印刷
定价：**112.00** 元
ISBN 978—7—112—19929—7
(29428)

编委会名单

目录 CONTENTS

序言 PREFACE

开放式建筑设计教学的新尝试
A BRAND-NEW ATTEMPT TO THE OPEN ARCHITECTURAL DESIGN TEACHING

庄惟敏 Zhuang Weimin

开放式建筑设计教学缘起于以下几点：第一，是与我们建筑学教育的特点有关，建筑学的学科特征从本质上来讲，是伴随着人类社会的进程而发展的，它是一个人居环境的营造过程，把这个过程中的教授与传承演变成为一个教化人的教育过程，就成为今天的建筑专业教育。师徒相传是建筑教育最本质也是最朴素的方法。无论是中国的鲁班，还是美国西塔里埃森的赖特，他们的专业传授中不仅有知识的传承，更重要的是作为一个建筑师的身体力行的影响。由于我们学制时间的限制，怎么样能使学生在有限的时间内，尽快变成一个对建筑学专业有理解、对未来自身定位有明确的方向、尽快进入职业状态的职业建筑师，这一点是很重要的。

第二，是学校教育层面的思考。在清华大学第24 次教学讨论会的开幕式上，校领导明确指出，改革开放30年来清华的教育是值得反思的：一是优秀的学生能进来但不能保证他们优秀地出去，或者说有没有做到更优秀；二是学生在清华有没有培养出自己对专业的志趣。志趣没有，就看不到未来作为一名职业建筑师的状态。环顾我们周边的新一代建筑师，他们在建筑实践的第一线，无论是实践作品还是思想理念都非常精彩，已经具有了很大的社会影响和业界赞誉，他们的执业状态是可以影响年轻人的。这一点对学生们来说更重要。这些中青年建筑师有各自的建筑理想，有积极的生活态度，作为一名活跃的建筑师，他们会有很多的感悟，但同时他们又身体力行，让他们来教授学生们，让学生们感受他们对建筑的热爱和执着，远远胜过书本里的说教，这种直观形象的感受是最感人的。

第三，是从职业角度的思考。我本人从2005 年开始到现在一直在UIA职业实践委员会工作，这么长时间以来特别明显地感到，我们现在的职业教育和国际上还有很大的差距，这个差距不是说我们课程设置怎么样，也不是说老师有怎么样的问题，最主要的是我们缺乏一种职业精神。我们的本科和硕士研究生教育是职业教育，我们培养的建筑学的学生是要拿专业学位的，是为了以后成为职业建筑师的。但针对这一目标，即我们的学生是否毕业以后能够达到一个职业建筑师的起码要求，我还是担心的。因为，按照以往的教学模式，作为职业建筑师的基本职业素养和职业精神，往往是在学校里学不到的。所以必须要有人手把手地教，这些人是需要经历过市场的，经历过很严酷竞争的，经历过和社会面对面、和业主面对面、和批评家面对面交流的。在他们身上反映出来的职业素养，其实就是最好的职业建筑师的表率，就是最好的职业精神的表现，他们的职业精神传授就是最好的教科书。

基于以上三点，寻求开放式教学的缘由变得愈发充分。正好有这样一个契机，2013年的北京市优秀建筑设计评审，北规委邀请了除资深专家和院士之外的一批中青年建筑师，他们有一些是民营建筑设计企业的总建筑师，有些是独立建筑师事务所的领衔建筑师。这些建筑师都在创作的第一线，都有敏锐的专业思考，同时还有很好的专业背景和国际视野，都有海外留学的经历。他们的参与对北京市优秀建筑的评优在学术层面起到了绝对的推动作用。在那次评审会后，我就更加坚定了这个想法，要聘请他们作为清华设计教学的导师，用他们的亲身实践和职业精神影响到学生。经院务会研究决定，由学院主管教学的副院长单军教授组织院教学办制定计划，第一批确定了15位中青年建筑师作为三年级设计课的导师。然而情况并不顺利，按照清华大学的一般做法，是没有“设计导师”这一聘任头衔的。于是，我们向学校专门陈述我们的理由和想法，讲述国外“评图导师”的先例。在我们向陈吉宁校长汇报设想之后，得到了校长的肯定，校人事和教务部门也对此给予大力的支持。15 位导师的证书是由清华大学颁发的“清华大学建筑学设计导师”聘书。这件事情就这样开始了。

最初开始时，我们对这件事的目的和意义并没有领会得太深，但在后来的教学和若干次的教学讨论的过程中，发现有两方面情况：一是这些设计导师们都很有荣誉感和责任感，很有激情，愿意用他们的经验去教导学生；二是，这些建筑师毕业于中国现有的教学体制，大部分又受过西洋教育，有宽广的视野，他们有自己的判断，这个判断是对当下我国建筑教育的判断，很多人直言不讳地对当下的建筑教育提出了批评，许多是以他们亲身经历的感悟而提出来的。这样我更觉得这件事情是有意义的，它不是一个简单的review，或者final review，而是希望他们参照自己以往的学习经验，在新的过程中教会学生去思考。同时，我也希望这些建筑师的引进可以触动我们自己的老师，我们的老师还有一些是纸上谈兵的东西，或者就理论说理论的东西。能盖出好房子的老师在业界是公认的好老师，至少他有实践经验，如果他的建筑又得了奖，那么他在学生们的心目中也是备受瞩目的，这就是设计导师的力量。

我们没有像其他学校那样设立“大师班”，而是在一个年级全面铺开，全面铺开的优越性在于：首先让学生们全盘面对一个真实建筑师的舞台，这些建筑师是与国际建筑师在同一个舞台上对话的，这一点让学生们很兴奋；其次是这些建筑师可以把他们想的东西讲出来，这种讲述，对自身而言也是一种互动；第三也是比较关键的，我们希望尝试一种教学方式的变化，这种变化的出发点，都是希望我们能有一个相对务实的、同时又很有效率的方法，来启发和教育学生对专业学习的志趣。

设计导师们会把自己作为职业建筑师的想法和体验灌输给这些年轻的学生，使学生们强烈地感受到来自职业建筑师的信念和追求。这一点恰恰是我们当下设计课教学中最缺乏的，也是学生们最需要的。各位导师出不一样的题目，都紧扣了当下建筑设计的热点，以及建筑教育的关键点，而学生们自由选择导师和题目的过程，也反映了我们学生们的价值取向，能够大胆地按照自己的选择去表达。

这件事另外一个附带的效果是仪式感。仪式感对建筑学的教学是很重要的，国外很多建筑院校的评图是相当有仪式感的，我们需要用仪式感去调动学生的主观能动性和成就感。这场秀一定要做，是在学生们面临市场和社会选择的情况下秀自己，同时也是秀老师。仪式感在公开评图那天相当起作用，因为建筑本身就是带有很强自我表现的科学和艺术的综合体，是一种品质的体现，综合气质的展现。

开放教学已经进行快两年了，设计导师们的激情不减，学生们的热情不减。每一次的教学研讨和学期初的教学准备会我们都会听到导师们对教学的思考和再认识，也一次次地被这些设计导师们的职业精神所感动。每一次的教学评图，无论是中期还是期末，我们都会听到学生们充满激情的反馈，也一次次地感受到学生们对这种教学模式的喜爱。我们为此感到欣慰。

希望我们的同学们可以更积极地参与其中。我们期待着明年。

The idea of open architectural design teaching originated for the following causes: first of all, it is related to the characteristic of architecture education. The discipline develops along with the development of human society since the beginning, which is a building process of living environment. Teaching and passing on knowledge during this particular process forms today' s architecture education. The most substantial and simple teaching method is the apprenticeship. Being either Luban from China or Frank Lloyd Wright known for Taliesin West, the passing on of his expertise includes not just the passing on of knowledge but to influence as an architect figure. Due to the limited length of school education, how to provide students with qualified education that would ensure they develop into professional architects with deep understanding of architecture expertise and clear vision of future in such a limited period of time is a vital theme.

The second source of the idea is, the reflection on college education. On the opening ceremony of the 24th teaching forum of Tsinghua University, the university leader has pointed out clearly that the education of Tsinghua University, for three decades since the Founding and Reform, is worthy of reflection: on one hand, the top students entered with no guarantee of their excellence when graduating, or to say not exceeding; on the other hand, whether Tsinghua has nurtured the interest of the students in their major or not was unclear. Without interest, one could not see himself as a professional architect in the future. Retrospectively, architects of the new generation stand in the front-line of architecture practice. Being brilliant both in theory and practice, their practice of work and their minds will influence the society and form their reputation. Their working status could impact the younger generation. This is of great importance to the students. All of the established architects or young architects have their own architectural dream as well as positive living attitude. As an active architect, they must have much feeling yet practice by himself. Allow them to teach the students and encourage the students to search for their enthusiasm and persistence on architecture prevail over the textbook lessons. The vivid and live experience is the most touching.

Thirdly, thinking from the occupational perspective. Since 2005, I have been working at UIA Professional Practice Commission. For a long time, I have noticed a very obvious fall out between the level of our domestic professional education and the international level. The gap exists in neither the course setting nor the teachers, but rather the lack of professional spirit. Our undergraduate and graduate education is parts of professional education. The students we nurtured are to get professional degree and become architects. I hold a little concern toward whether our students could live up to the requirements for a professional architect. As for the previous teaching mode, the basic professional spirit and virtues could not be taught at school. There must be hand by hand tutoring by those who have survived from the fierce competition in the market and experienced the face-to-face communication with the clients and the critics. The professional virtues in these people are the best examples as professional architects.

Based on the three aforementioned aspects, there are sufficient reasons to pursue open teaching. There appears to be a great opportunity then when the 2013 Beijing Best Architecture Design appraisal invited some newly established architects and young emerging architects besides senior experts and academicians. They are the chief architects from private architecture design enterprises or founding architects from independent firms. All of them are from the front-line of innovative creation with clear visions, excellent expertise background and global perspectives rooted from their overseas education experience. Their participation in the appraisal definitely propels the academia. After the appraisal, I am convinced to hire them as teaching instructors of architecture design for Tsinghua to influence the students with their practical and professional spirit. The faculty commissioned the deputy dean Pr. Shanjun, who is in charge of teaching, to organize and set up the course. A first batch of 15 newly established architects and young emerging architects were to become tutor for junior-year teaching course. However, the process is not so smooth. According to the normal practice of Tsinghua University, there is no such title as "design tutor". Thus, we specially stated our reasons and thoughts to the university and the precedent of "review tutor" adopted abroad. Our report to President Chen Jining received approval and great support from the HR and Teaching Department. The certificates of Tsinghua Architecture Design Tutor for our 15 tutors was issued by Tsinghua University. Thus the whole story began.

At the very beginning, we haven't fully grasped its purpose and significance, but during the post course review and discussion, we noticed the two situations: First, these architects carried great sense of honor and responsibility, who are passionate and willing to teach students with the extent of their capacity; Second, though the architects are educated under the typical teaching system in China, many of them also have the experience of studying abroad with broad horizon and independent judgment. Their critics are made against the current architecture education, and even blunt criticism from their personal experience. I believe this is significant. It is not a simple review or final review, but that they want to refer to their previous study experience and teach the students to think during the learning process. Meanwhile, I also hope that the introduction of the architect could influence us as teachers, who talk in theory with no practice. The best teacher is the one who could build great houses. At least he should have some practice experience. If his building was publicly rewarded, then he became a mentor among students. This is the power of architect tutor.

Unlike the "Master' s Course" in other universities, we offer equal opportunity for all students in the same year to participate in the course taught by architecture masters. The advantage of this widespread studio is that: first students are confronted with influential practicing of architects, who is on the same platform as international architects, which the students are thrilled about; secondly, the architects could talked and rephrase their thoughts and ideas which become an revision for themselves; last but not least, we hope to try a different teaching mode. This change is our wish to have a more practical and yet efficient method to raise the interest of students and educate them towards professional study. Architect tutor would pour their professional ideas and experience into the young students bring about the strong belief and dream as a professional architect. This is what we lack the most in our current design teaching courses and what the student need the most. Each tutor has a different topic, firmly focusing on the current hot issue of architecture design and the key points of architecture education. Whilst the free selection of tutor and topic, the student' s choice of selecting reflects the value orientation of our students and encourage them to boldly express themselves based on their own choice.

Another effect is the sense of ceremony, which is very important to architecture teaching. In many architecture colleges overseas, the review process is very ceremonial. We need the sense of ceremony to activate the students' subjective initiative and sense of accomplishment. The show must go on. Particularly, the students could display themselves and be confronted by the selection by market and society, and demonstrate to the teachers as well. The sense of ceremony worked really well on the Public Review Day, since architecture is an integration of science and art with strong need for self-expression. It is also a demonstration of the overall ability.

It has been almost two years since we introduced open teaching. The design tutors are passionate as ever and so do the students. For each teaching discussion and the teaching preparation at the beginning of the term, I would listen to the tutors about their thoughts and reform my understanding of teaching. Repeatedly, I was moved by the professional spirit of the design tutors. Every time at the design review, either mid-term or final, we would receive warm feedback from our students. Over and over, we see the students are affected during this teaching mode. We are genuinely gratified with what we have accomplished.

Hope my students could participate more actively.

Looking forward to the coming year.

社会住宅

马岩松

MAD 建筑事务所创始人、合伙人

马岩松

MAD 建筑事务所创始人、合伙人

教育背景

1994年 – 1999年
北京建筑大学建筑与城市规划学院建筑学 学士
2000年 – 2002年
美国耶鲁大学建筑学院建筑学 硕士

工作经历

2004年至今
MAD建筑事务所创始人、合伙人

主要论著

Mad Dinner [M]. Actar出版社, 2007
《疯狂兔子》[J]. a+u, 2009/12
10*10/3[M]. Phaidon出版社, 2009
Douglas Murphy. "It' s a MAD World" [J]. ICON, 2011/04
Edwin Heathcote. "Conquering the West" [J]. 英国金融时报, 2011/03
Matthew Allen. 封面报道 "An Empathetic Twist" [J]. Domus, 2012/11
马岩松. Bright City, Blue Kingfisher出版社, 2012
Ma Yansong From (Global) Modernity to (Local) Tradition, Actar出版社 & FUNDACIÓN ICO，2012
马岩松.《山水城市》，广西师大出版社理想国, 2014
马岩松. Shanshui City, Lars Müller出版社, 2015
马岩松.《鱼缸》，中国建筑工业出版社, 2015
MAD WORKS, Phaidon出版社, 2016
Isabelle Priest. 封面报道 "Stepping Forward, Looking Back" [J], Riba Journal, 2015/12
Alexandra Seno. 封面报道 "Dangerous Cures" [J]. Architectural Record, 2015/12
Sara Banti. 封面报道 "A New Harmony between Art and Nature" [J]. Abitare, 2015/12
Harry den Hartog. 封面报道 "Taking Nature to the Next Level" [J]. Mark, 2015/12
马传洋. 封面报道《马岩松MAD》[J]. PPaper, 2016/01
封面报道《哈尔滨大剧院》[J]. Domus中文版, 2016/02
封面报道《哈尔滨大剧院》[J]. 世界建筑, 2016/02
方振宁. 封面报道《建筑与自然和音乐迂回》[J]. 时代建筑, 2016/03
杨志疆.《第二自然的山水重构》[J]. 建筑学报, 2016/06

设计获奖

2008 – 全球最具影响力20位青年设计师, ICON
2009 – 全球最具创造力10人, Fast Company
2011 – 国际名誉会员, 英国皇家建筑学会
2011 – 中国最具创造力10公司, Fast Company
2012 – 黄山太平湖公寓: 2012年度十佳概念建筑, Designboom
2012 – "假山" : 最佳住宅群建筑, 世界地产奖
2012 – 梦露大厦: 年度建筑, ArchDaily
2012 – 梦露大厦: 美洲地区高层建筑最佳奖, CT BUH（高层建筑与人居环境委员会）
2013 – 梦露大厦: 2012全球最佳摩天楼, EMPORIS
2014 – 2014年度世界青年领袖, 世界经济论坛
2014 – 全球商界最具创造力100人, Fast Company
2014 – 鄂尔多斯博物馆: 最佳 "建筑 – 金属" 奖, 世界建筑新闻奖
2014 – 南京证大喜玛拉雅中心: 2014年度十佳住宅, Designboom
2014 – 朝阳公园广场: 2014年度十佳高层建筑, Designboom
2015 – 威尔士大道8600: 设计概念奖, 洛杉矶建筑奖
2015 – 康莱德酒店: Beyond LA奖, 洛杉矶建筑奖
2015 – 全球设计权力榜, Interni杂志
2015 – 哈尔滨大剧院: 2015年度十佳艺术中心, Architectural Record
2015 – 哈尔滨大剧院: 2015年度最瞩目建筑, Wired
2016 – 哈尔滨大剧院: 年度文化建筑, ArchDaily
2016 – 哈尔滨大剧院: 最佳 "表演空间" 奖, 2016世界建筑新闻奖
2016 – 哈尔滨大剧院: Beyond LA奖, 洛杉矶建筑奖

代表作品

加拿大梦露大厦（图1）、哈尔滨大剧院（图2）、鄂尔多斯博物馆（图3）、日本四叶草之家（图4）、芝加哥卢卡斯叙事艺术博物馆（图5）、中国木雕博物馆（图6）、北海 "假山"（图7）、胡同泡泡32号（图8）、黄山太平湖公寓（图9）、北京朝阳公园广场（图10）、南京证大喜玛拉雅中心（图11）

Diesel

社会住宅的社会性

三年级建筑设计（6）设计任务书

指导教师：马岩松
助理教师：齐子樱　何晓康

社会住宅的社会性

在词典中，对于社会性的定义是生物作为集体活动中的个体、或作为社会的一员而活动时所表现出的有利于集体和社会发展的特性。人的社会性是人不能脱离社会而孤立生存的属性，主要包括利他性、协作性、依赖性以及更加高级的自觉性等。换言之，在城市里我们没有办法一个人生活，城市是属于人群的。人们在城市里一同创造价值与活动，一同享受城市的便利与活力。

住宅是建筑世界中的重要项目，它反映了当代的人文活动与社会趋势。其中，社会住宅更是占有关键的地位。社会住宅(social housing)，在欧洲亦称之为“社会出租住宅”（Social Rented Housing)，是指政府直接兴建、补助兴建或民间拥有之适合居住房屋，以低于市场租金或免费出租给收入较低的家户或特殊的弱势对象的住宅，美国称之为affordable housing、日本称公营住宅、香港称公共屋邨（简称公屋）、新加坡与马来西亚称组合房屋（简称组屋）等，台湾又因为年代法源和地方政府政策施行而有不同名称之社会住宅，例如平价住宅、出租国宅、公营住宅、青年住宅、劳工住宅等。

社会住宅通常都具有容积率高、体量大、造价要求严苛、户型简单高效等特性要求，并且需要形成一定的社区性质，并体现社会公平，以及设计均好性。新的社会需要新的住宅形式，与城市产生尊重人的居住尊严和人们生活精神需求的对话。高容积率的城市居住社区的规划和建筑设计，对于一个建筑师来说永远是非常具有挑战性的项目课题。它的设计难度会随着建筑师对建筑、社会形态等因素理解，以及其自身对设计质量的要求和成熟度不断提高而增加。现在在中国并没有真正意义上的社会住宅。在城市化进程中以及社会形态不断演变以及信息化工业革命中，人的社会性不断退化。住宅建筑设计不断向实用主义和城市化进程妥协，丧失了社会性和对人居住尊严和精神追求的尊重。我认为，社会住宅的建筑设计要超越坚固、实用、美观，希望能够通过这学期的课程，与学生们一起从居住的精神理想出发检验中国城市住宅同质化的症结，畅想并展现新一代的中国居住。

课程安排

本学期课程分为两个阶段，均分组进行，2~3人一组。

第一阶段为社会住宅研究，以现代全世界社会住宅建筑为基本案例，重点研究北欧以及日本，搜集材料、分享研究成果、共同讨论，进行深刻分析，提出并且不断清晰、加强自己心目中理想的住宅、社区的印象。

第二阶段为设计，以事务所正在进行的北京保障房项目为具体案例，利用实际项目的各项指标，以及实际地形、周边环境，让学生们在一定的局限性中，体会对于第一阶段研究成果的应用。

课程要求

学生能够体会到一个完整的大体量项目的设计过程。勘查现场并进行分析和研究，破题立意，处理较为复杂的流线、功能和户型要求以及平面排布，对于立面、材质、光照、规范等等细节问题的考量。最后形成正式的汇报文件，达到一定的概念深度，组织与各界公开汇报。

在设计过程中，学生需要严格遵守业主给出的指标及相应住宅规范，充分考虑以成本为指向的可实施性，以小组为单位体验职业建筑事务所的工作流程和方式，以及团队协作，包括最终汇报。设计结束后，同学们可以拿着设计成果与学期初始阶段提出的理想住宅、社区相比对，检验设计过程中对于建筑师情感表达的追求和坚持。

在当下建筑专业教育中，要求建筑师在处理实际项目中拥有很强的理性思维能力，并且利用标准逻辑思维方式来解决问题。希望学生们可以带入自己对理想居住环境和社区的细腻情感，以感性表达为最本源的出发点并且坚持，在面对真实项目的设计处理中，体会到将理性思维和逻辑方式当成工具使用，而非被其使用。

教学安排

由于课程设计量和内容较大,希望同学们开学之前就能查阅一些社会住宅经典项目案例。

1~3周，研究。

4~8周，设计。

重访光辉城市 / REVISITING THE RADIANT CITY

项目选址：北京百子湾

项目类型：集合住宅，公租房

建筑面积：30 万 m^2

用地面积：10 万 m^2

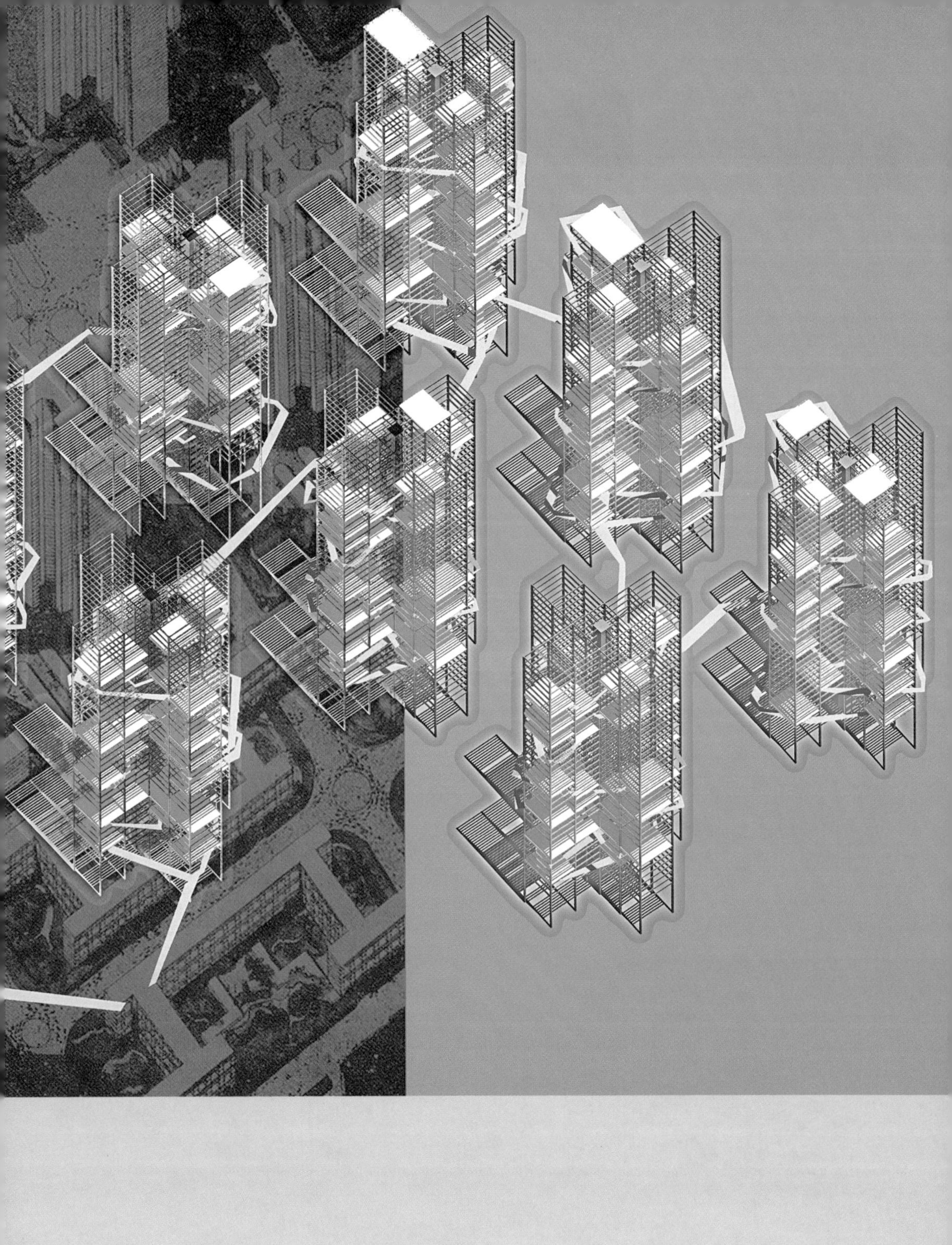

方案设计：张嘉琦
指导教师：马岩松
完成时间：2015

方案设计：詹旭强
指导教师：马岩松
完成时间：2015

方案设计：胡曦阳
指导教师：马岩松
完成时间：2015

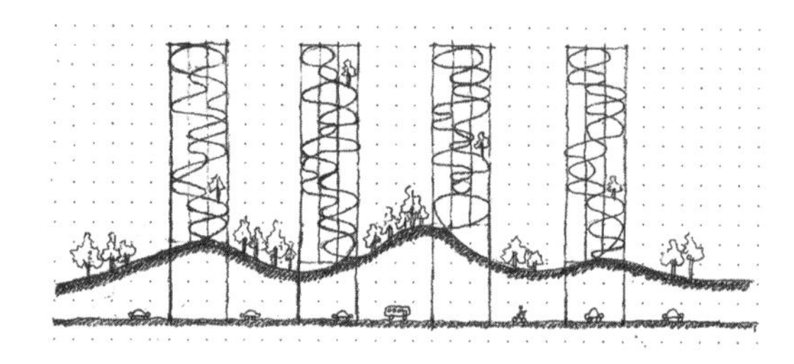

开篇：主要方案图。**本页上图：**概念草图。**本页中图：**十字塔楼变体。**本页下图：**概念立意。**对页上图：**轴测图。**对页下图（从上至下）：**各层平面图，立面拆解。

我们三个一直对建筑学科的外延，即其中"非建筑"的部分很感兴趣，同时也都想尝试一下研究性多于设计性的设计课程（这是我们在之前两年多的设计课中没有接触过的），所以这个题目的研究性和社会性是最吸引我们的地方。第一节课的一个"脑洞"让我们从历史视角切入（这也是社会学常用的研究方法），然后奇迹般地从第一节课磕磕绊绊地走到了最后。

勒·柯布西耶在将近一百年前提出的光辉城市理念在西方世界被证明无法可持续地实施（1972年Pruitt-Igoe社区的炸毁为这一理念执行了死刑），但百年之后却在万里之遥的中国落地生根并野蛮生长。放眼现在的北京以及中国其他城市，旧城居住区千百年来积淀和积累出的细腻而有机的肌理正在或已经被巨构的塔楼和空旷的场地侵蚀殆尽。现代主义的高

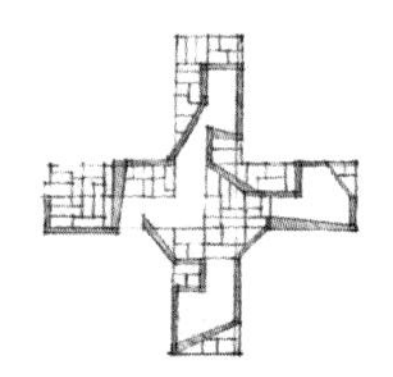

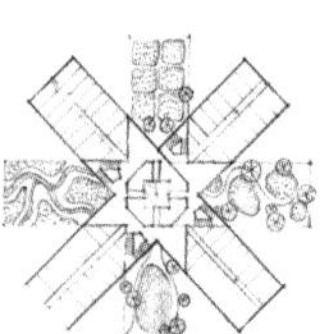

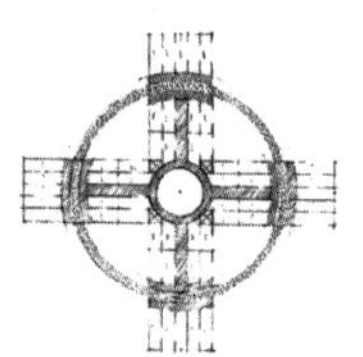

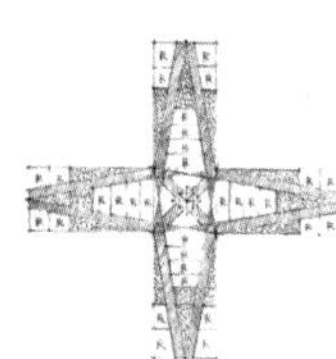

层塔楼带来了阳光和新鲜空气，但随之而来的是“人”的缺席——超人尺度和机械的重复扼杀了居民之间的社会交往与居住形态的多样性，住区的设计变成了关于金钱和效率的数学题。

为了对现在大多数国民所使用但却熟视无睹的居住形式进行尖锐的批判，我们试图在十万m^2的场地上借鉴现成品艺术的“戏仿”手法，在光辉城市原型的基础上用“一个动作”将其彻底打破，以期为当今中国的社会住宅带来真正能被居民所感知和使用的多样性的公共空间，并与社会与环境互动。

经过许多节课的探索（中间开过无数靠谱和不靠谱的“脑洞”），我们最终确立了用连续的坡道贯穿原来的十字塔楼结构的手法。坡道在楼间自由穿梭，穿过的开间就变成绿化的露台，这样若干条立体的绿道贯穿在整个住区之中，从超人尺度的空间结构中“掰碎”出一个个多样且近人尺度的社交空间。与此同时，我们认为原来光辉城市设想中对楼间空地巴洛克园林式的处理不过是设计师对自然的意淫，所以我们决定在楼间空地营造一层真正有山有水，植物自然生长的环境供居民真正地亲近自然，然后在“自然层”之下布置社区中心、停车场等大型社区公共空间。这样虽然在外观上基本保留了光辉城市十字塔楼的结构，但对居住单元间联系、多样小尺度社交空间和居民与自然亲近关系的追求却完全突破了原来方案的立意。

总之，这是一个解构多于建构，驳论多于立论的方案。

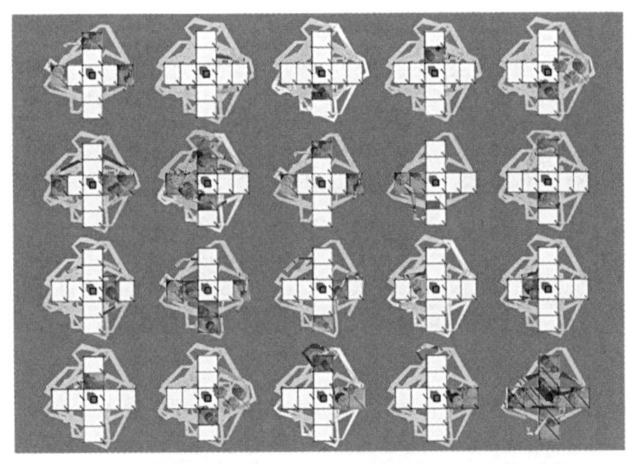

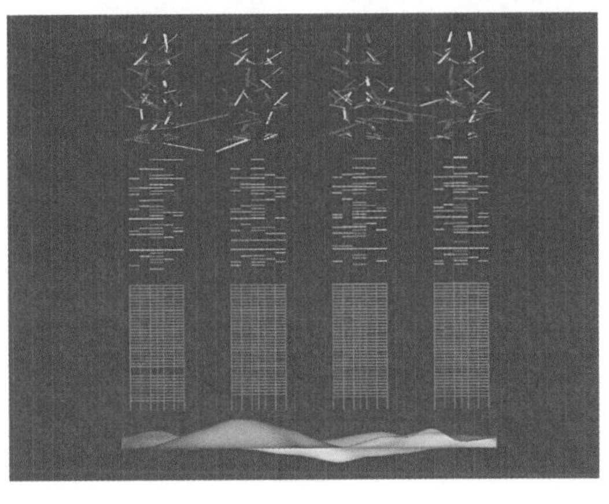

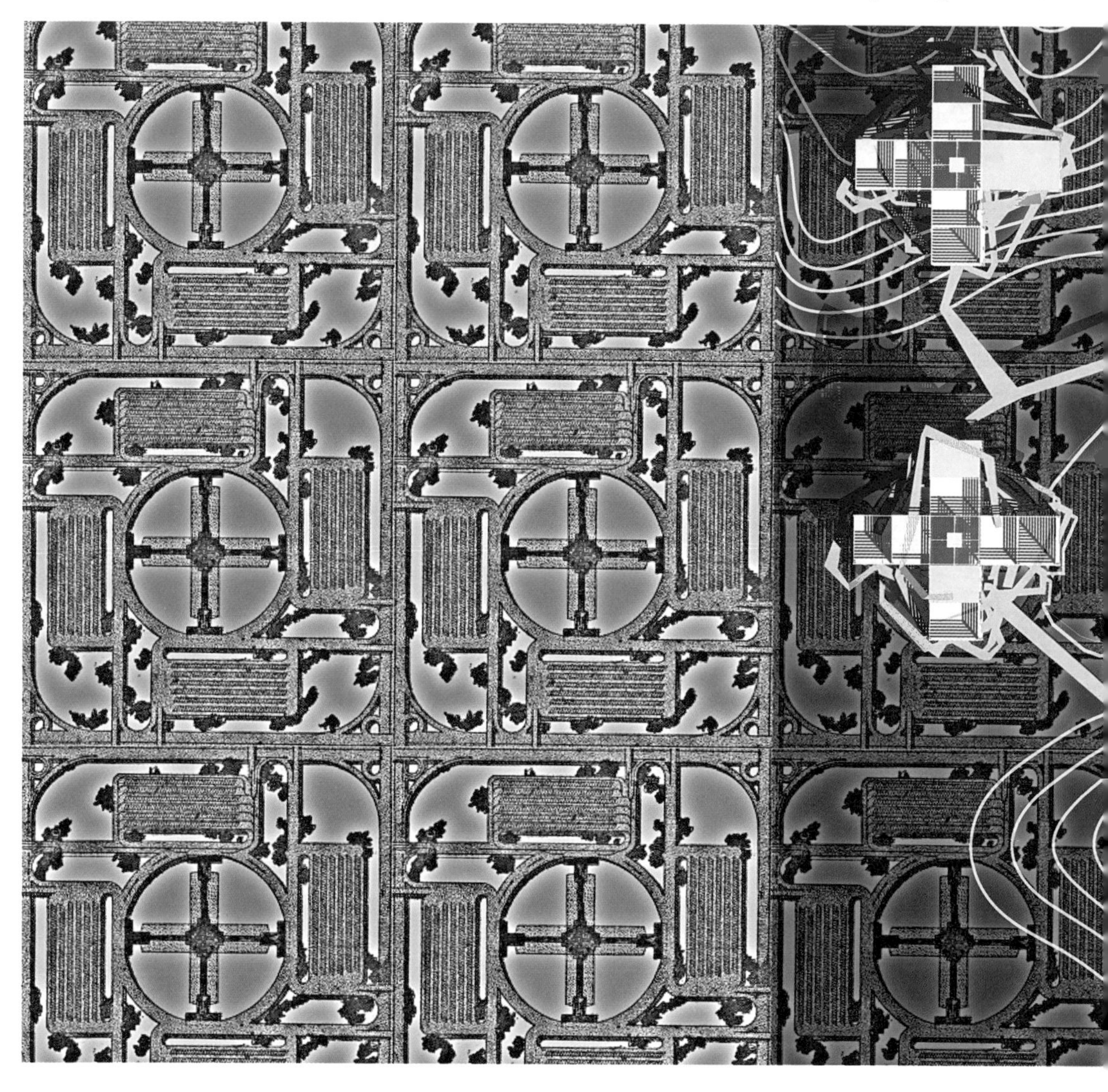

We have been very interested in the extension of architectural discipline, which is the part of "non-architecture", and we have also tried to attend the design courses that focus more on research than design (this is what we haven' t tried in the design courses in more than two years), so the research and sociality of this subject are most attractive to us. The imagination in the first lesson let us cut into it from the historical perspective (a commonly used method in sociology), and led us to the final like a miracle.

The concept of Radiant City proposed by Le Corbusier nearly a hundred years ago has been proved unable to be implemented sustainably in the western world (the blow-up of Pruitt-Igoe community in 1972 ended the concept). However, it is rooted in China, a million miles away, and grows wildly. Looking at Beijing and other cities in China, the delicate and organic texture that was accumulated from the residential area in the old city for thousands of years is, or has been, eroded by the mega towers and empty grounds. The modernist high-rise towers help bring sunshine and fresh air, but then causes the absence of "people". The XL scale and the mechanical repetition strangles the social interaction among the residents and the diversity of living forms, while the design of residential district becomes a mathematical question about money and efficiency.

In order to conduct sharp criticism to the living form that has been used by the majority of the population but neglected, we referred to the "parody" approach of readymade art and tried to break the prototype of Radiant City with "one action"

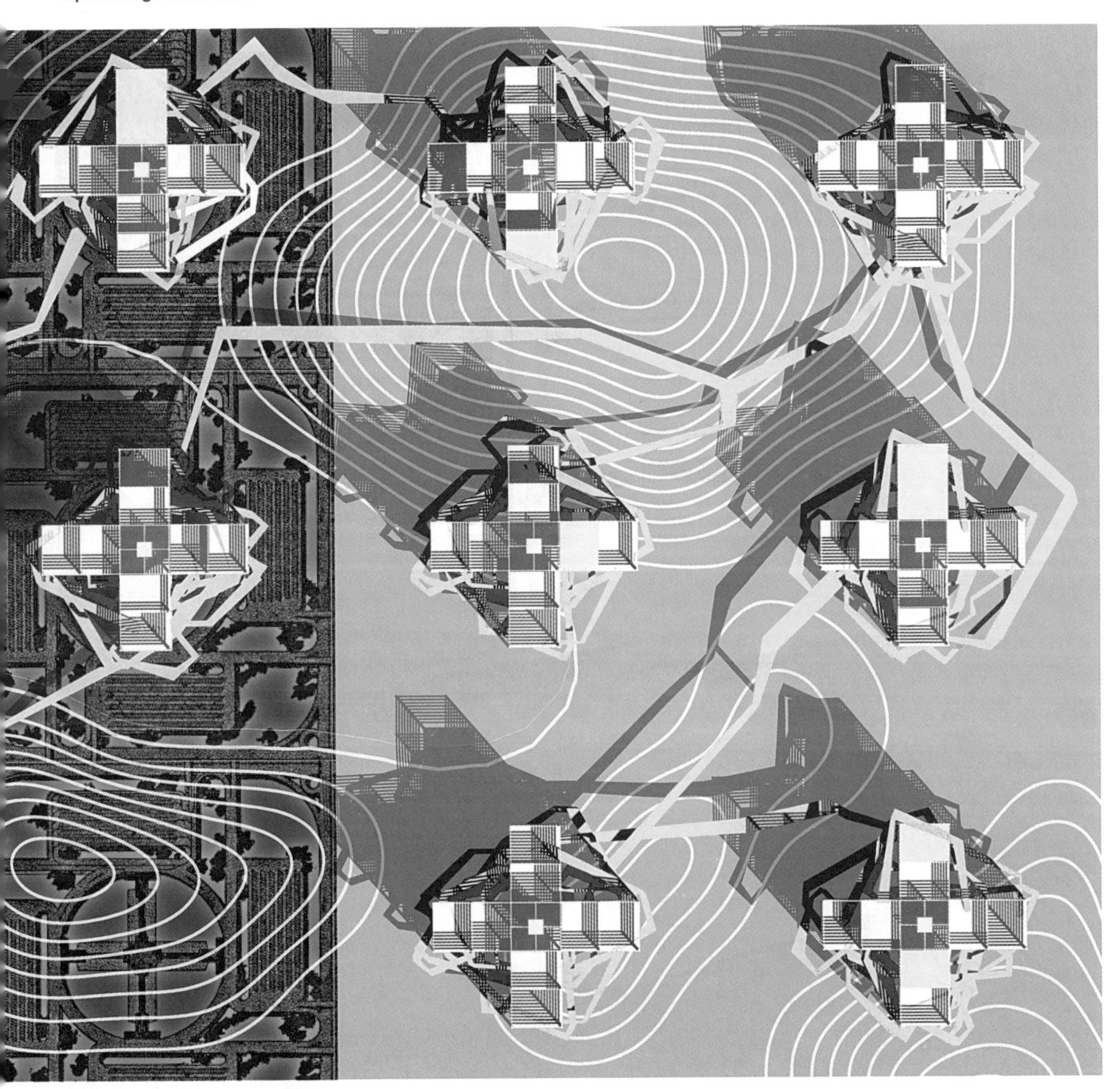

in the space of one hundred thousand square meters. The aim is to bring diversity, public space and social interaction to the social housing of contemporary China, which can be perceived and used by the residents.

After the explorations in many classes (through a myriad of reliable and unreliable imagination), we finally designed a continuous ramp throughout the original cross-tower structures. The ramp has free access into the floors, where the space along the way becomes green terrace. Thus, several three-dimensional green roads penetrate through the whole residential area, by breaking the XL spatial structure into many various and approachable scale of social spaces. At the same time, we believe that the introduction of Baroque garden into the open space between the towers in the original concept of Radiant City was just a daydream to the nature, so we decided to create an environment with true mountain and water in the space between towers, where the plants could grow naturally and the residents could really be close to nature. And we arranged the community center, parking lot and other large community public space below the "nature layer". So although the cross-tower structure of Radiant City is basically retained in appearance, the pursuit of the connection among the residential units, the diverse small scale social spaces and the close relationship between the residents and nature completely break through the original conception.

In short, this is a scheme with more deconstruction than construction and more refutation than proposition.

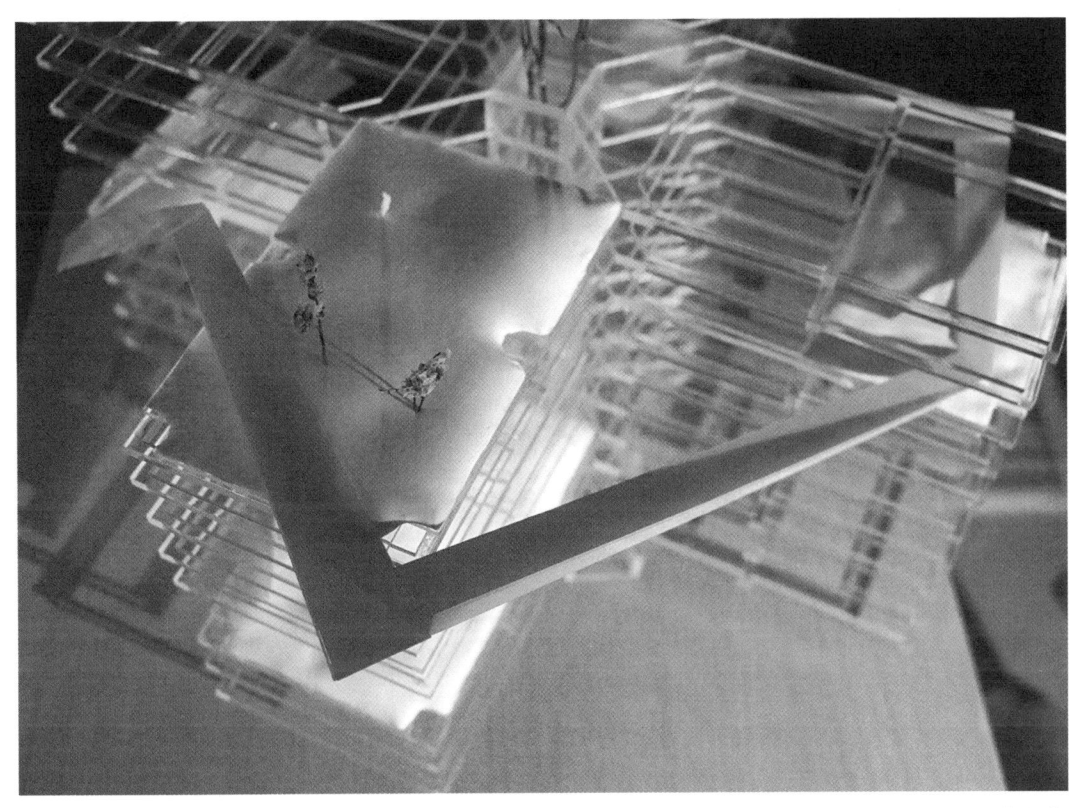

上页图: 平面图。**对页图:** 模型细节。
本页上图: 模型鸟瞰图。

教师点评

这组年轻人很大胆地质疑了柯布西耶年轻时提出的光明城市。他们很清楚我们今天所处的年代,我们的城市、建筑、文化还基本处于现代主义末期的功能主义和以构成为设计方法的阶段。今天,当新一代建筑师质疑人和自然在这样的城市中所处的地位时,必定会质疑它的源头,是什么造成今天千城一面的机械冰冷,反人性的城市?住宅是居住的机器么?

光明城市被当作一个反例,但其本身又是现代主义初期思想的经典作品,被这组年轻人用来研究、批判和进行改造。改造大师的经典作品是对时代的回应,是对经典延续到今天而成为一种普遍思路的反叛。他们的改造是对人性和自然的赞美,这是新的时代中住宅、尤其是社会住宅中的人所需要的。

Teacher's comments

The group young students boldly query the "Radiant City" theory put forward by Le Corbusier at his early age. They clearly know the age where we live. Our cities, buildings and cultures are basically at the end of Modernism (Functionalism) and at the stage of taking constitution as design method. Nowadays, when the new generation of architects query the positions of humans and nature in such a city, they will definitely query its source: What makes cities machinery, cold, uninhabitable and look alike? Is the house a machine for living in?

Radiant City is researched, criticized and transformed by these young people as a counter example, but itself is the classical work of Modernism at the preliminary stage. Transforming the great master's classical work is a response to the age, and a rebellion to the classical work that has become a common thinking nowadays. The transformation is a kind of praise to humanity and nature, which is required by people in houses, especially social housing, in the new age.

双城记/ A TALE OF TWO CITIES

项目选址：北京央视大楼原址
项目类型：媒体中心，城市地标
建筑面积：50 万 m^2
用地面积：20 万 m^2

方案设计：段然
指导教师：马岩松
完成时间：2015

方案设计：彭鹏
指导教师：马岩松
完成时间：2015

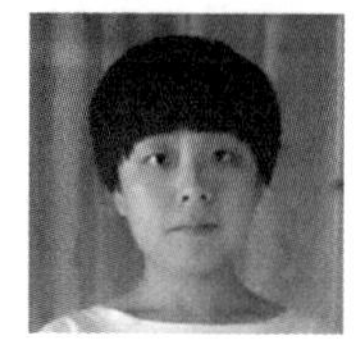

方案设计：石南菲
指导教师：马岩松
完成时间：2015

方案设计：温从爽
指导教师：马岩松
完成时间：2015

开篇图：剖面图。**本页图：**分层轴测图。**对页图：**总平面图。

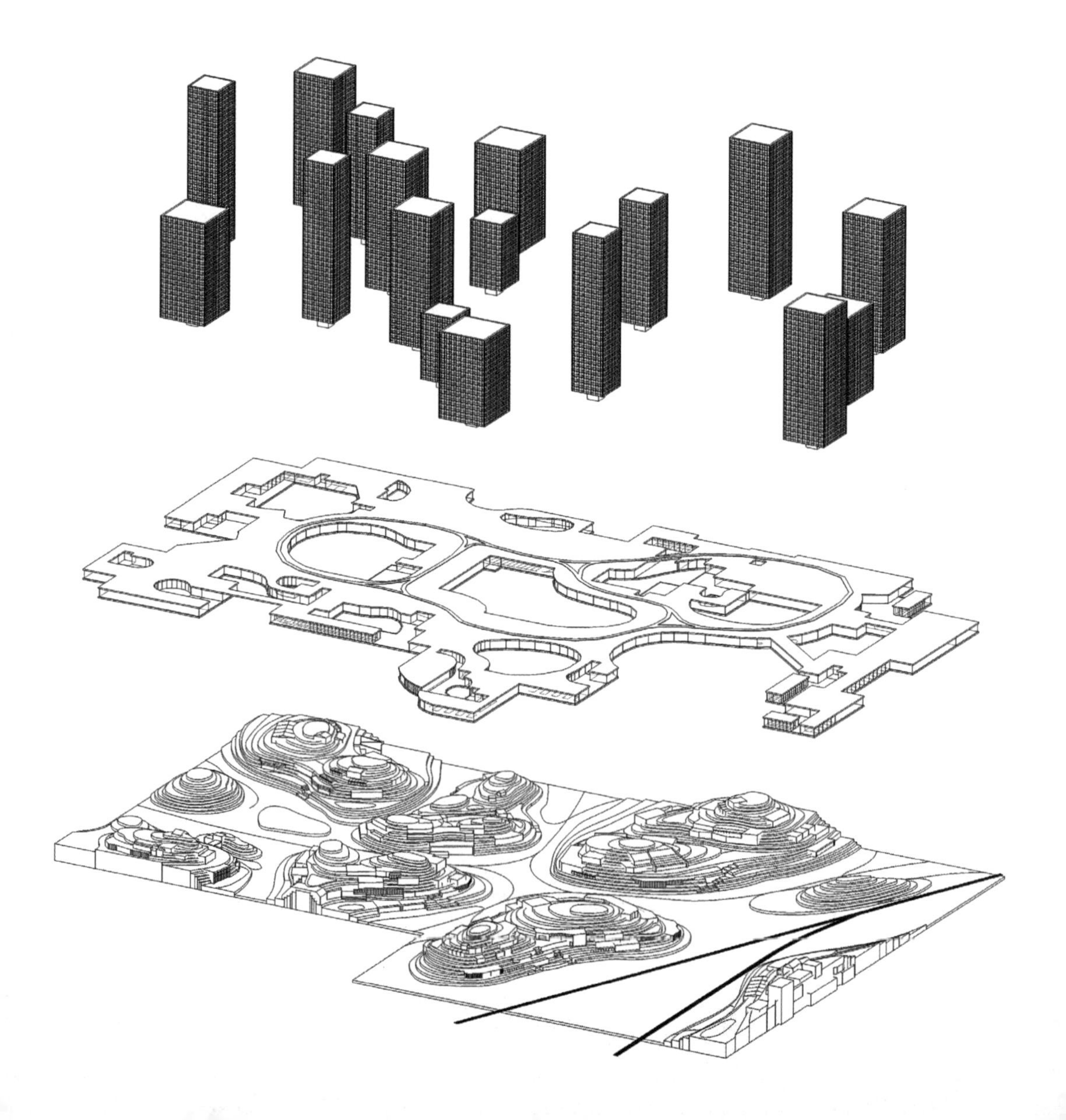

我们小组作品的名称为“双城记”，其概念是“分层”，将一个高密度的住宅社区架空，上层作为满足最基本居住、睡眠需求的住宅空间，解决“社会住宅”的高密度需求，下层则将土地还原给自然，将空间还原给社会活动，将建筑功能隐藏于山林之中，创造出可供居民进行社会活动和交流的空间。中间用一个经过设计的“板”状大空间分隔开来，可以进行锻炼、休闲娱乐和学习集会等多种社区活动，成为上下两层的连接部。这样上下两城，一动一静，一个为“北漂一族”提供了一个经济舒适的立足之地，另一个则是创造了一个绿野仙踪般的梦幻之城。

提到社会住宅，我们首先想到的是偏远的地段、低质量的建筑、局促的空间设计、高容积率、低绿化率等等。在概念设计阶段，我们提出了公共活动与私密空间分层而设和将自然纳入到城市当中的想法。我们从“分层”这个概念出发，将高效集约的塔楼架空在一个高空的板上，在下层则随心所欲地创造理想中的城市空间。上部的塔楼采用最简约的方筒形态，内部采用模数化的户型排布方式，每个户型都是一个内部空间可通过隔板调整的小单位，满足人最基本的睡眠、洗漱和简单备餐的空间需求。中间的“板”形态自由生成，为应对下面的光照需求，在开洞上进行了设计。板内空间作为“社区客厅”，成为整个社区的人们交流活动的重要空间，内部设置小型会客厅、可出租的个人studio、预约研讨间、图书馆、网吧、餐厅、集会场所和一些室外庭院，为居住在其中的年轻一族提供充足的学习条件和丰富的精神生活。板的上部则是“社区花园”，成为社区的锻炼场，设置长跑道、球场、健身区并沿跑道布置花田、藤架长廊、树阵花园等景观。下部的城市，采用覆土建筑、半地下建筑、种植屋面等做法，将建筑藏匿于山林之中，以有机自然的形态、景观将建筑掩盖，使人虽处于热闹的城市综合体，却仿佛置身于丛林之中。地段的东侧原有一条废弃铁路线，我们将其保留作为一个景观要素形成集约绿地，作为社区活动中心，同时也在山谷等地形吸引点处布置了林荫大道、芦苇荡、游泳池等景观。在使用功能上，利用山体内部的大空间布置影剧院等公共建筑，沿道路则有餐饮、商业、办公等功能，同时结合山体设置景观平台、上山步道等。我们希望即使在高密度的住区中，也有丰富的交流空间，有美好的环境。

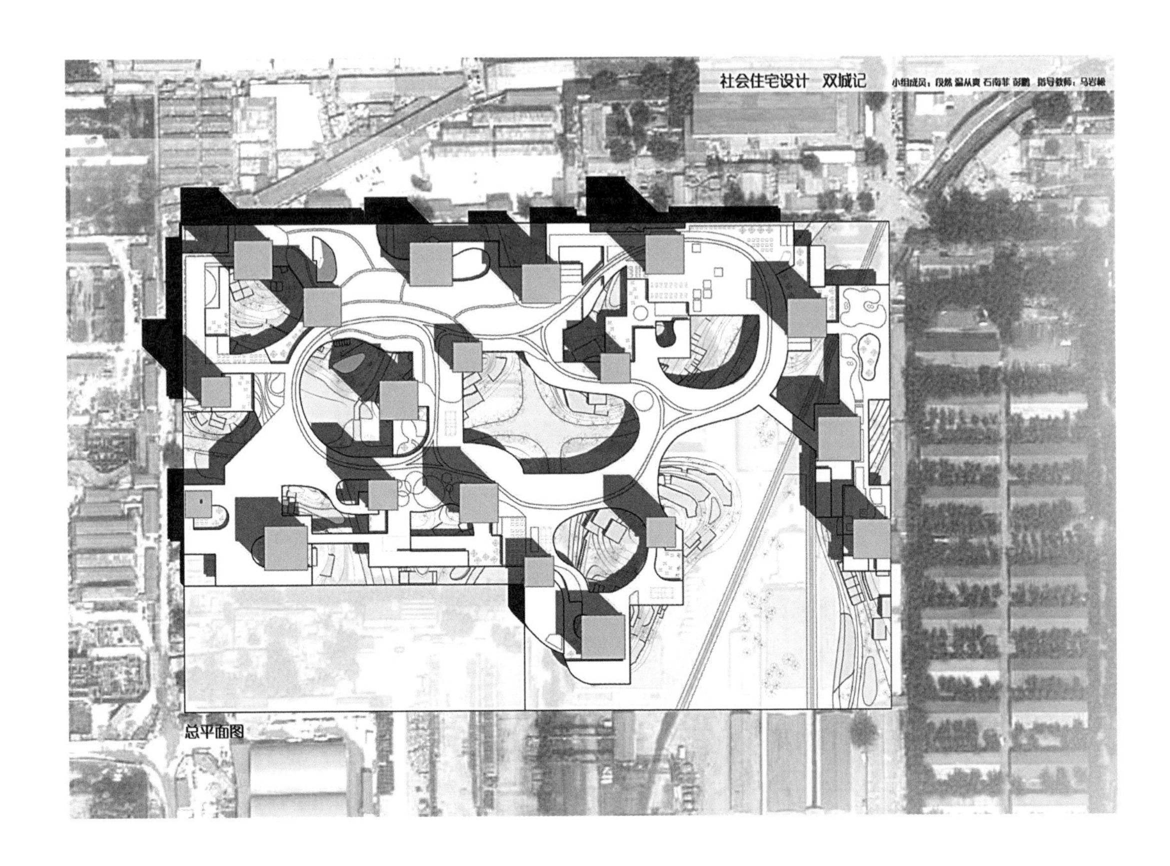

总平面图

The name of our work is "A Tale of Two Cities", and its concept is "layered". The high-density residential community was designed to be elevated. The upper layer was used as the residential space that met the most basic living and sleeping needs to solve the demand of the high density "social housing", while the lower layer was returned to the nature, restored the space to social activities and hided the building functions in the mountains, so as to create a space for residents to carry out social activities. In the middle there was a designed "board" where people could carry out multiple community activities for exercise, recreation and entertainment, learning, assembly, etc., becoming the connection of upper and lower layers. For the upper and lower cities, one is dynamic, the other is static; one provides a comfortable economic foothold for "drifters in Beijing", and the other creates a city of dreams. When mentioning social housing, we would first think of the buildings in remote areas with low quality, cramped space, high volume rate and low rate of greening, and so on. In the conceptual design phase, we proposed two ideas, a layered setting of public activities and private spaces, and the integration of nature into the city. We started from the concept of "layered" to elevate the high efficient tower over the board high above the ground, creating a free ideal city space in the lower level. The upper tower building is in the most simple cube shape. The inside adopts the modular arrangement and each household is a small unit. The interior space can be adjusted by the partition board and meet the most basic space requirements of sleeping, washing and simple preparation for meal. The form of the "board" is generated freely, with the opening holes designed to deal with the illumination requirements in the lower level. The space inside the board, serving as "community living room" , has become an important space for the communication of people in the whole community. The small parlor, personal studio for rental, seminar room, library, internet bar, restaurant, meeting place and some outdoor courtyards are set inside, which provides adequate learning conditions and rich spiritual life for young people who live there. The upper part of the board is community garden, which has become the exercise field, the long running track, court and the fitness area. Flower field, pergola corridor, tree array gardens and other landscapes are arranged along the runway. Adopting the design of earth sheltered architecture, semi underground building and planting roof, the lower part of the city hides the buildings in the mountains, and covers the buildings with natural organic forms and landscapes, so that though

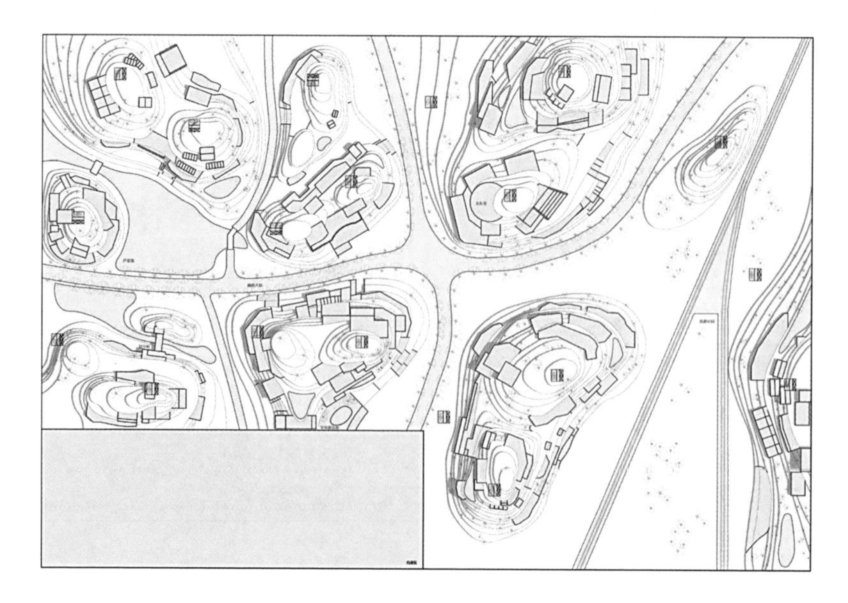

people are in the busy urban complex, they feel they are enveloped in the jungle. The eastern side of the site was originally a disused railway line, so we retained it as a landscape element to form an intensive green space, which became the center of the community activities. In addition, the boulevards, reed marshes, swimming pools and other landscapes are arranged in the valley and other attractive terrains. In functions, public buildings such as the theaters are arranged by means of the large space inside the mountains, providing catering, commercial, office and other functions along the road. Meanwhile, landscape platforms, mountain trails, etc., are arranged in combination with the mountains. We hope that even in the high dense settlements, there are plenty of communication space and beautiful environment.

对页图：平面一底层。**本页图（从上至下）：**平面一板内。平面一板上。

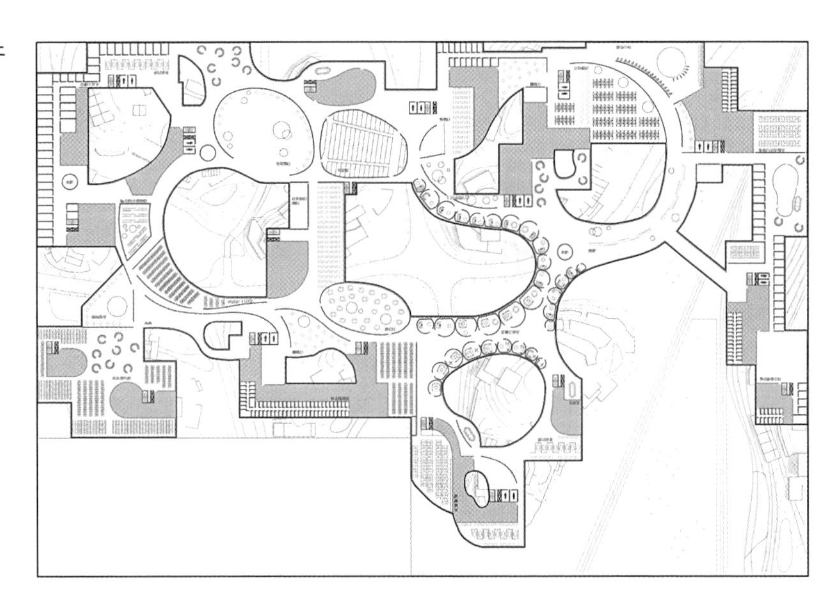

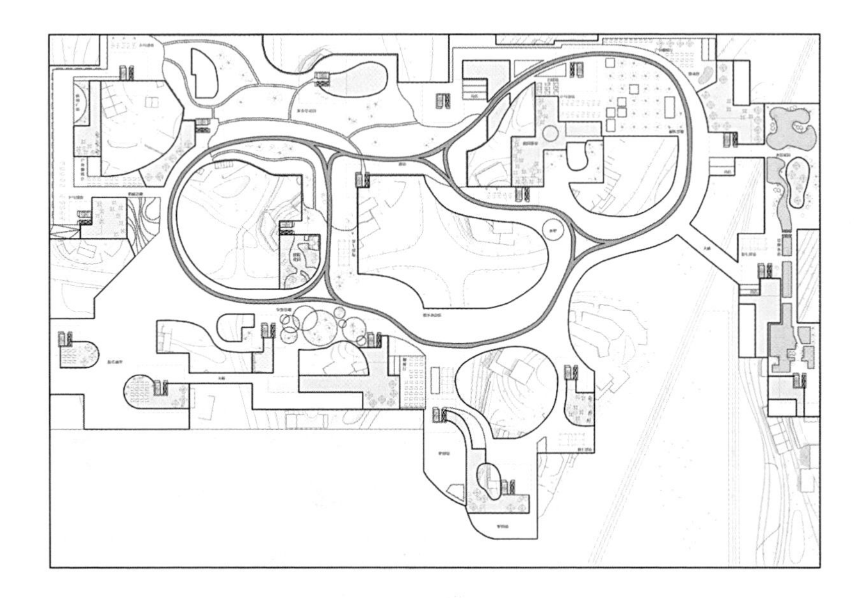

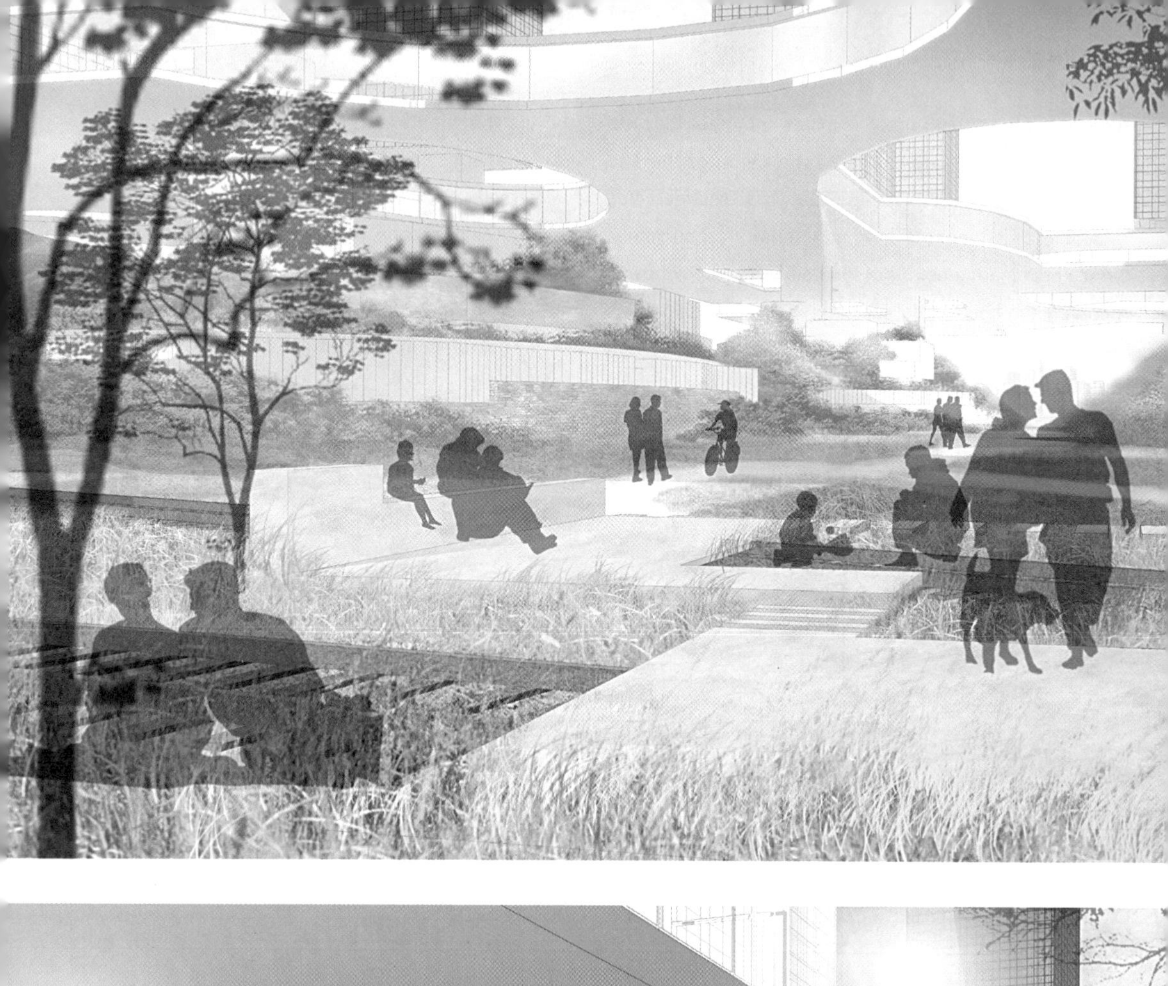

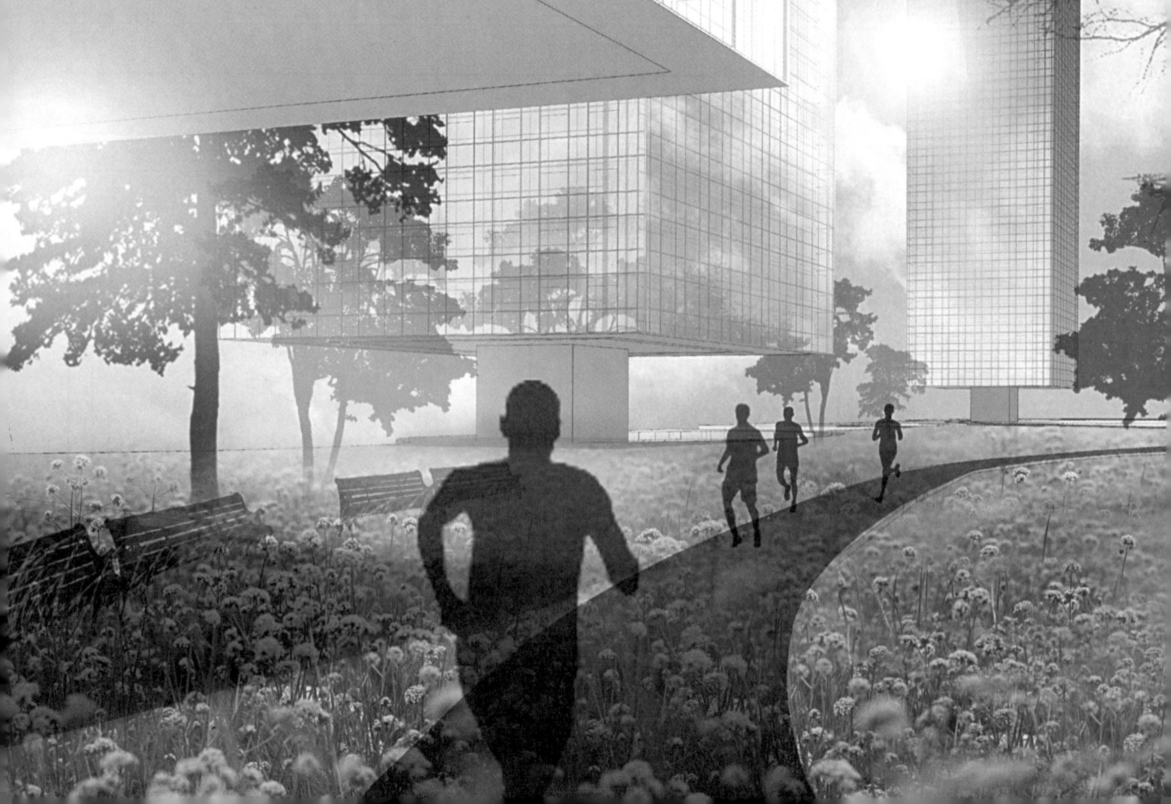

教师点评

这是一个立体城市的微缩版。立体城市在现代主义城市初期被反复提出，并始终被认为是未来城市的样貌。这个作品看起来是一个马上就可以实施而又充满社会理想的独特作品。将社交空间从每一个住宅单元剥离出来集中设置在空中花园下方，作为社区住宅，它对促进人与人的交流合作有着积极的意义。同时我们不禁要质问，为什么这么明显可行、对社区和环境有益的设计出现在学校中年轻学生的笔下，但中国数以千万计的城市社区和住宅却又那么千篇一律？现实中的城市究竟包含了那些不可抗拒的负面力量，社会住宅的社会性理想如何实现？它不是一个乌托邦，也不再是早期提出的立体城市那样的具科幻感的象征性图像，它的社会现实感、可实现性奠定了作为社会住宅的理想跟现实的结合。

Teacher's comments

It is a micro edition of the three-dimensional city. The three-dimensional city was repeatedly put forward at the preliminary stage of modernistic city and considered as the appearance of future cities. It is just like a unique work, which can be put into effect immediately and is full of social ideals. It separates the social space from each dwelling unit and intensively sets them under the hanging garden. As a residential community, it has positive significance for promoting the communication and cooperation among people. Meanwhile, we can't help querying that why the obviously feasible design being beneficial to the community and environment is made by young students in university and why tens of millions of China's communities and residences are still at low levels and all alike. How many irresistible negative forces are contained in real cities? How to realize the social ideal of social housing? It's neither a utopia nor the science-fiction symbolic images like the three-dimensional city put forward at the early stage. Its sense of social reality and realizability determine the combination of the ideal of social housing and the reality.

跨页图（从上至下）：景观—铁路公园。景观—花田。**本页图：**林荫道。

点子 手段

王昀

方体空间工作室创始人、主持建筑师

王昀

方体空间工作室（www.fronti.cn）
创始人、主持建筑师

教育背景

1981年 – 1985年
北京建筑工程学院建筑系 建筑学 学士
1992年 – 1995年
日本东京大学 工学 硕士
1995年 – 1999年
日本东京大学 工学 博士

工作经历

2001年至今
执教于北京大学
2002年至今
方体空间工作室 创始人、主持建筑师
2013年至今
北京建筑大学建筑设计艺术研究中心 主任、
现代建筑研究所主持人
2013年 – 2015年
清华大学建筑学院 设计导师

主要论著

王昀.《传统聚落结构中的空间概念》[TU]，中国建筑工业出版社，2009
王昀.《向世界聚落学习》[TU]，积木，2010
方体空间工作室编著.《当代建筑师系列——王昀》[TU]，中国建筑工业出版社，2012
王昀.《音乐与建筑》[TU]，中国电力出版社，2015
王昀.《绘画与建筑》[TU]，中国电力出版社，2016
王昀.《建筑与废物》[TU]，中国电力出版社，2016
王昀.《建筑与书法》[TU]，中国电力出版社，2014
王昀.《建筑与园林》[TU]，中国电力出版社，2015
王昀.《中国园林》[TU]，中国电力出版社，2014
王昀.《密斯的隐思》[TU]，中国建筑工业出版社，2016
王昀.《聚落平面图中的绘画》[TU]，中国电力出版社，2016

设计获奖

1993年 – 日本《新建筑》第20回日新工业建筑设计竞赛获二等奖
1994年 – 日本《新建筑》第4回S×L建筑设计竞赛获一等奖

代表作品

善美办公楼门厅增建（图1）、60m²极小城市（图2）、石景山财政局培训中心（图3）、庐师山庄住宅A+B（图4）、庐师山庄会所（图5）、百子湾中学（图6）、百子湾幼儿园（图7）、杭州西溪湿地艺术村H地块会所（图8）

点子·手段·空间研究

三年级建筑设计（6）设计任务书

指导教师：王昀
助理教师：赵冠男 张捍平

建筑是什么？

建筑不是文学，不是诗歌，不是绘画，不是雕塑，不是电影，不是舞蹈，不是服装、不是……。

建筑就是建筑。

建筑不是建筑之外的任何事情。

建筑有属于自己的语言系统。

点子

对于建筑设计而言，“点子”是重要的，“点子”就是想法，是构思，是设计师在自己进行设计时的所有知性的综合体现。当然“点子”本身也包含着设计师对于建筑功能的全部理解。

手段

“点子”的确是重要的，但是所有人都会有“点子”。

这个世界实际上不缺少“点子”和馊主意。

“手段”是使得点子得以转化成为被设计的“对象物”的通道。

建筑设计的“手段”是建筑师的专属。

课程目标

解决“点子”和“手段”的关系问题。关注“空间研究”是本课程的教学目标。

课程描述

根据既定的教学大纲和教学内容的要求，进行相应的题目设定。

重要的教学内容不是设计的对象物，而是设计本身。

成果

相关的表现图纸和模型

注意事项

本教学内容仅仅在课堂上的时间是不够的，需要同学们花费大量的业余时间和精力。

请大家利用假期时间锻炼好身体、并尽可能提前安排本来计划安排在下学期前半学期预计要做的工作。

艺术行廊/ ART GALLERY

项目选址：北京市石景山游乐场东北部地区的一片湿地中

项目类型：艺术行廊

层高：9m　**建筑面积：**6500m²

用地面积：5100m²

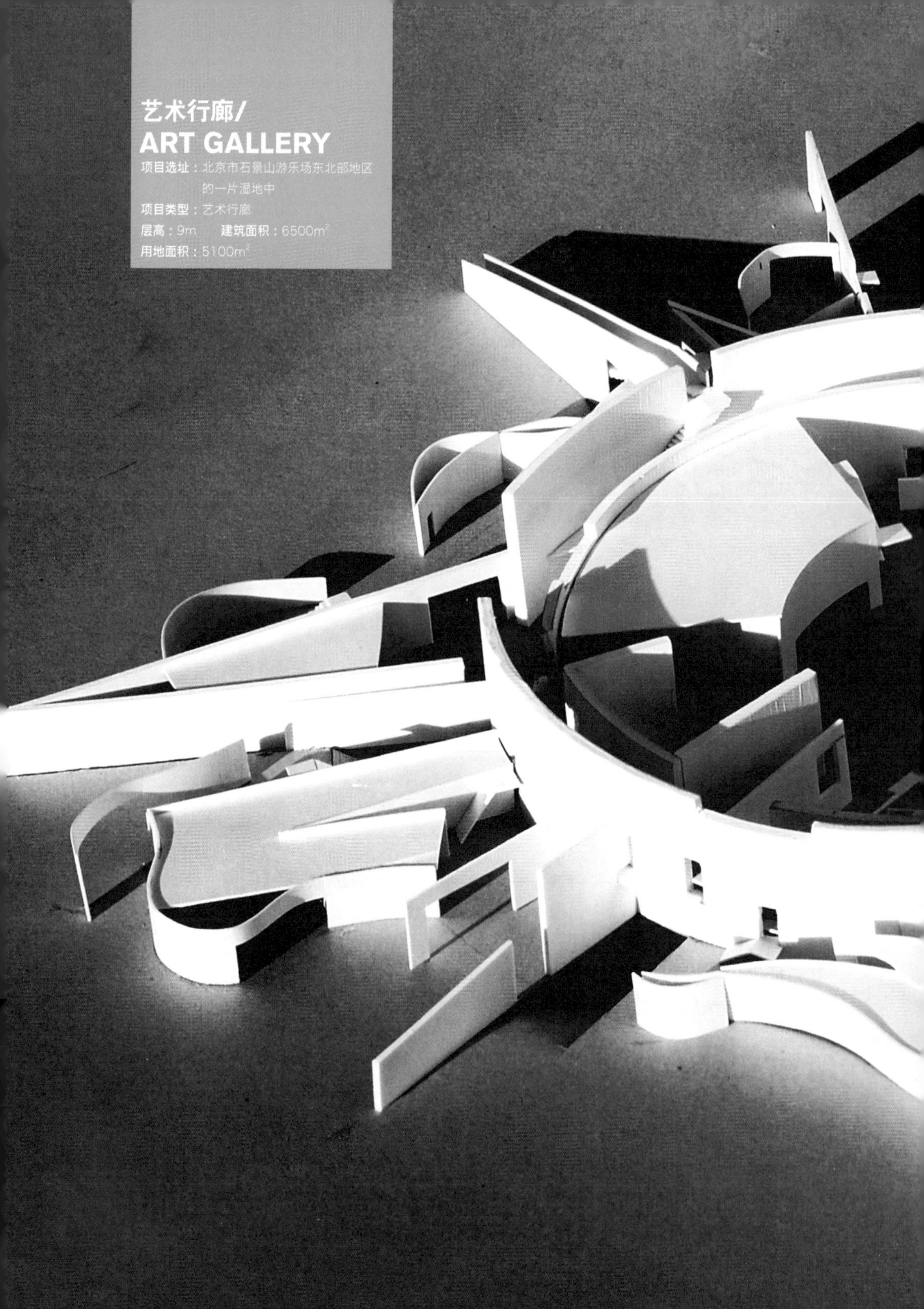

方案设计：丁惟迟
指导教师：王昀
完成时间：2015

开篇：模型。**本页图：**概念。**对页上图：**功能分区图。**对页下图：**轴测图。

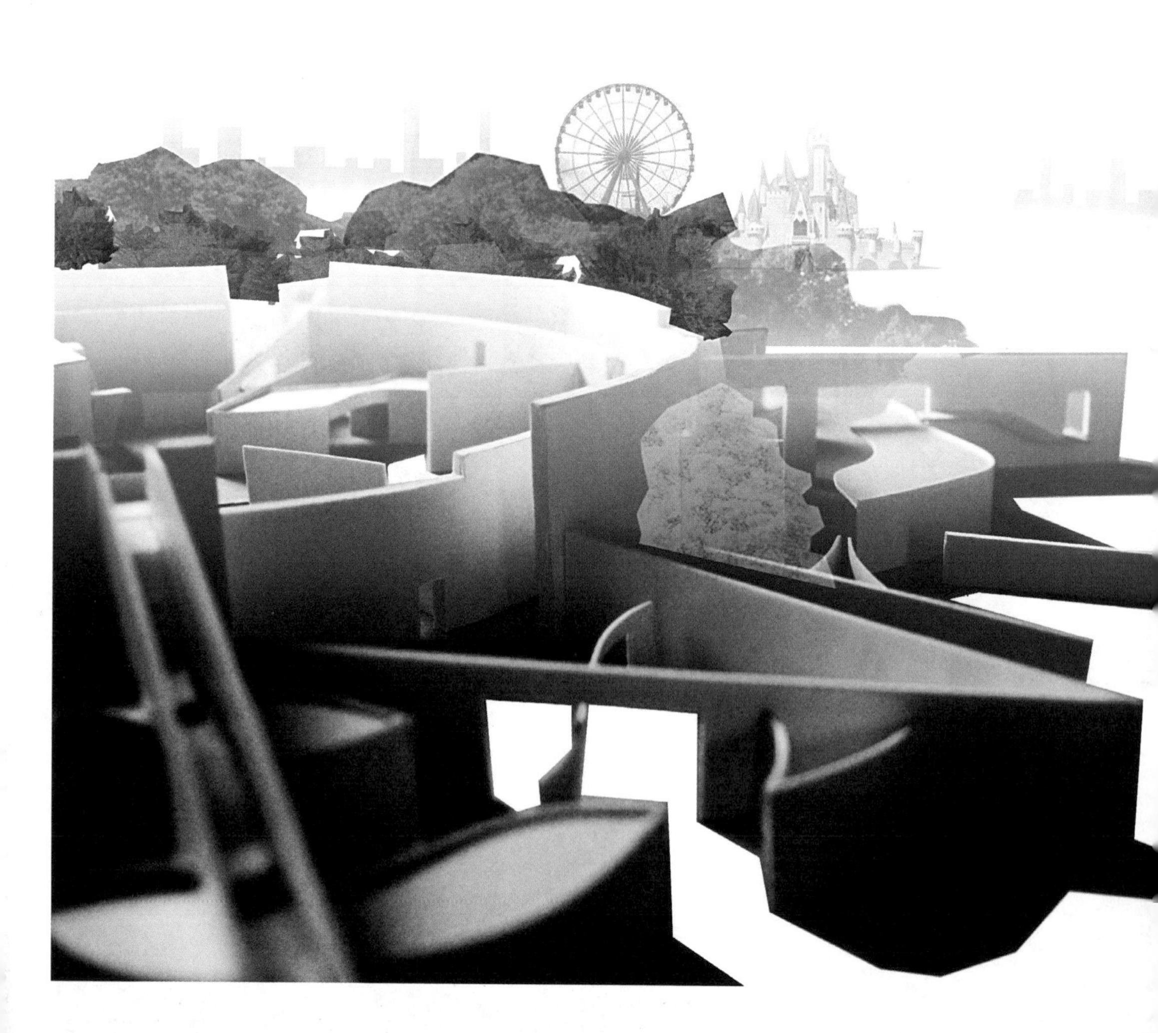

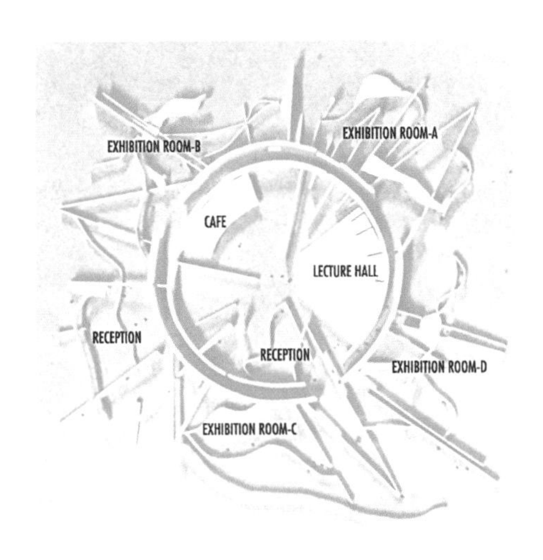

该地段位于石景山游乐园附近，各色的游乐设施以及周围的建筑都为其环境增加了一丝不可诉说的游乐感觉，当得知石景山游乐园要拆除时，又有了一丝悲伤的感觉，希望向这个为我们童年回忆增色的地方致敬。

概念当中选取了钟表的概念，其中在一个循环的时间轴中穿行正好符合了追求的艺术走廊的气质，也与当地的场域气质不谋而合。

当一段时光成为永恒，它为人们带来的记忆是不可磨灭的，而钟表带来的气质也正是一种无尽的永恒留念，在其中循环，感慨人间变化，世事无常。

而这座行廊矗立在那里，又像是破碎的摩天轮轰然倒塌，平拍在地上，以破碎而崭新的姿态迎接新的游人。

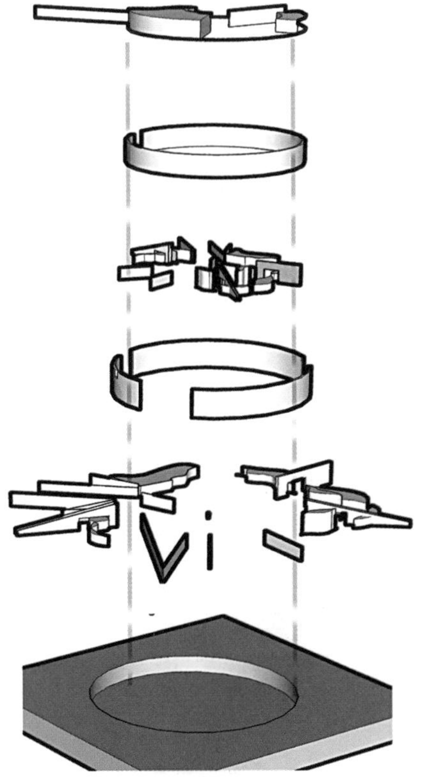

内圈的区域被核心空间所围，多为报告厅，图书广场，花园等公共性强的区域。

内墙，厚600，将公共性质的空间包裹其中，并有无数小“孔”使人在空中穿梭。

核心通道，为时间轴的主线，与内外圈与地下空间相连。有各种楼梯与通道穿过。

外墙，厚600，将展览性质的空间隔在外面，并有无数小“孔”和地下楼梯使人在空中穿梭。

展览空间，在外墙之外，有丰富的游线组织，并有楼梯，地下通道等多种手段进入内部。

地下空间，有餐厅和停车场以及办公区域，组成，是增加深厚感的产物也为地上的交通流线提供了缓冲的机会。

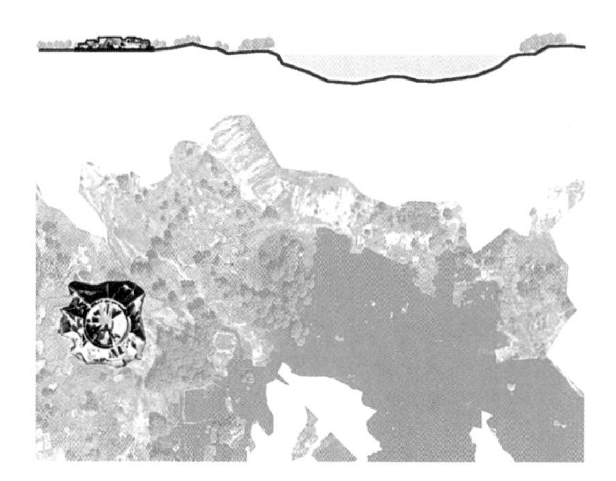

本页上图：总平面图。**本页下图**：中间的通道。**跨页**：内部广场。**对页上图**：剖面图。

The site is near the Shijingshan Amusement Park. Various recreation facilities and surrounding buildings add a relaxing atmosphere to the environment. When I learned that the Shijingshan Amusement Park would be dismantled, I felt a bit of sadness and paid tribute to the place that added happiness to the childhood memories.
The clock concept was selected. Passing in a cycling time axis is exactly in line with the pursuit of the art gallery characteristic and also coincidence with the genius loci.
The memory is indelible when it becomes eternal. In the same way, the temperament of a clock is also an endless eternal memory. Its cycle makes people sigh with emotion of the changing and inconstant world.
The gallery stands there as a broken Ferris wheel collapses suddenly, smoothly patting on the ground, and welcomes new visitors with a broken and new attitude.

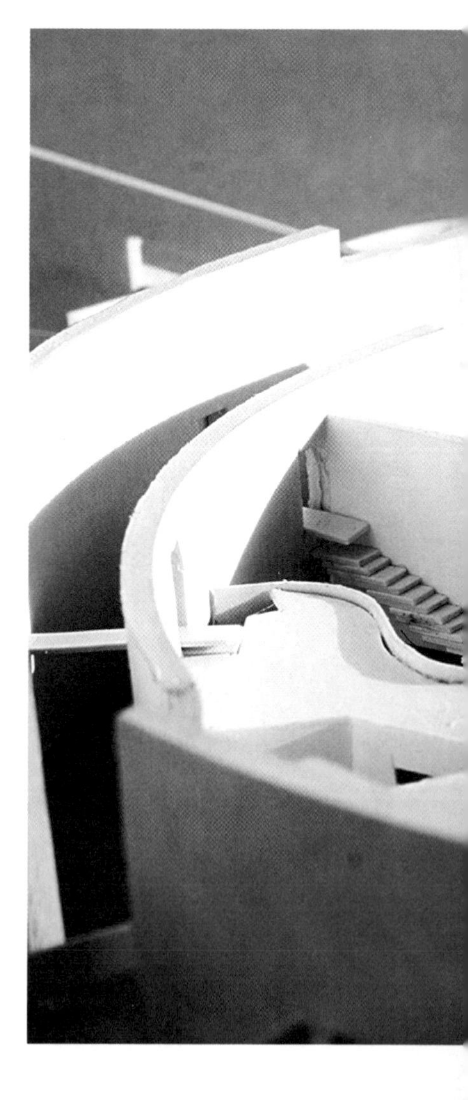

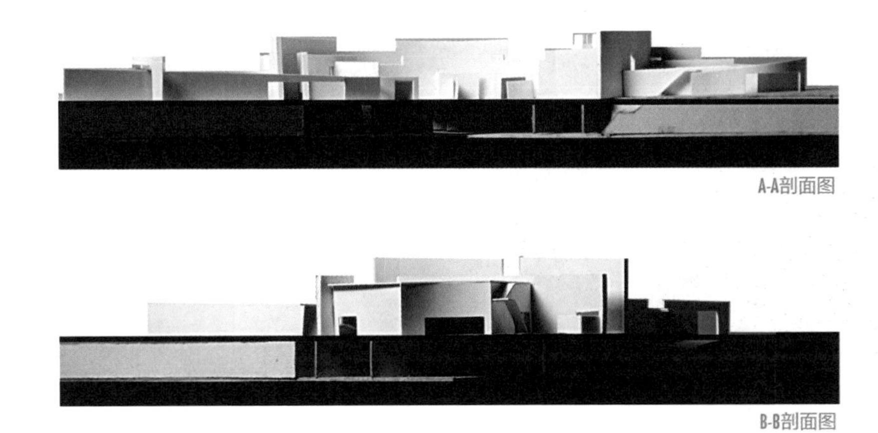

A-A剖面图

B-B剖面图

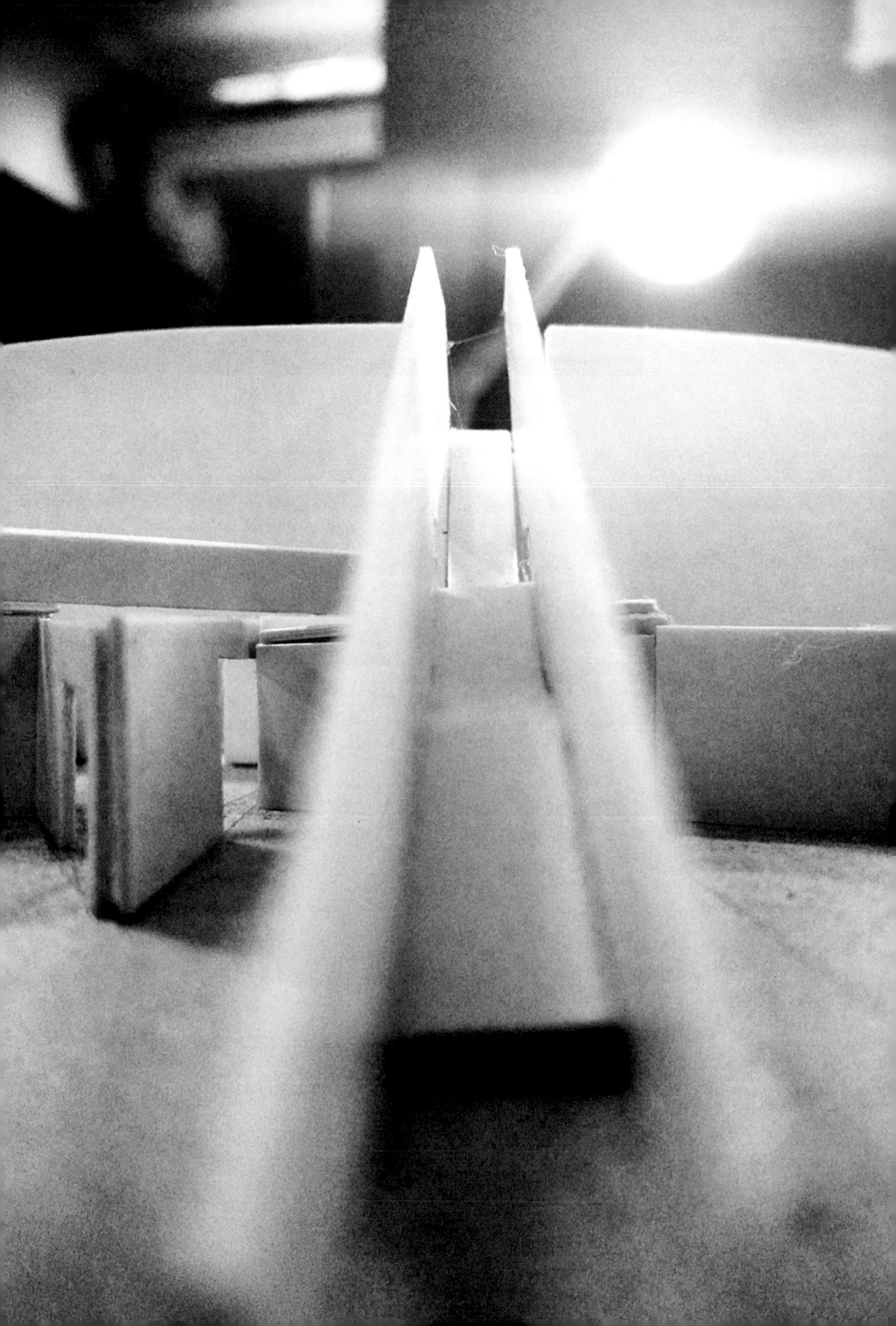

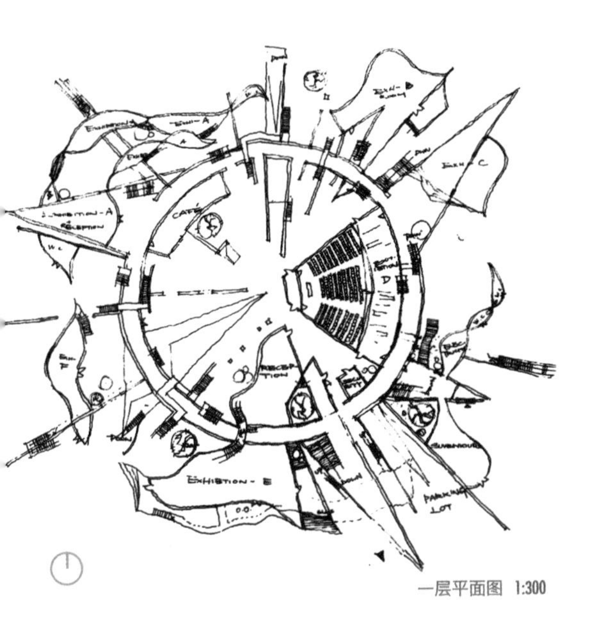

一层平面图 1:300

教师点评

丁惟迟同学的石景山游乐园纪念艺术行廊，空间形态富于乐感及魅力。向心性和模糊处理的建筑边缘与周边的环境获得了巧妙地协调。两道圆形围墙所形成的“街道”构成了内外空间过度的同时，有机地组合了功能。同时飘逸和碎片化的墙体消解了建筑的体量，形成了建筑的特征。

Teacher's comments

In the Art gallery of Shijingshan Amusement Park by Ding Weichi, the spatial form is full of musicality and charm. The centrality and the ambiguity of architectural edges provide the best coordination with the surrounding environment. The “street” formed by the two rounded enclosures not only constitutes the transition of inside and outside spaces, but also organically organizes the functions. Meanwhile, the elegant and fragmented walls lighten the building volumes, forming the features of the building.

对页: 三层的楼梯。**本页上图:** 平面图手图。**本页下图:** 外部入口。

建筑与音乐/ ARCHITECTURE AND MUSIC

项目选址：北京市 798 艺术区

项目类型：跨界艺术家村落——居住、创作、展览

建筑面积：6,931m² **用地面积：**19,972m²

占地面积：7,779m² **容积率：**0.35

方案设计：叶雪粲
指导教师：王昀
完成时间：2015

开篇：主要方案图。**本页：**概念分析图。**对页：**主要方案图。

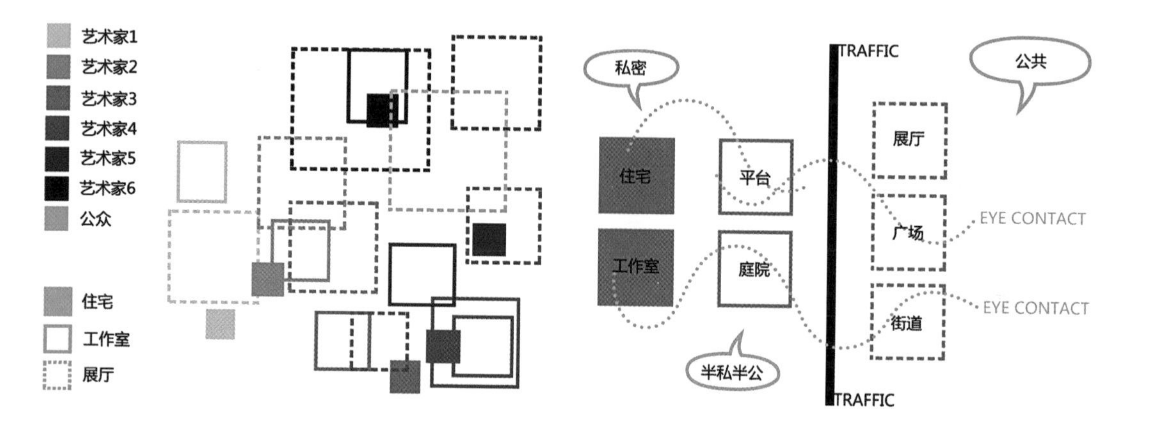

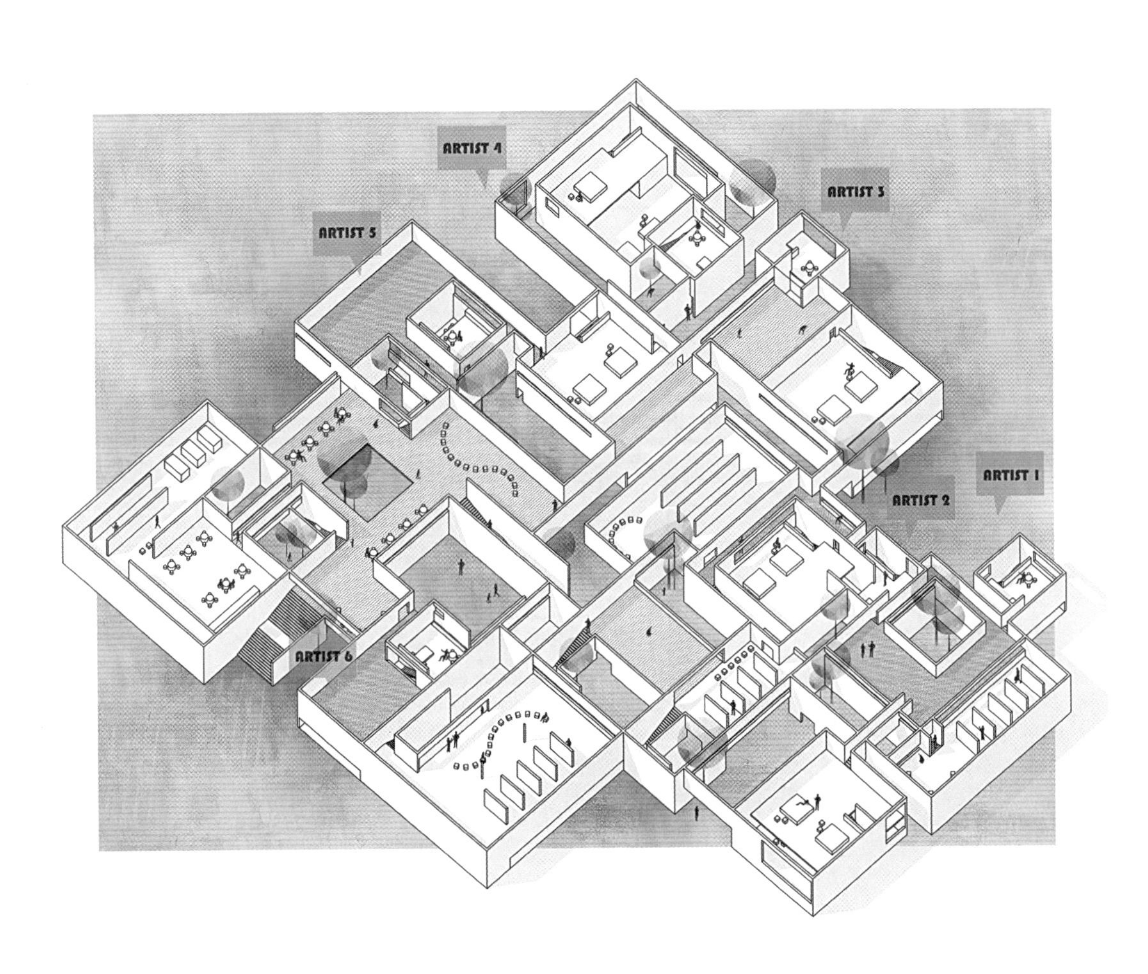

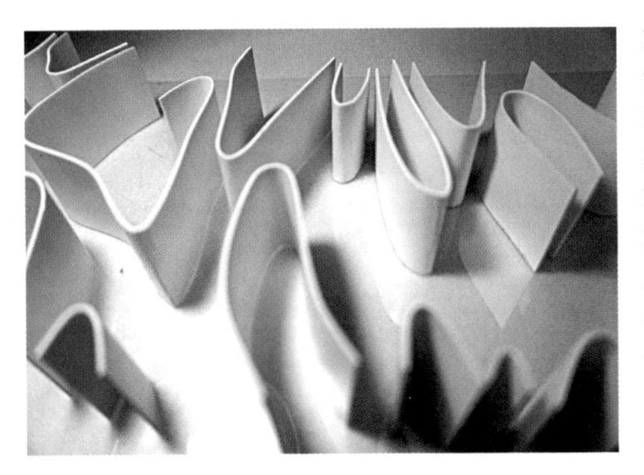

建筑一定是“功能决定形式”吗？能否是“形式启发功能”？……建筑学习的过程中，我们不断被灌输了一些好像必须去遵守的规则，但最终是需要我们有勇气去打破的。这一次，对于建筑空间，我们以一种“无为”的状态，尝试去“发现”既有的，而非野心勃勃地想着“生造”出些什么。我们一开始将功能、场地、文化、经济、结构考虑都架空，纯粹地去世界里“发现”形式，再以建筑师的眼与手将其转译为建筑空间，在“读解”形式以后，赋予其功能与结构等因素。我从“音乐”领域出发，从一张乐谱中转译出空间，在保证建筑学“本质要求”的前提下，尽可能保存原谱的客观秩序美，得到了这样一个跨界艺术家聚落。

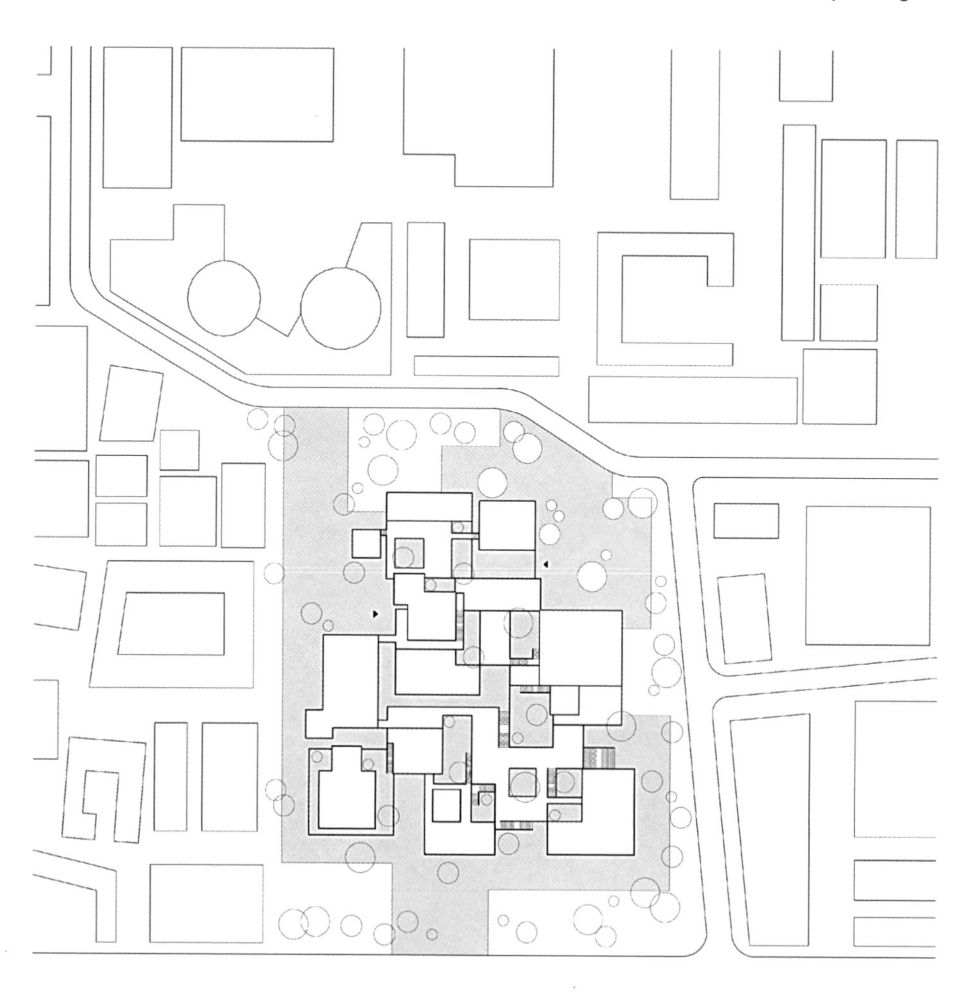

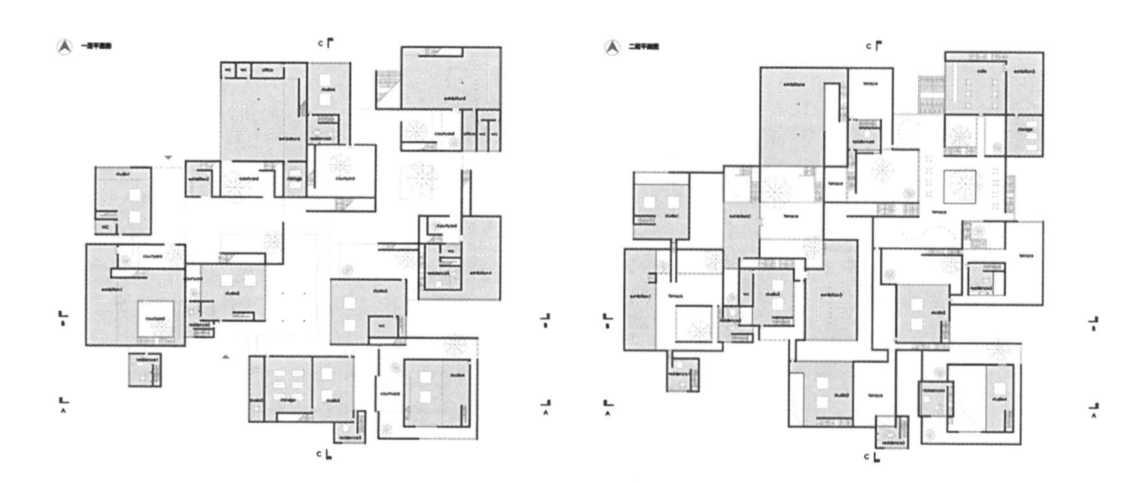
一层平面图
二层平面图

Does form certainly follows function in architecture? Can it be "form decides function"? ... In the course of architecture learning, we are constantly being infused with rules to be obeyed, but in the end, we need the courage to break them. As for space, with a kind of state of "inaction", we tried to "find" the existing things rather than "coin" things with ambition. We considered none about the function, space, culture, economy and structure, but purely "found" the form in the world, and then translated it into the construction space with the architect's eyes and hands. After the "interpretation" of form, it is given its function, structure and other factors. I started from the field of "music" to translate the space from a piece of music score, saved the objectivity, order and beauty of original music score on the premise of ensuring the "essential requirements" of architecture, thus obtaining such a boundary-crossing artist settlement.

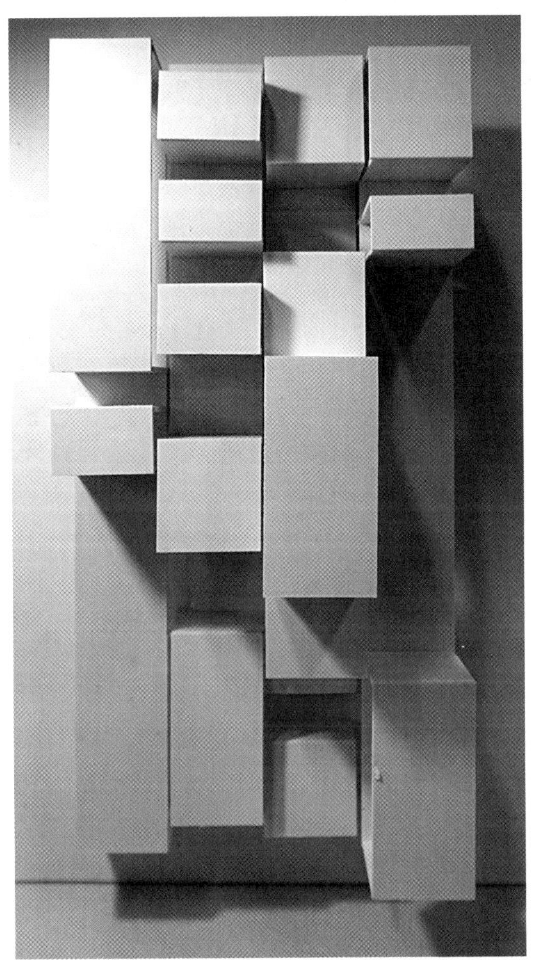

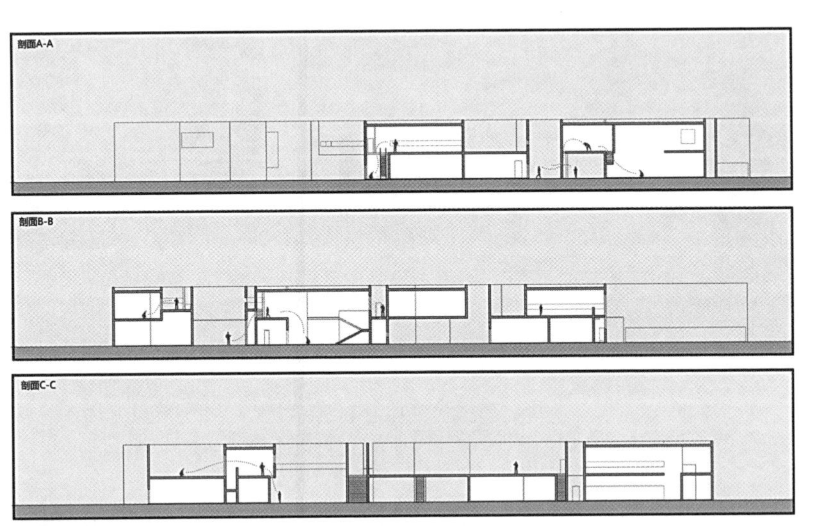

对页上图：总平面图。**对页下图：**平面图。**本页上图：**模型照片。**本页下图：**剖面图。

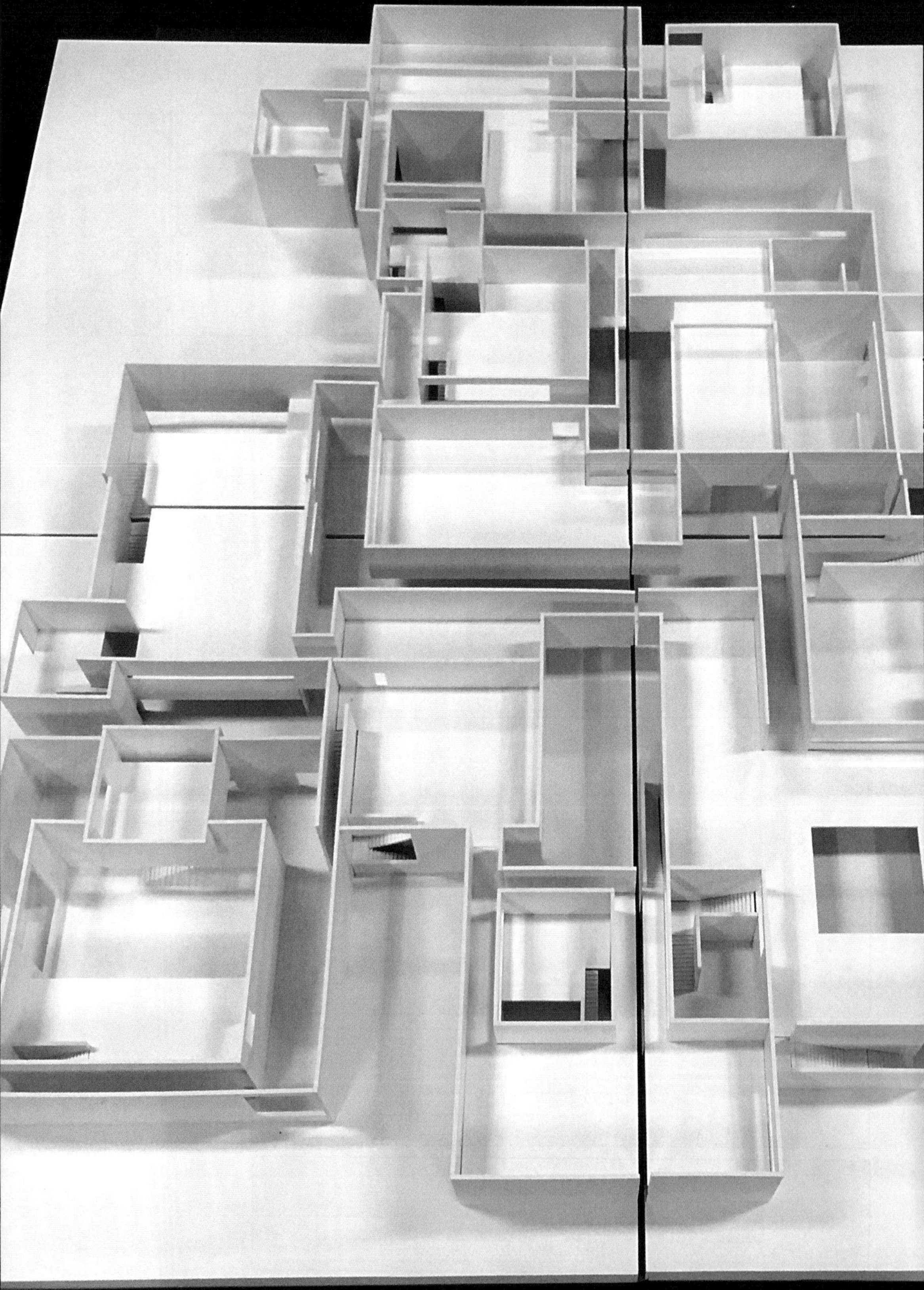

教师点评

叶雪粲同学的艺术家工坊将空间研究过程中所发现的空间“交集”特征进行抽取、放大，构成了富于聚落感的艺术家村落。建筑整体使用了彼此相连、跨越、视线交汇、阶路连接等运作手段，触发了群落整体的迷宫特质，简洁的处理给建筑带来了清醇的感觉。

Teacher's comments

Artist Workshop by Ye Xuecan extracts and magnifies spatial "intersection" features discovered in the process of space research, constituting the artist village full of sense of settlement. The integral building adopts the operating methods of mutual connection, crossing, sight intersection, step connection, etc., and triggers the maze traits of the whole community. The simple processing endows the building with a mellow feeling.

跨页：模型照片。**本页下图：**模型照片局部。

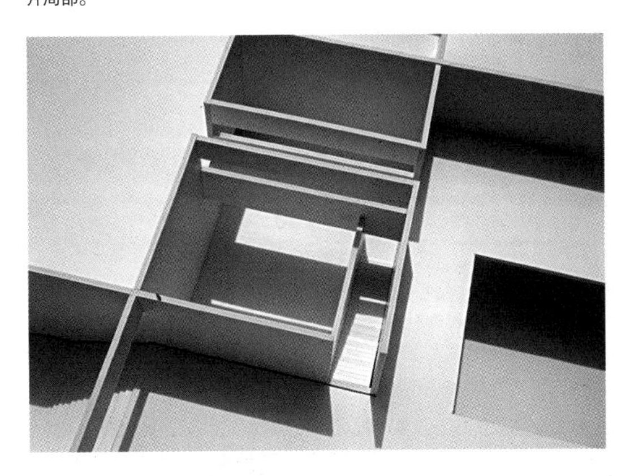

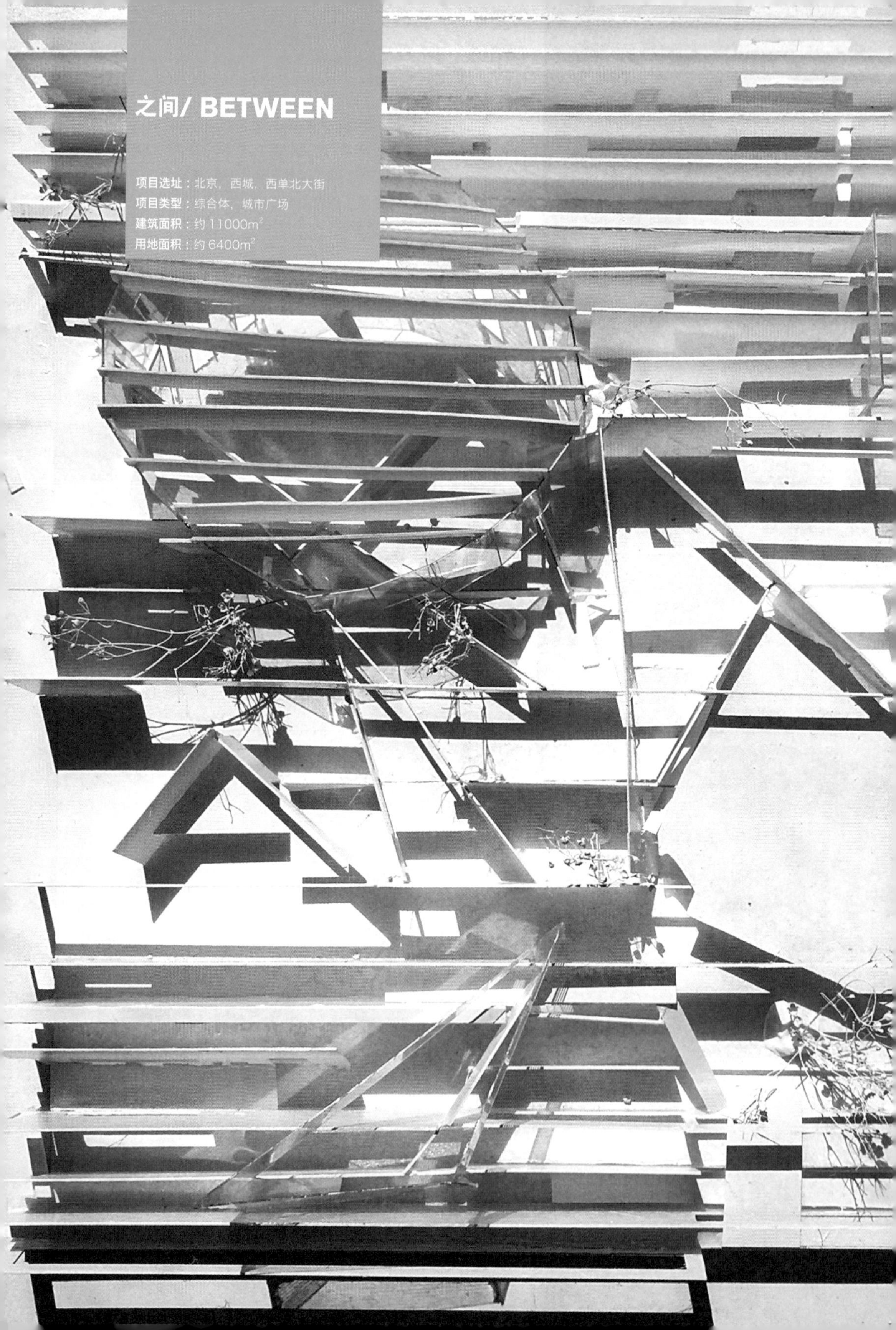

之间/ BETWEEN

项目选址： 北京，西城，西单北大街

项目类型： 综合体，城市广场

建筑面积： 约 11000m^2

用地面积： 约 6400m^2

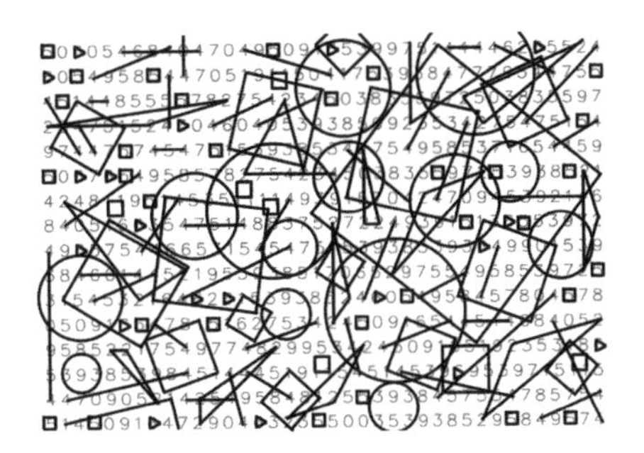

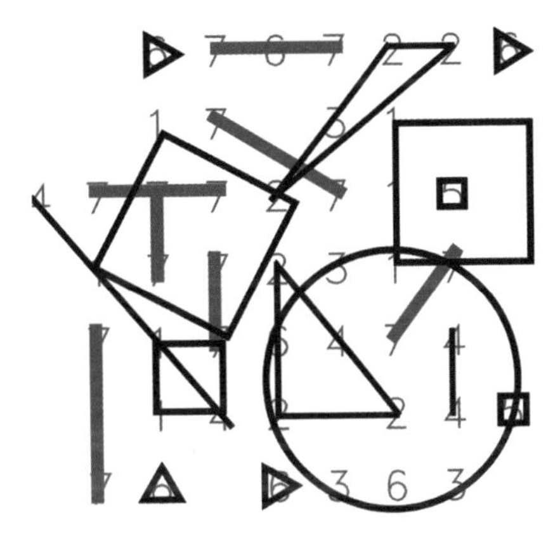

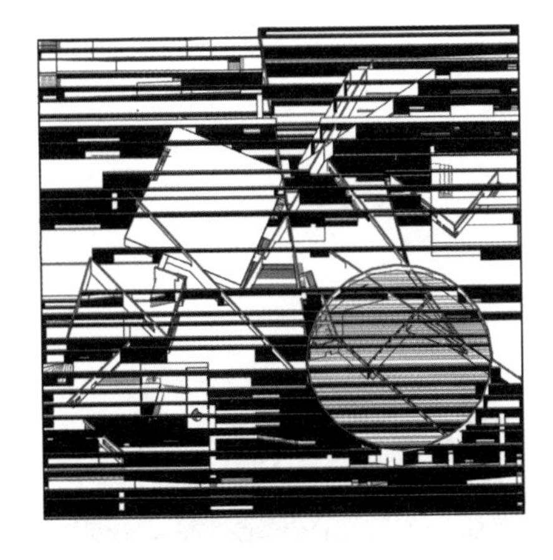

方案设计：陈梓瑜
指导教师：王昀
完成时间：2015

开篇：模型整体。**本页上图：**平面图。**本页下图：**剖面图。**对页：**主要方案图。

底层平面图

西单北大街自20世纪80年代开始的商业规划和改造持续至今，形成了大尺度的商业立面与小尺度的胡同民居相互割裂却又共生的现状。购物游览的来访者还是周边的居民对相对有质量的休憩和聚集的空间的需求没能在片区中得到满足，同时考虑到附近不乏文教区域但缺乏文化艺术的中心，选择综合小型艺术商店、阅览室、画廊等功能，将西单北大街仅存的一片处于尺度交界上的空地改造为边界开敞的城市广场。

方案利用平行墙体的形式在沿干道的立面上以完整的界面协调周边商业立面的尺度，而面向次级干道以小尺度的间距呈现出通透开敞的印象。墙间的尺度呼应周边逐渐被拆除消失的胡同，希望被一个完整的形象接纳，同时唤起记忆和熟悉感。人们在墙面间自由但清晰的穿行中，体验空间并与他人相遇。

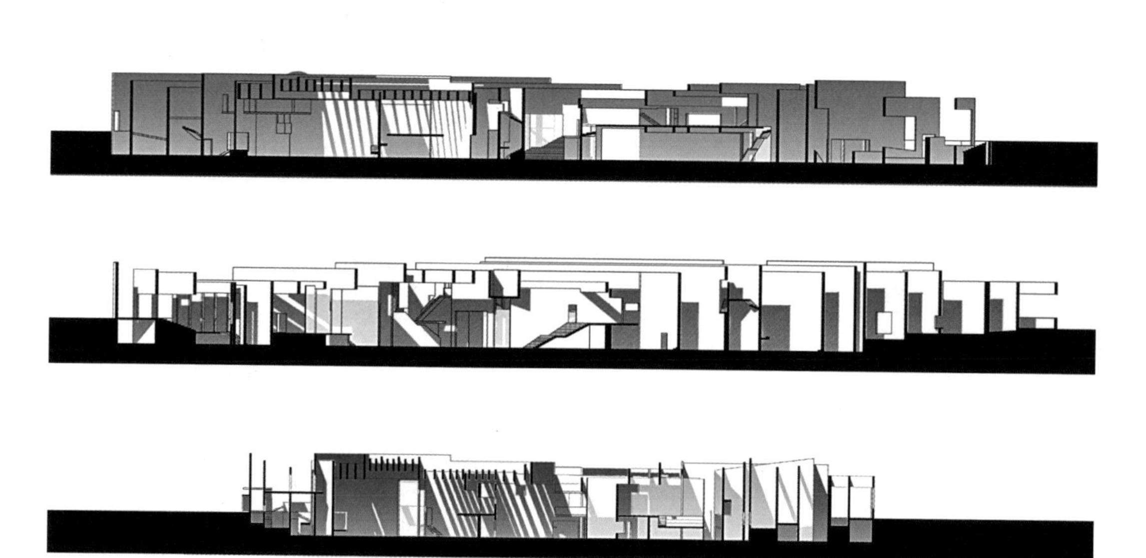

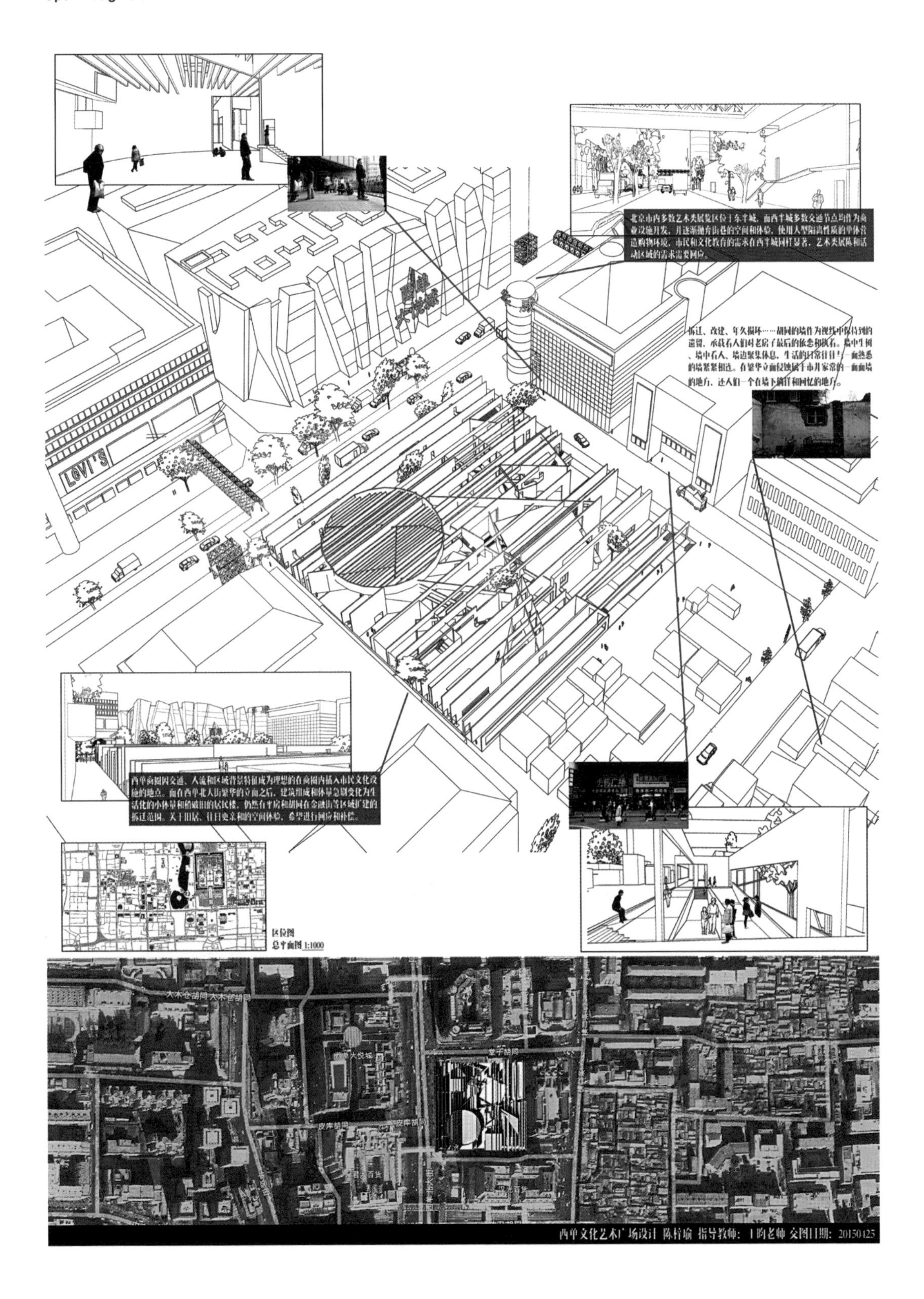
北京市内多数艺术类展览区位于东半城，而西半城多数交通节点均作为商业设施开发，并逐渐抛弃街巷的空间和体验，使用大型隔离性质的单体营造购物环境。市民和文化教育的需求在西半城同样显著，艺术类展陈和活动区域的需求需要回应。
拆迁、改建、年久损坏……胡同的墙作为视线中保持到的遗留，承载着人们对老房子最后的依恋和执着。墙中生树、墙中看人、墙边聚集休息，生活的日常往往与一面熟悉的墙紧紧相连。在繁华立面侵蚀属于市井家常的一面面墙的地方，还人们一个在墙下徜徉和回忆的地方。
LEVI'S
西单大悦城
西单商圈因交通、人流和区域背景特征成为理想的在商圈内插入市民文化设施的地点。而在西单北大街繁华的立面之后，建筑组成和体量急剧变化为生活化的小体量和稍破旧的居民楼，仍然有平房和胡同在金融街等区域扩建的拆迁范围。关于旧居、往日更亲和的空间体验，希望进行回应和补偿。
区位图
总平面图 1:1000
大木仓胡同
西单大悦城
皮库胡同
君太百货
汉光百货
西单文化艺术广场设计 陈梓瑜 指导教师：王昀老师 交图日期：20150425

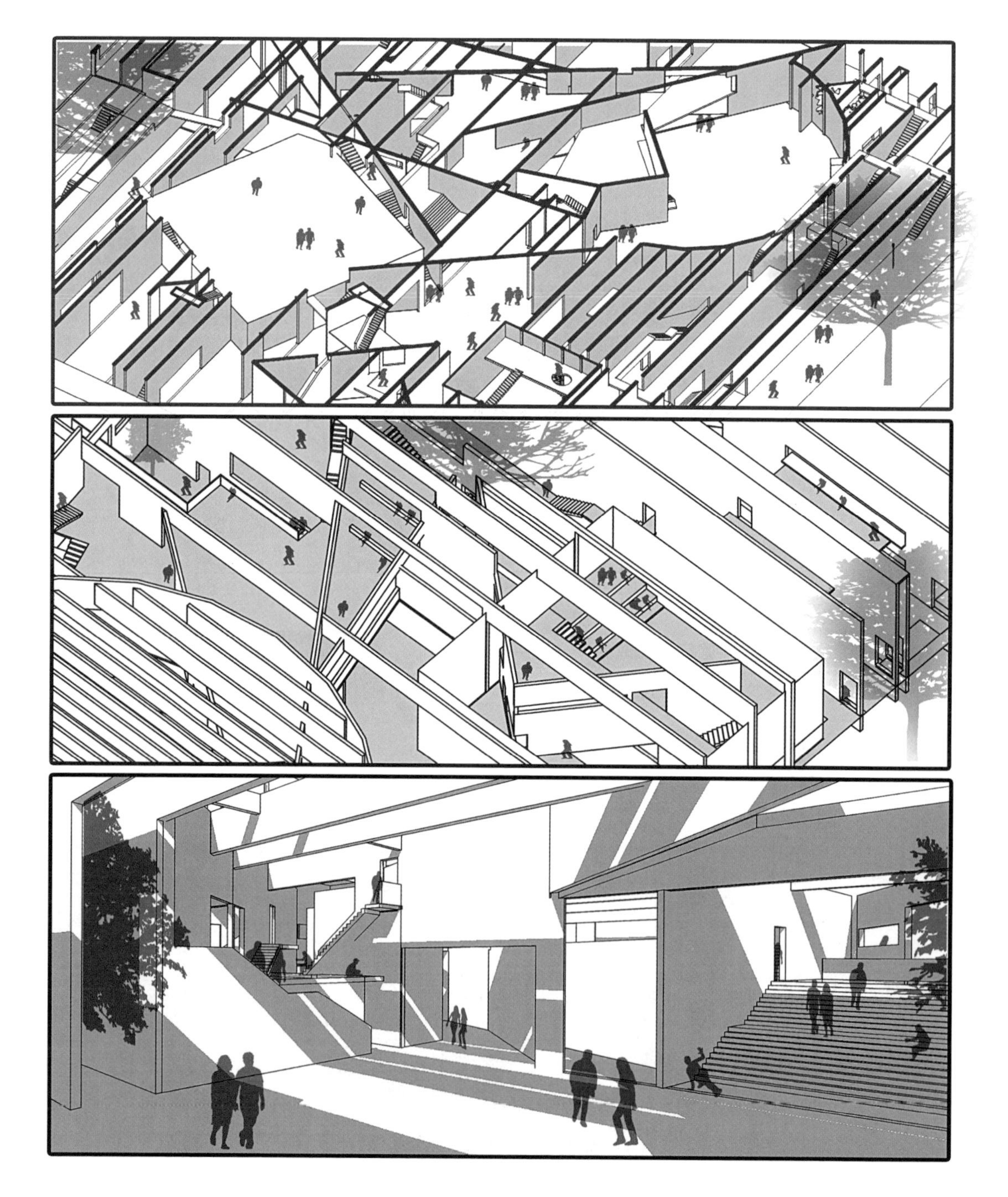

The business planning and transformation of Xidan North Street have existed till now since 1980s, which formed the status of symbiosis of fragmented space of large scale commercial elevation and closed commercial facilities along the street and the small scale alley houses behind. The demands of shopping visitors and the surrounding residents for the relative qualified rest and gathering space could not be satisfied in the area. In addition, though there are some cultural and educational areas, there is a lack of culture and art centers nearby. Taking the fact into account, I intended to transform the only space on the boundary of Xidan North Street into an open city square through various small art shops, reading rooms, galleries, etc.

The entire surrounding commercial elevation is adjusted in scale with facade along the main line through the form of parallel wall. However, for secondary roads, the small-scale space presents an open and transparent impression. Spaces between the walls echo the scale of the surrounding alleys that are gradually removed, hoping that they can arouse the memory and familiarity by the complete image. People use, experience and meet in free walk between the walls.

本页及对页：场景图。

教师点评

陈梓瑜同学的西单文化艺术广场对于当下的城市空间进行再度的加密处理，空间的层级叠加与异质形体的导入形成了该设计的空间语言特征，层级的切分式形态与叠加造成一种拼贴感觉。表达出分节的城市特征。

Teacher's comments

Xidan Culture & Art Square by Chen Ziyu suggests a further densification for the current urban space. The overlap of spatial scale and the introduction of heterogeneous form produce spatial language features of the design. The segmented form and overlapping of level of the space cause a sense of collage, expressing the segmental urban characteristics.

跨页及本页：透视图。

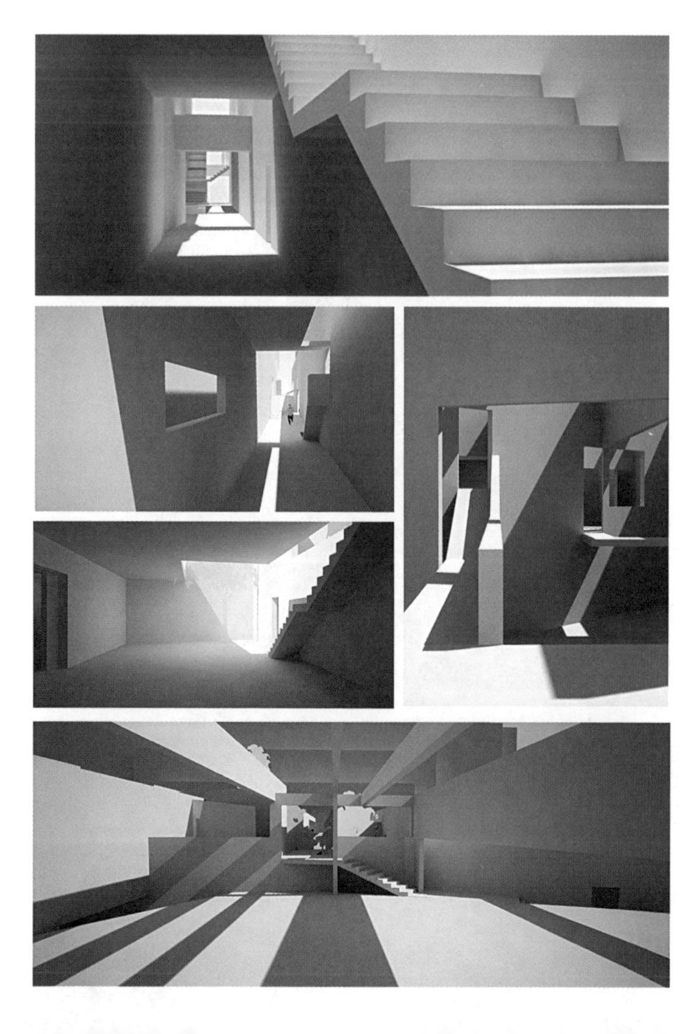

未来工作室1-9号/ 9 HOUSES

项目选址： 雨儿胡同，北京市

项目类型： 艺术家工作室

建筑面积： $1338m^2$

用地面积： $841m^2$

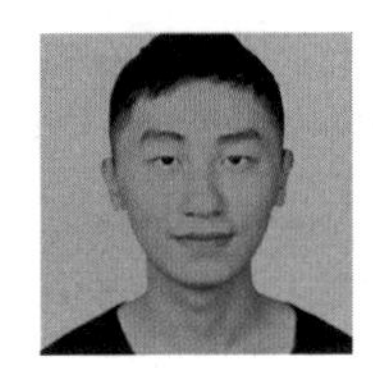

方案设计： 谢志乐
指导教师： 王昀
完成时间： 2015

本设计从一种常见的单元模式出发，通过一定程度的扭转和随意线条的干扰，让原本单调的空间呈现一种互相咬合的状态。与此同时，原本复制并列的单元将由此变得各异，进出方式、平台位置以及自然光的形象等都随之改变。彼此不同的单元相互契合，为“个人”与“他”的转换提供了条件。更进一步，在创造了单元之间特异性的同时，希望能保留其原本可以任意组合的特征。于是产生了一些不同条件下的组合形式，启发了更多的使用可能。

开篇： 主要方案图。**本页下图（从上至下）：** 生成分析。功能分析。**对页上图：** 总平面图。**对页下图（从左至右）：** 一层平面图。二层平面图。

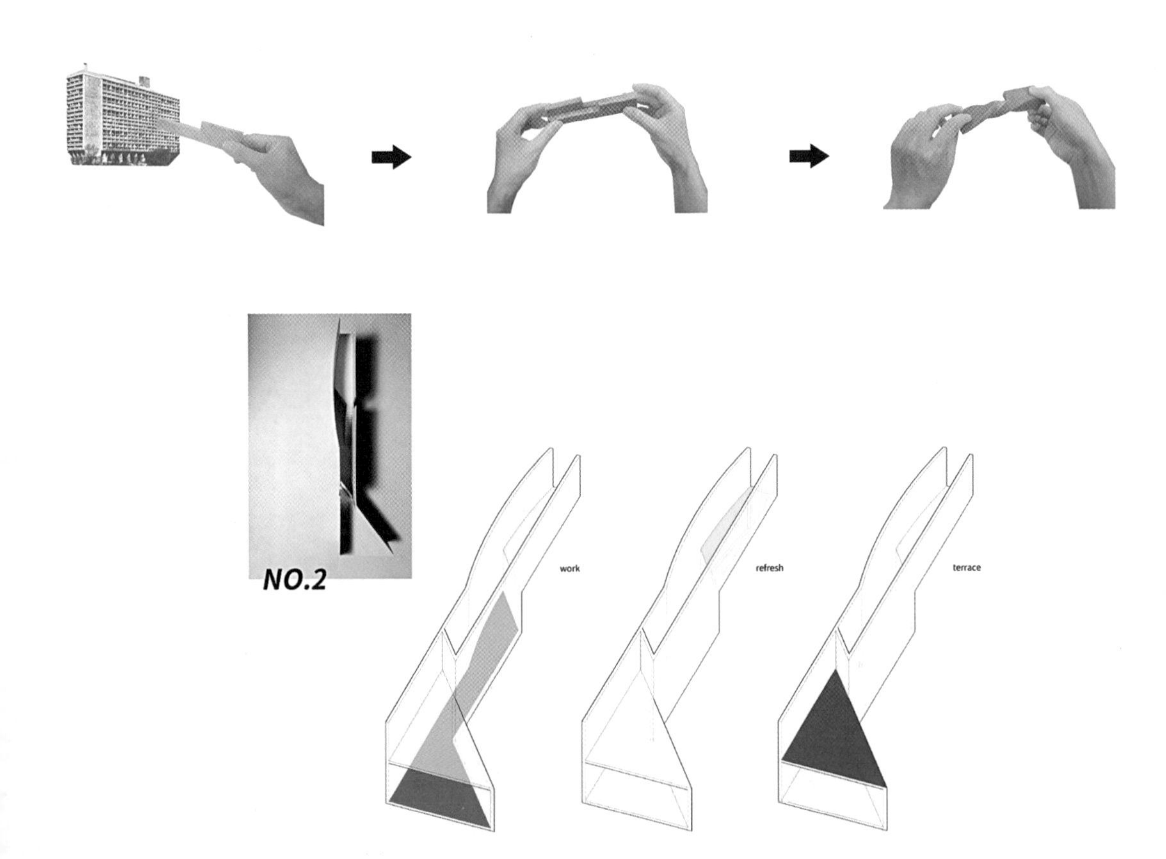

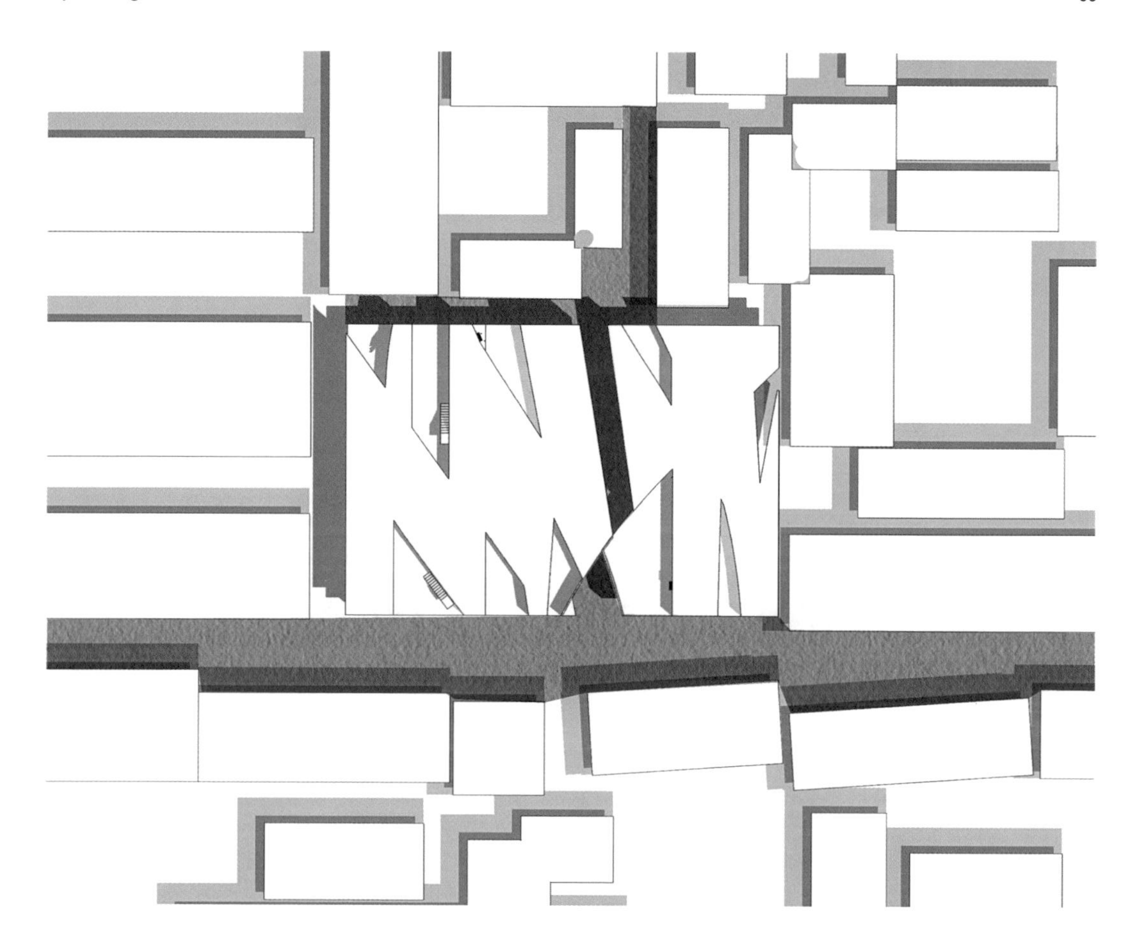

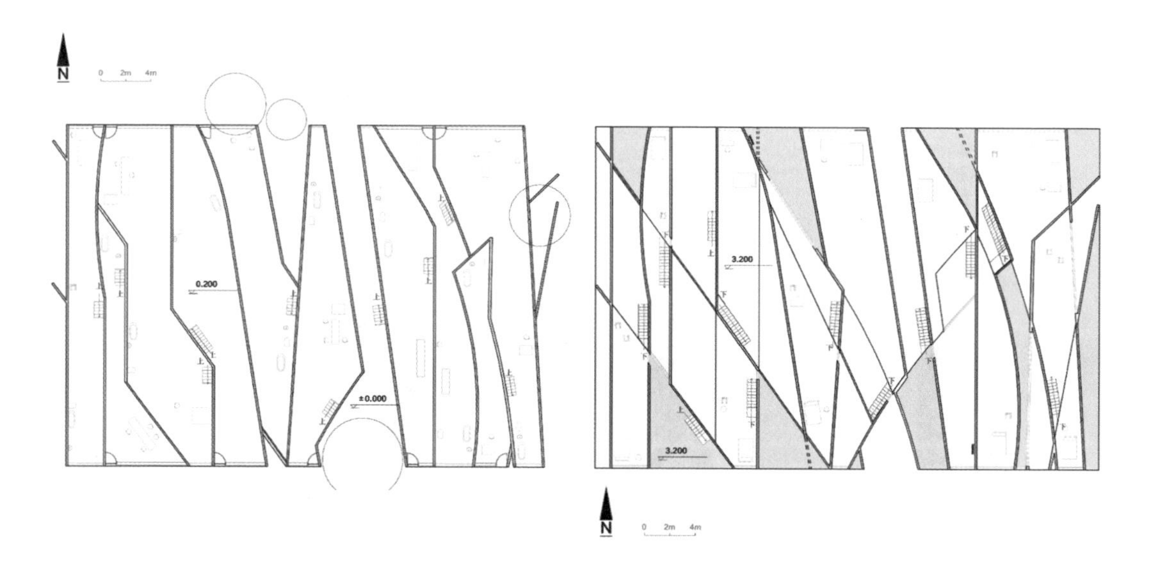
N
0 2m 4m
0.200
±0.000
3.200
3.200
N
0 2m 4m

This design, starting from a common unit model, presents a state of mutual integration from the original monotonous space through a certain degree of torsion and interference of random line. At the same time, the originally copied parallel units will thus become different, the in and out method, position of the platform and the image of natural light will be changed. Different units will fit each other, which provides the conditions for the conversion of the "individual" and "others". Further, while creating the specificity between the units, it's hoped that the features to be combined randomly can be retained. Thus the combination forms under a number of different conditions are produced, inspiring more use possibilities.

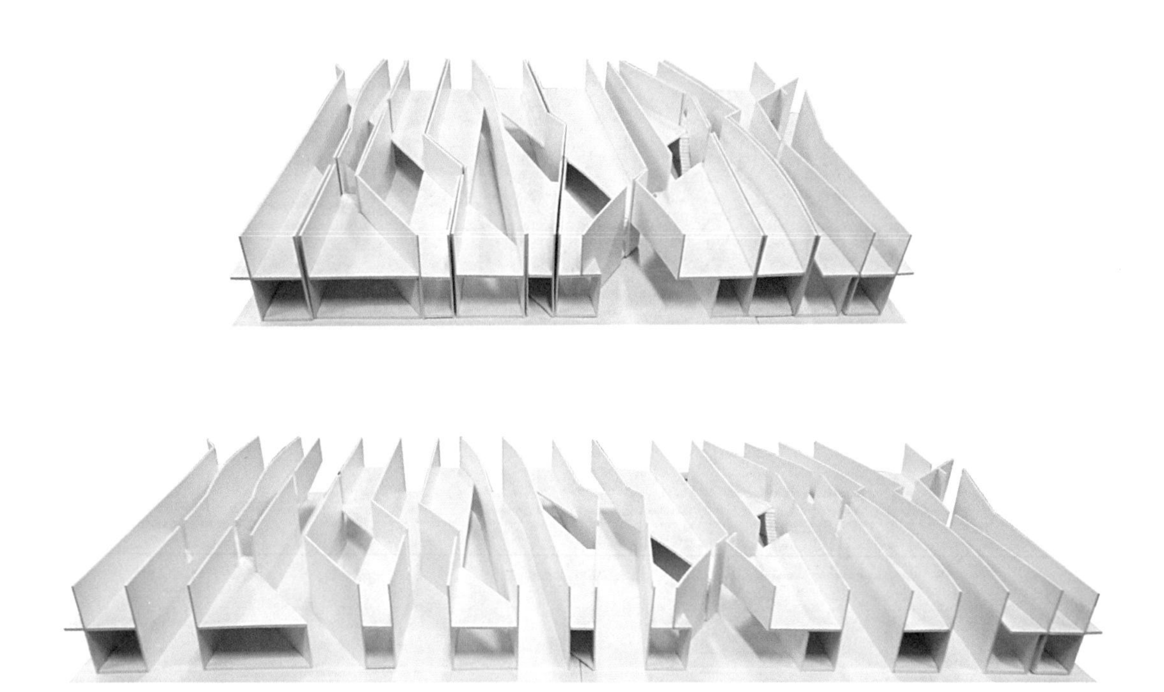

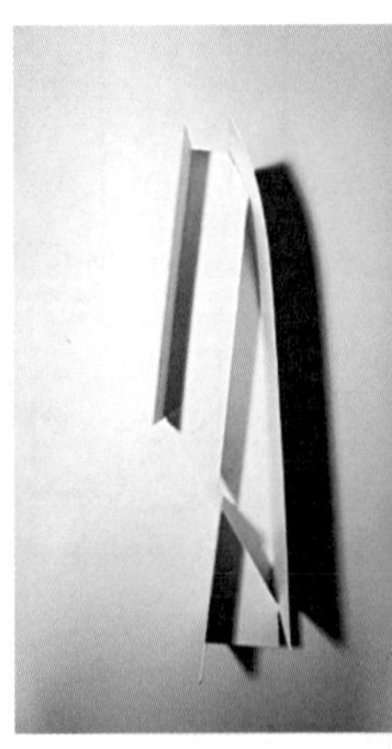

对页左图：模型拆解分析。**本页右图**：模型局部人视效果。**下图跨页**：空间单元模型。

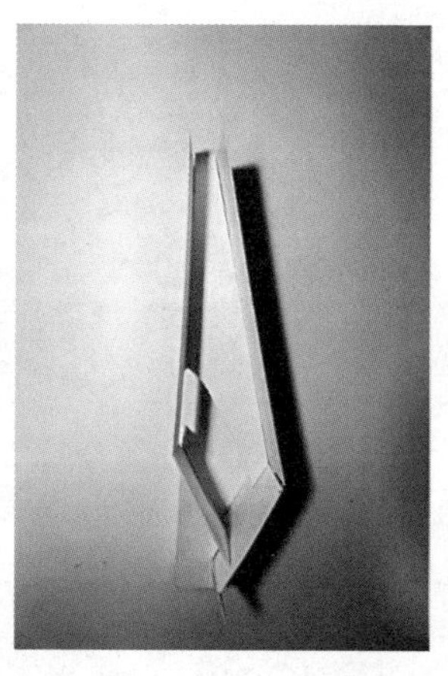

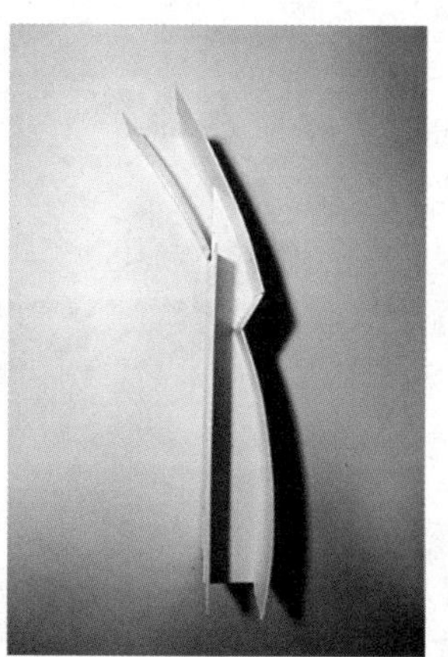

教师点评

谢志乐同学的设计从偶然中提取出空间构成的因子，通过一系列的组合形成了一系列可以被使用的空间形态。设计仔细地对从偶然中获得的空间进行读解，寻找出每一个个性空间所呈现的可以被使用的期待和特征，并赋予各自恰如其分的使用功能。可能性是该方案的特征。

Teacher's comments

Xie Zhile's design extracts factors of space composition accidentally, and forms a series of available spatial forms through a series of combination. The design interprets the space accidentally obtained, to find out possible assumptions and features shown by each individual space, and provides each with appropriate function. Possibility is the feature of this scheme.

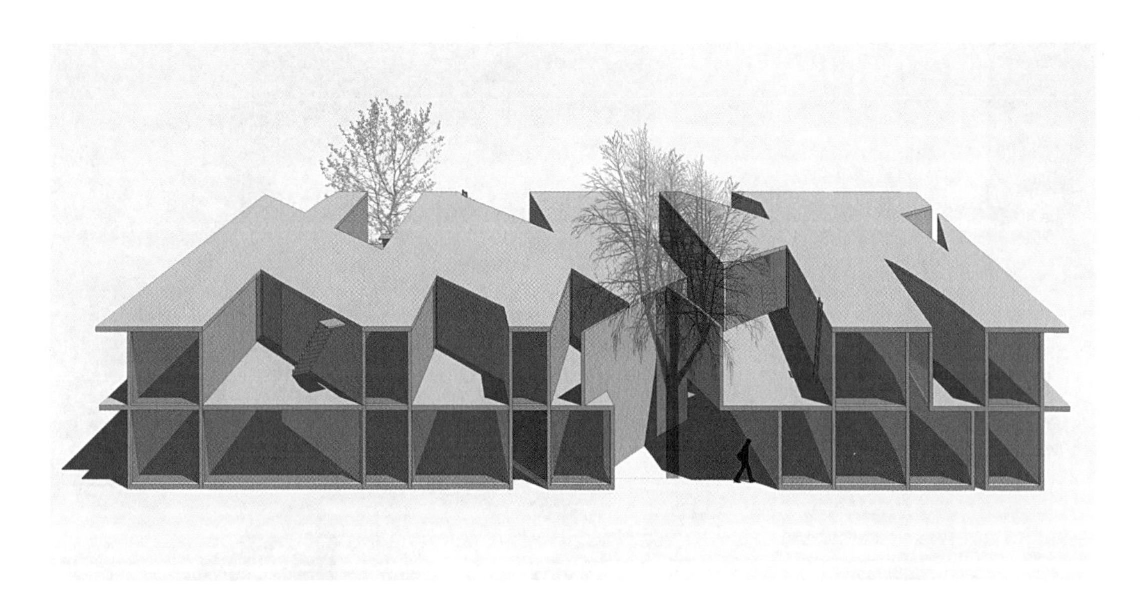

对页左上：模型室内光影效果。**跨页**：轴测意向图。**本页右上**：鸟瞰意向图；**本页右下**：模型局部人视效果。

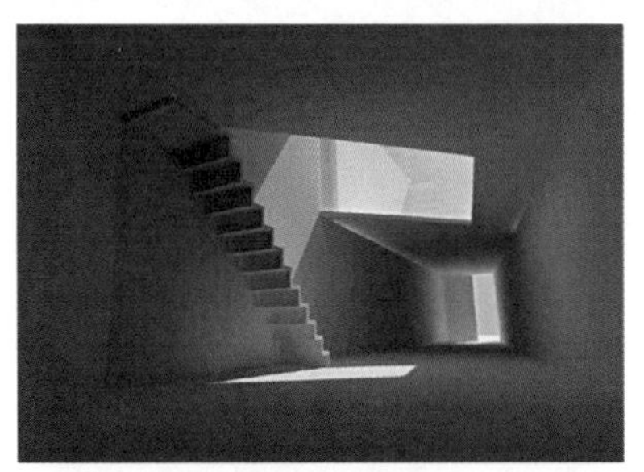

破碎/BROKEN

项目选址：畅春园公园
项目类型：大学生、居民和艺术家的开放聚落
建筑面积：$5623m^2$
用地面积：$8000m^2$

方案设计：杜光瑜
指导教师：王昀
完成时间：2015

项目位于与北京大学学生公寓有一墙之隔的畅春园公园内。传统而言，大学所象征的精英教育同普通百姓常有距离。而在新的时代里，大学应该以更开放的姿态融入社会。项目试图打破围墙，以形式破碎的缓冲空间连结居民公园和大学公寓，创造一个为居民、大学生和艺术家所共同享有的开放聚落，促进交流、理解和体悟，并希望以此为各个年龄段的人们提供共同成长的生命历程。

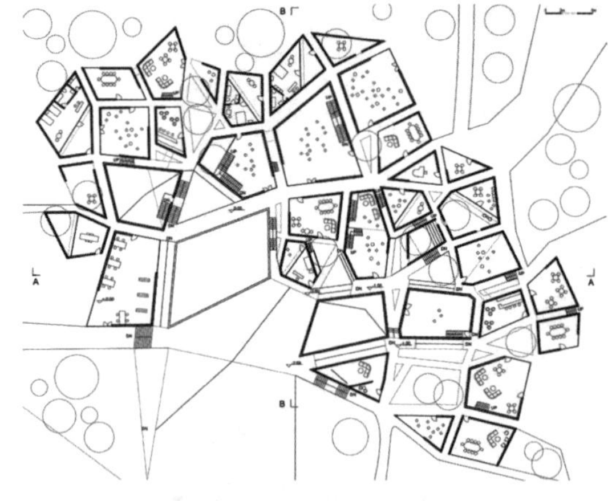

悬圃/ HANGING GARDEN

项目选址：浙江省舟山市普陀区东极镇庙子湖岛

项目类型：东极岛居民文化娱乐中心

建筑面积：4500m²

用地面积：8000m²

方案设计：杜京良

指导教师：王昀

完成时间：2015

通过对大地艺术的研究抽象出的曲线，借助带电粒子在磁场中的运动原理，反推出磁场，即生成了多样化复合性的空间。空间之间在保持一定的模块化的同时又产生了内在相互联系的韵律感，同时也对东极岛的山势做出了回应。

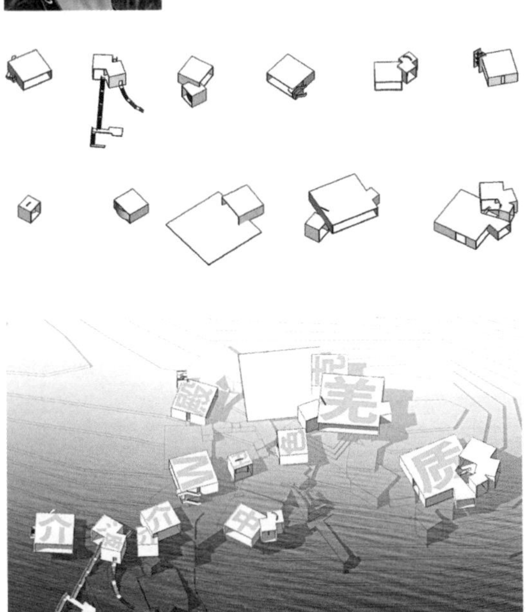

消解方格网/ DIGESTION GRID

项目选址：陕西省西安市雁塔南路
项目类型：市民活动中心
建筑面积：7850m²
用地面积：14000m²

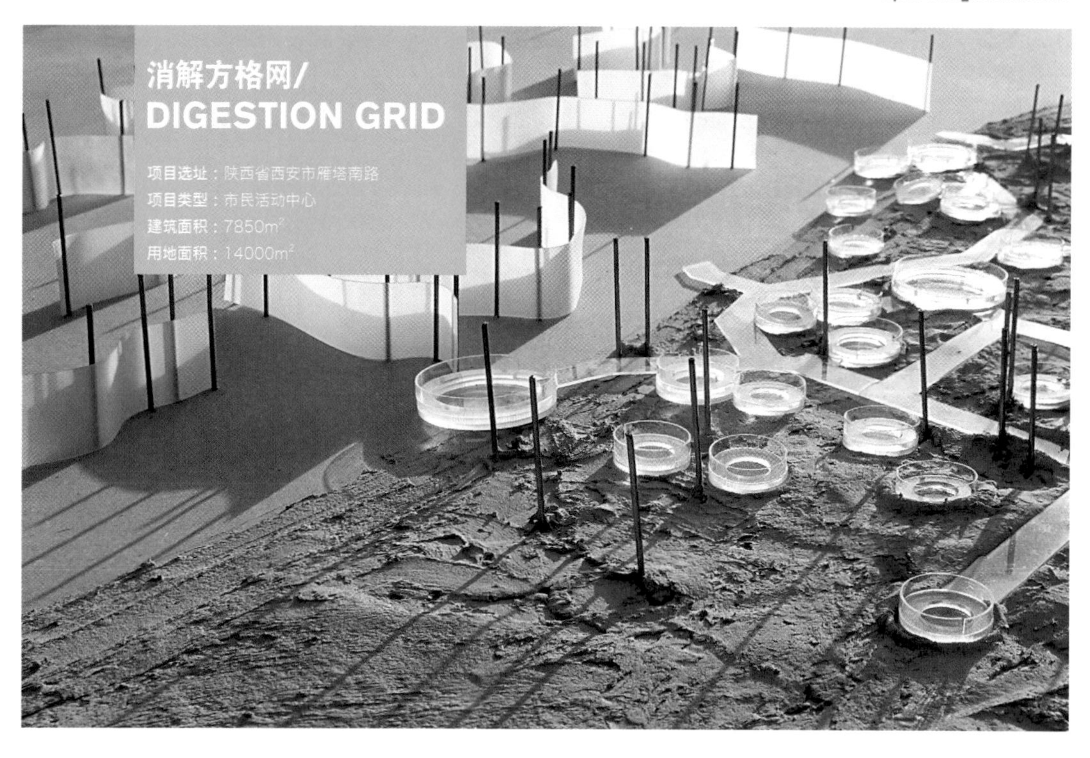

方案设计：高钧怡
指导教师：王昀
完成时间：2015

西安市具有严整的方格网格局，极具历史传统，但缺少一丝灵动。此设计在大雁塔轴线旁拟建一个市民综合活动中心，以自由的形式，引入阳光与植物，消解方格网；提供多功能活动场所，激活城市空间。

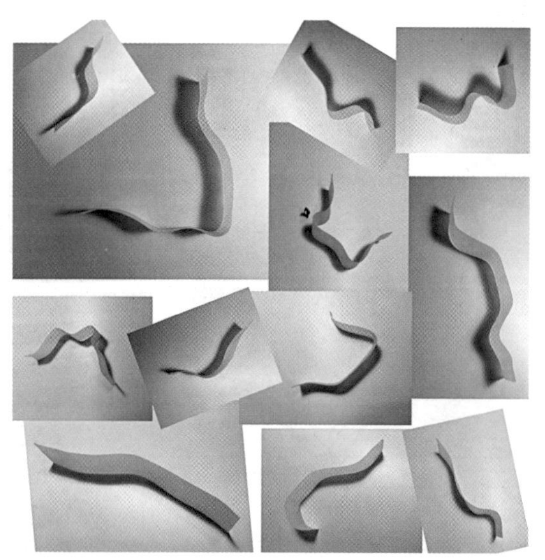

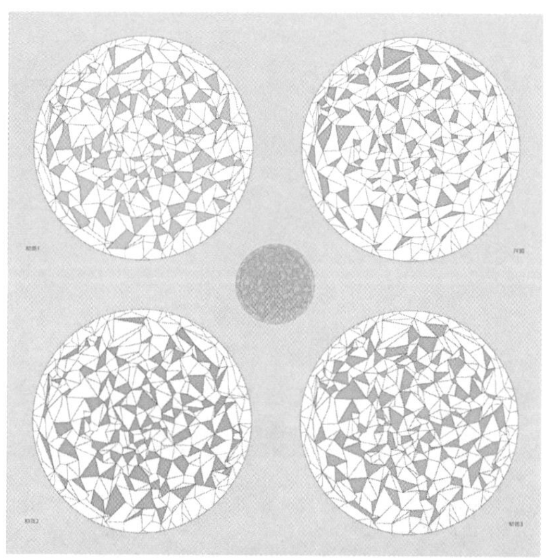

VORONOI村/ VORONOI VILLAGE

项目选址：清华大学紫荆操场南侧，紫荆路北
项目类型：复合式艺术家住宅及工作室
用地面积：4000m²

方案设计：侯兰清
指导教师：王昀
完成时间：2015

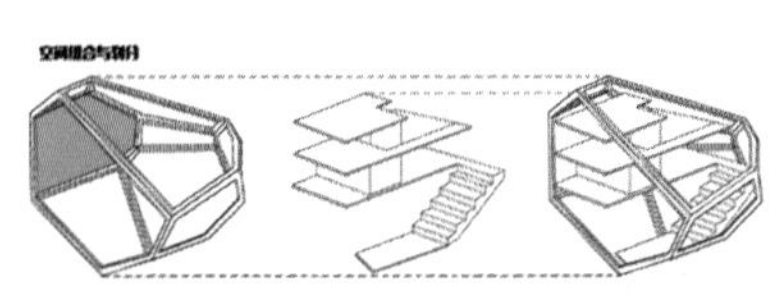

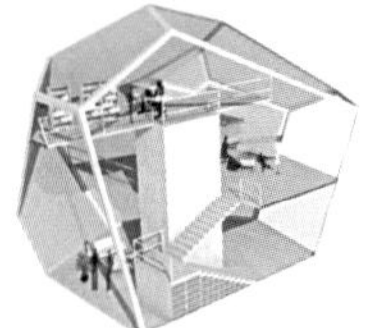

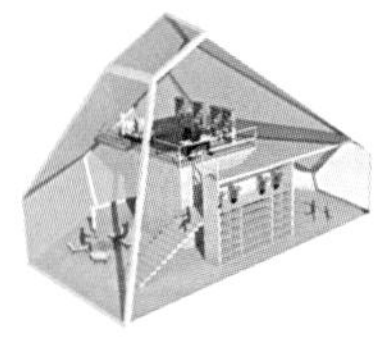

对聚落形态进行研究得到空间散点图；
以3D voronoi的算法划分空间，形成独特的空间单元；
保留单元独特形态，在水平向的楼板上进行处理，使之契合于空间状态，而在纵向上以楼梯进行联系，每个单元空间围绕中心体布置。

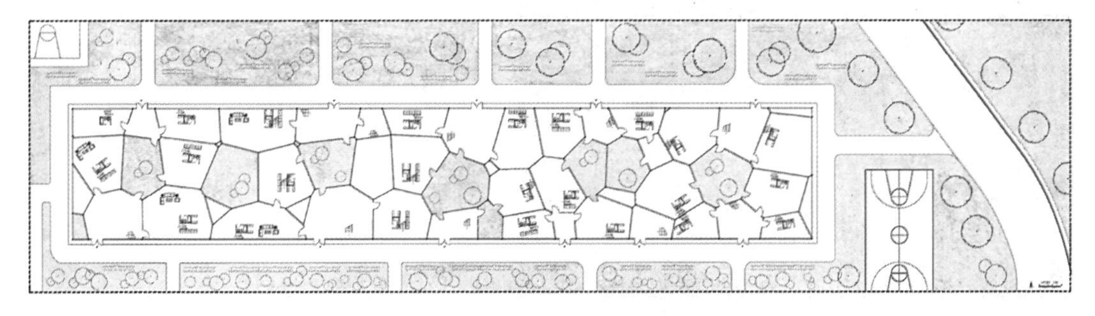

光店/THE LIGHT SHOP

项目选址：王府井大街
项目类型：临时商业空间
建筑面积：2049m^2
用地面积：2856m^2

方案设计：吴之恒
指导教师：王昀
完成时间：2015

本设计为一次空间形式探索，试图在王府井大街厚重的建筑物之间营造一个轻质而半透明的商业空间。The Light Shop 有两种含义：一是材质和建造上的light（轻质），二是意指光线的light。设计灵感来自对纸的折痕的光影变化的研究。

长弄堂博物馆群/LONG LANE MUSEUM

项目选址：江苏省用直古镇
项目类型：博物馆群
建筑面积：6125m²
场地面积：960m²

方案设计：徐逸
指导教师：王昀
完成时间：2015

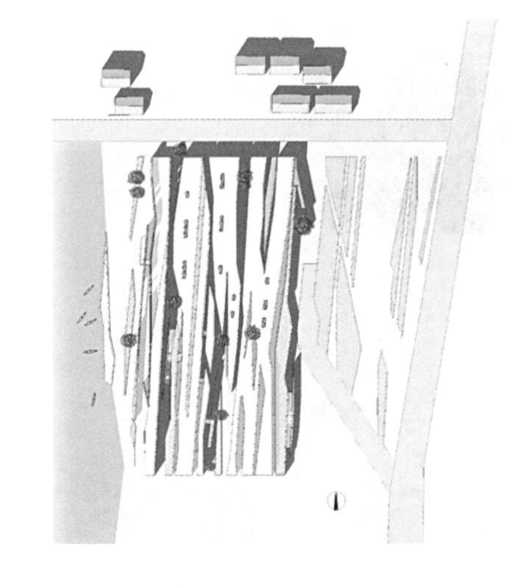

在人工智能迅猛发展的今日，是否有可能借由计算预测城市的发展肌理、由此生成建筑？本案将古镇的一个现存的界面转译为数字语言，通过元胞自动机110进行数十万次迭代生长，其过程本质上等同于城市在历史过程中的更新，由此得到的图像作为建筑的肌理依据。

小径交叉的花园/ A GARDEN WITH FORKING PATH

项目选址：清华大学清芬园西侧绿地，
项目功能：学生活动中心，
建筑面积：1280m²
用地面积：4690m²

方案设计：杨隽然
指导教师：王昀
完成时间：2015

清华是一所强调集体意志与规范的学校，校园内的建筑和规划也无不体现出这种严谨规整的氛围。然而，从与同学们交流所得的印象却是，大家渴望一个自由探讨或是独处的空间。因而这是一个反叛性的设计。一方面是功能的自由性，没有墙作为限定，只有拥有展墙的流动空间。另一方面是形体的自由性，模仿景观步道，将校园内的无效绿地立体化。本设计通过建筑的介入，使得人更充分地体验场所与空间，塑造立体庭院，改变无效绿地的现状。

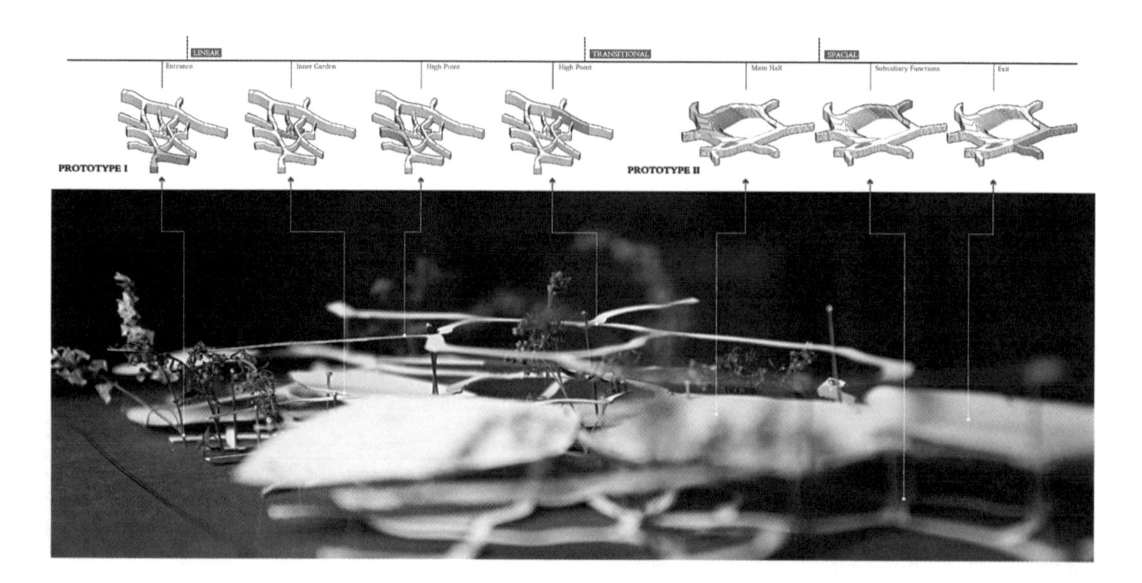

延庆白河峡谷度假中心/ BAIHE VALLEY VACATION CENTER

项目选址：北京延庆白河峡谷

项目类型：度假中心

建筑面积：4800m²

用地面积：7500m²

方案设计：周桐

指导教师：王昀

完成时间：2015

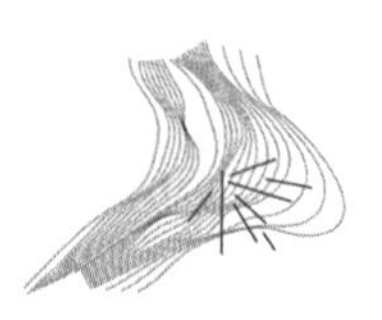

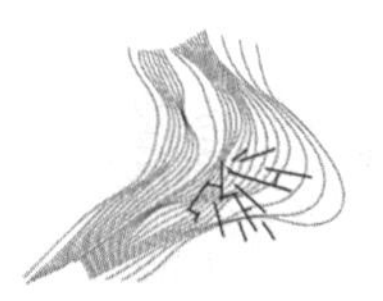

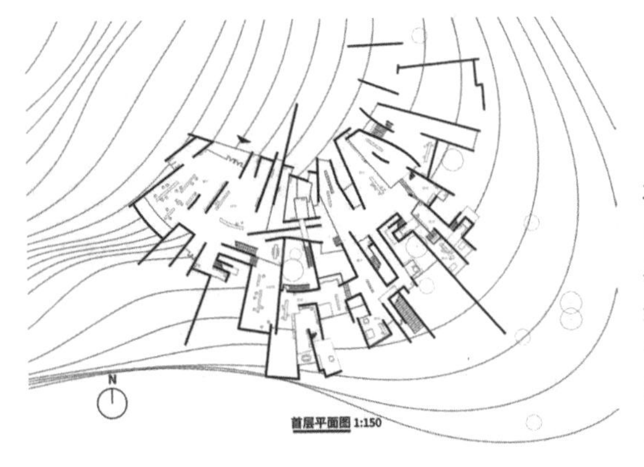

首层平面图 1:150

本方案在八周的前四周经过了研究、发现与寻找，通过对自然中的原型进行提取和操作，得到了空间的基本格局，再为这一具有迷宫特质与神秘气息的空间寻找合适的地段与功能。最终项目选址为风景秀丽的北京延庆百合峡谷，并将功能定位于较为灵活的度假中心，希望蜿蜒曲折、高低错落的空间如森林一般嵌入地景，与场地紧密结合在一起，并希望使用者在空间中寻找与发现，得到意料之外的空间感受。

芍里人家/ CHINESE PEOPLE

项目选址：北京市朝阳区芍药居地铁站西
项目类型：商业街区
建筑面积：6650m²
用地面积：19580m²

方案设计：李明玺
指导教师：王昀
完成时间：2015

我们的城市存在许多碎片空间，包括高楼与高楼之间的空白地带，无人问津的袖珍广场，街角围合的寂寥院落……本设计旨在唤醒城市的碎片空间，利用富有活力的建筑及设施，加以合适的功能，为城市的“细枝末节”带来新的可能性。方案在芍药居城铁附近的商住混合用地上，整理出一片碎片空间，用灵活的手法，营造出一片微商业街区，供小成本商贩在此营业。

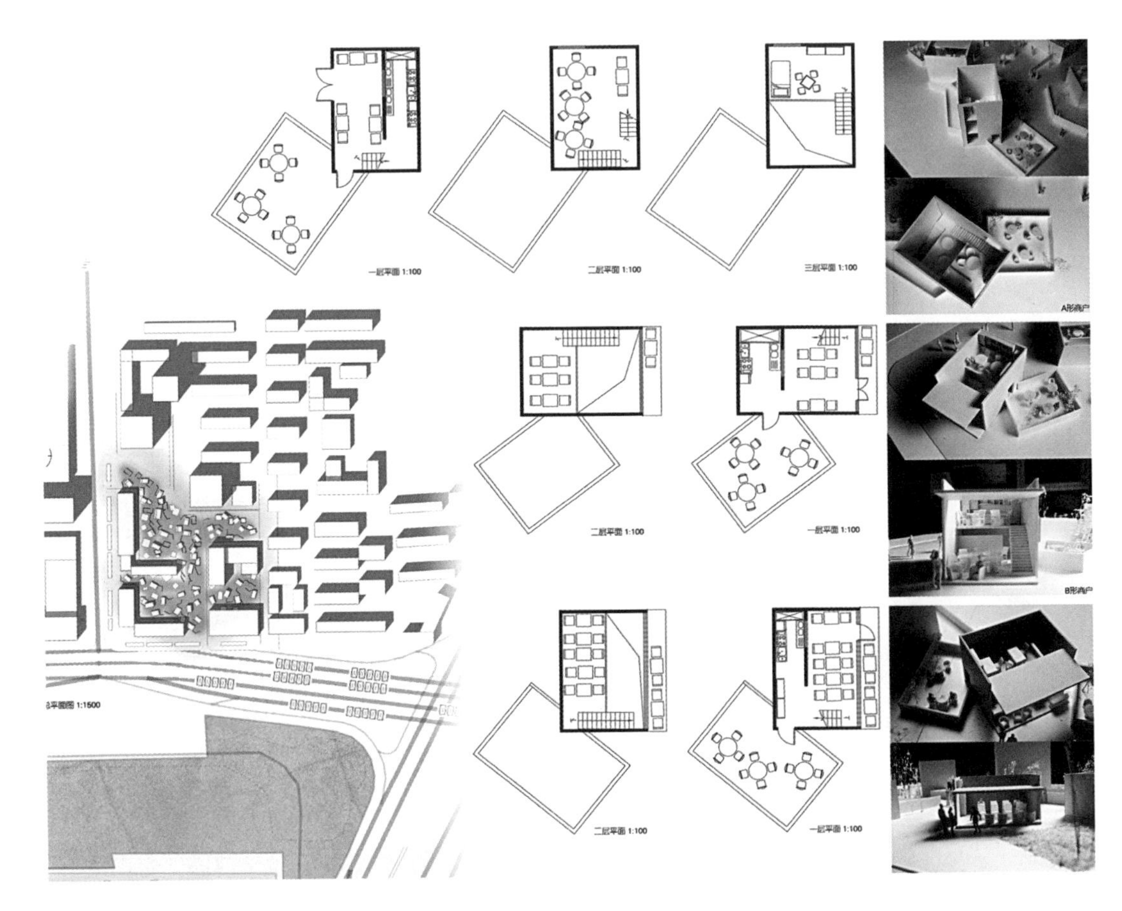
一层平面 1:100
二层平面 1:100
三层平面 1:100
A形商户
二层平面 1:100
一层平面 1:100
B形商户
平面图 1:1500
二层平面 1:100
一层平面 1:100

年轻人

王辉

URBANUS 都市实践建筑设计事务所
创始合伙人，主持建筑师

微城市

王辉

URBANUS 都市实践建筑设计事务所
创始合伙人，主持建筑师

教育背景

1985年 – 1990年
清华大学建筑学院　学士
1990年 – 1993年
清华大学建筑学院　硕士
1995年 – 1997年
美国迈阿密大学建筑系　硕士

工作经历

1993年 – 1995年
中央工艺美术学院环境艺术系 教师
1997年 – 2001年
纽约Gensler 、Gary Edward Handel等事务所
建筑师
1999年至今
URBANUS都市实践建筑设计事务所 创始合伙人，
主持建筑师

主要论著

王辉，范凌《十谈十写》，上海：同济大学出版社，2016
《唐山市城市展览馆》，天津:天津大学出版社，2009
《都市实践》，北京：中国建筑工业出版社，2012
《URBANUS都市实践》北京：中国建筑工业出版社，2007

设计获奖

2016 – 威尼斯双年展 穿越中国——中国理想家 杰出贡献奖
2015 – 城建集团杯·第八届中国威海国际建筑设计大奖赛优秀奖
2014 – 北京国际设计周 设计“为城市更新”服务——推动智慧城市建设的优秀项目
2013 – 北京国际设计周 最受大众欢迎项目
2013 – 首都第十九届城市规划建筑设计方案优秀方案奖
2010 – 第三届美国《建筑实录》“好设计创造好效益”中国奖 年度奖及最佳公共建筑奖
2009 – 中国建筑学会建筑创作大奖
2008 – 第四届WA奖 佳作奖
2008 – 第二届美国《商业周刊/建筑实录》“好设计创造好效益”中国奖　最佳公共建筑奖

代表作品

深圳规划大厦（图1）、唐山博物馆（图2、图4）、唐山城市展览馆（图3）、北京白云观云起时珍宝花园（图5）、北京前门西河沿街小蚂蚁皮影剧场（图6）、上海嘉定“现厂”创意园（图7）、山西芮城五龙庙环境整治（图8、图9）

1
2
3
4
5
6
7
8
9

年轻人的微城市

三年级建筑设计（6）设计任务书

指导教师：王辉
助理教师：唐康硕　张淼

课题预设

随着当代大都市核心区的土地价值攀升，年轻人越来越被排挤到城市的边缘：不仅仅是他们无力在城市中心找到租金合适的居所，也很难找到其创业的办公地点。而另一方面，城市越是发展，城市人口越是面临老龄问题，就越需要年轻人口来使用城市的服务设施，并对城市提供积极的服务。为年轻人在城里找块可能的地变成一个课题。

在北京，分布在三环与四环之间的公交场站，成为完成这一课题的一种可能。通过调研发现，城市内现有的公共交通场站占据了大量的城市用地，同时，现状中的一些场站在空间上也无法与城市公共空间、城市社区生活发生联系，而是在城市中犹如飞地般的存在。目前，这些场站基本上是空地，其地上和地下的空间资源尚未被充分开发。这些空间的开发，不仅仅会带来土地的收益，以弥补公交的财政投入，以及稳定和扩大公交事业，还能带来城市更新的一种新的契机。

因此，公共交通场站的综合再利用还承载着一系列现阶段所亟需的城市功能，以达到完善土地价值、集约用地和促进城市社区活力的目标。在城市区域由功能的单一性逐渐演变为多元化时，对城市存量中现有的公共交通场站的综合利用也可以做到“一地多用”和“一站多用”的原则：从单一的交通场站功能到与城市、社区需求相结合的多层次综合功能；从单一的车场形态到可以容纳交通、年轻人创业和居住的复合空间形态。

从城市发展的时间轴上，这种契机往回看，是如何用这一新的城市填空机会，完善与车场毗邻的城市住区的公共配套；这种契机往前看，则更有挑战意义，它提出了一个未来年轻人的微城市模式，使年轻人有可能在这里居住和创业，并用一种新型的建筑类型来激活城市。

本设计课程提出“微城市”的课题，既从形态和功能出发，让一座建筑体现出小型城市的感觉。

课程目标

基于这样的课题想象，这个八周的课程，将通过从大到小、从虚到实的循序渐进的方式，培养学生多种设计技能：

1. 案例研究的技能：

每一个课程步骤都结合一定的案例研究，以及研究表述。学生通过对每一步骤中给出的相关设计案例的学习分析，总结出其设计特点、空间特殊性和空间与功能的辩证关系。

2. 做任务书的技能：

当城市从增量发展走向存量挖掘时，这一代学生的未来职业生涯更面临一种需要设计师去发明、发现设计任务的处境。因此，非常有必要锻炼学生对设计任务的主动提议和对设计任务书的编写制定能力。

3. 从整体到局部的方法统一性的技能：

从整体空间、结构、组成形态以及建筑类型学的方式入手，培养学生从整体性思考出发的设计方法，以及在不同尺度对空间设计方法的思维一致性。

4. 从社会学角度进行批判性思考的技能：

通过对城市和社区中年轻社会阶层的观察，从社会学角度思考针对这一人群对居住和创业的实际功能、空间的真实需求及其在城市、建筑空间上可能的投射，并因此产生对传统建筑设计过程和建筑空间的批判性反思。

5. 构想未来的技能。

课程安排

1. 基地建造：1.1 选址；1.2 上盖的地形地貌设计；1.2 上盖的结构设计。

2. 功能设想：2.1 案例研究；2.2 任务书评价。

3. 微城市设计：3.1 微城市的形态；3.2 微城市的几何性；3.3 微城市的社会性。

4. 单体设计：4.1 从剖面引发的设计；4.2 从功能图示引发的设计；4.3 从内向外引发的立面设计。

5. 局部设计（可选项）：5.1 局部功能单元的设计；5.2 局部场地的设计；5.3 局部细部及材料的设计。

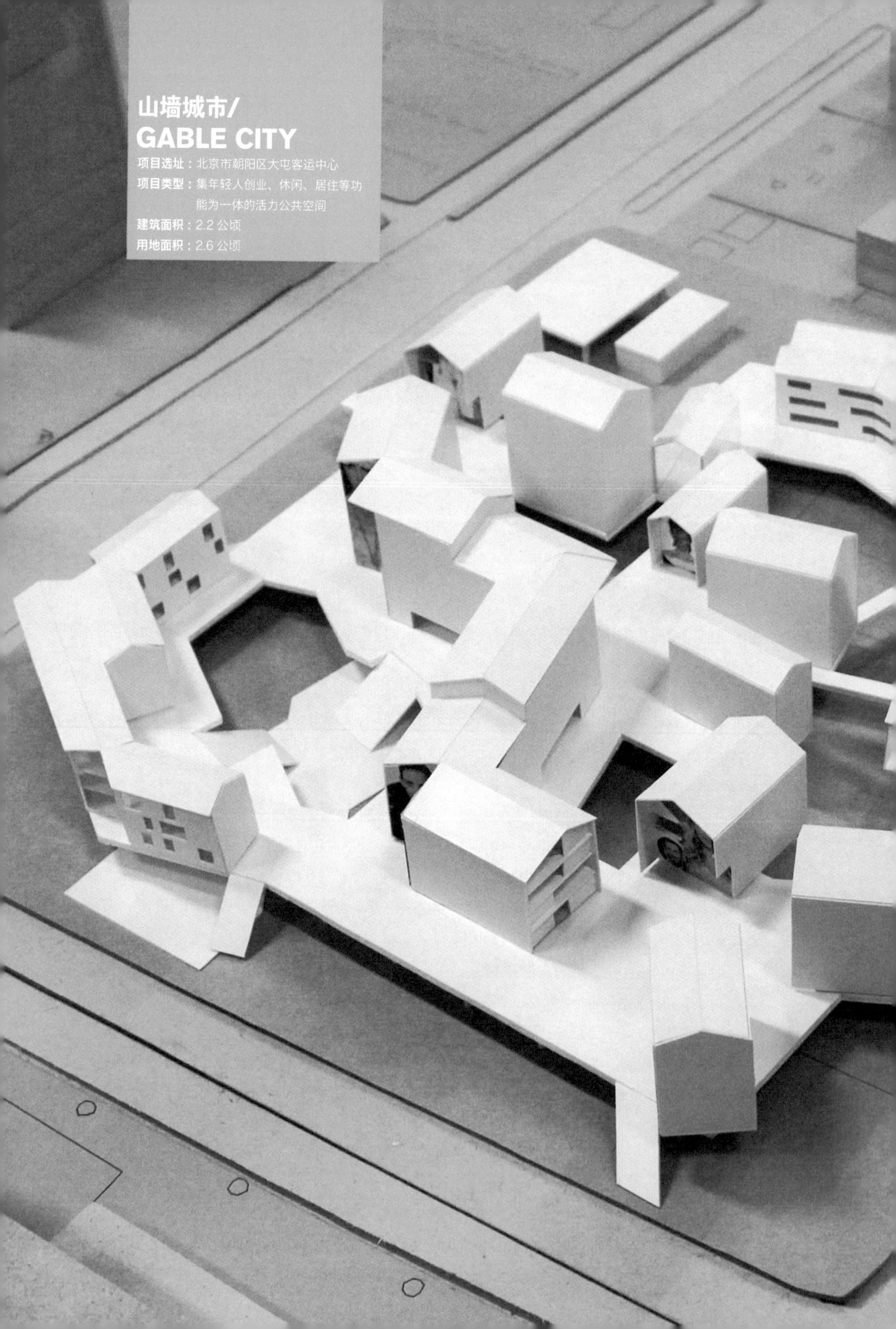

山墙城市/
GABLE CITY

项目选址：北京市朝阳区大屯客运中心

项目类型：集年轻人创业、休闲、居住等功能为一体的活力公共空间

建筑面积：2.2 公顷

用地面积：2.6 公顷

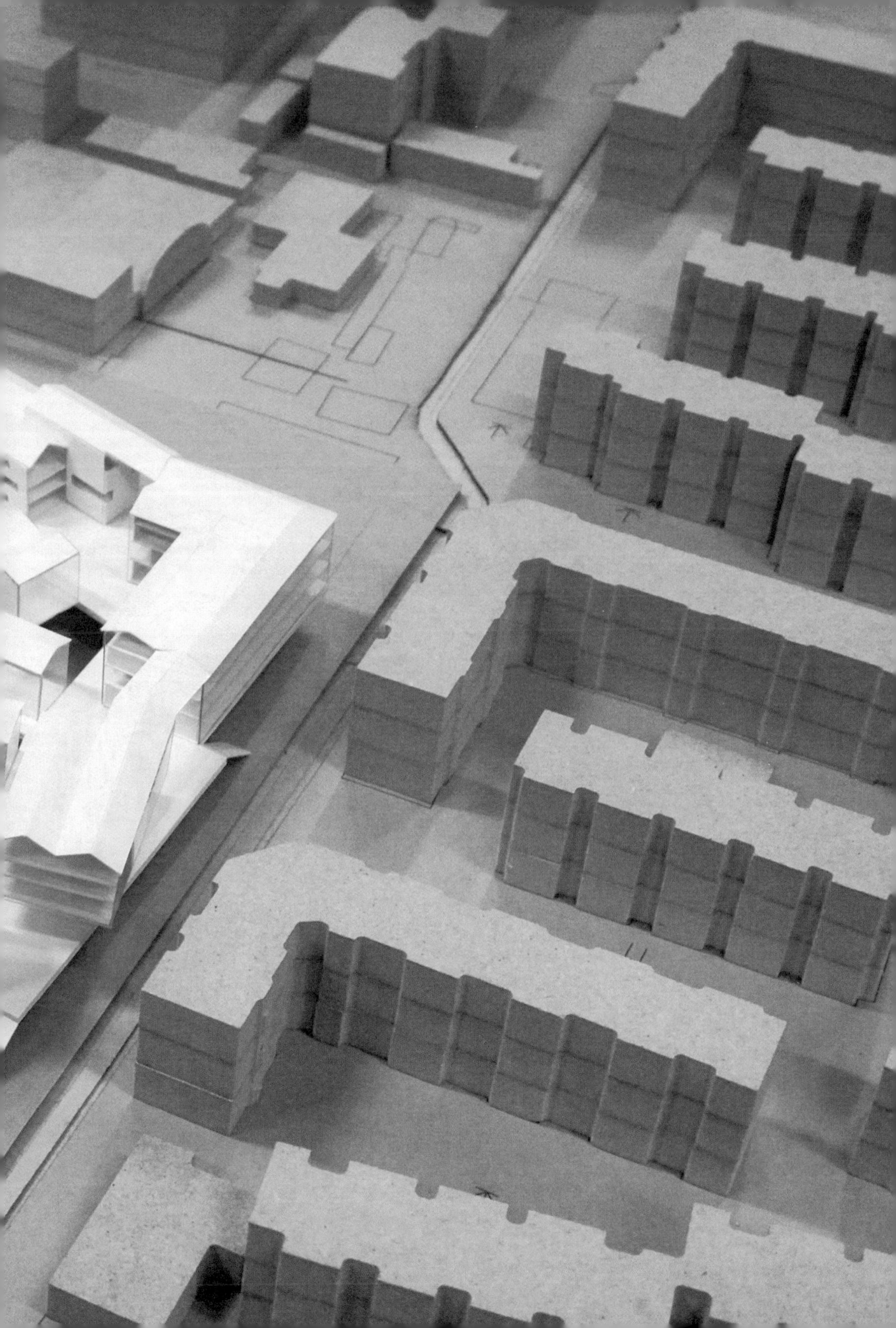

方案设计：刘倩君
指导教师：王辉
完成时间：2015

我将想象画中“平凡”的生活状态转换为建筑语言——对朴素的楼梯元素予以放大，连接不同功能的楼层与建筑，使之成为年轻人生活的载体，期待着建筑的使用者们在不同场景的转换过程中在此发生反应。

山墙城市最终以一种直白的、形而下的手段为“年轻人的微城市”作了一解。挖掘平素的生活片段，未尝不是空间创新的一种姿态。

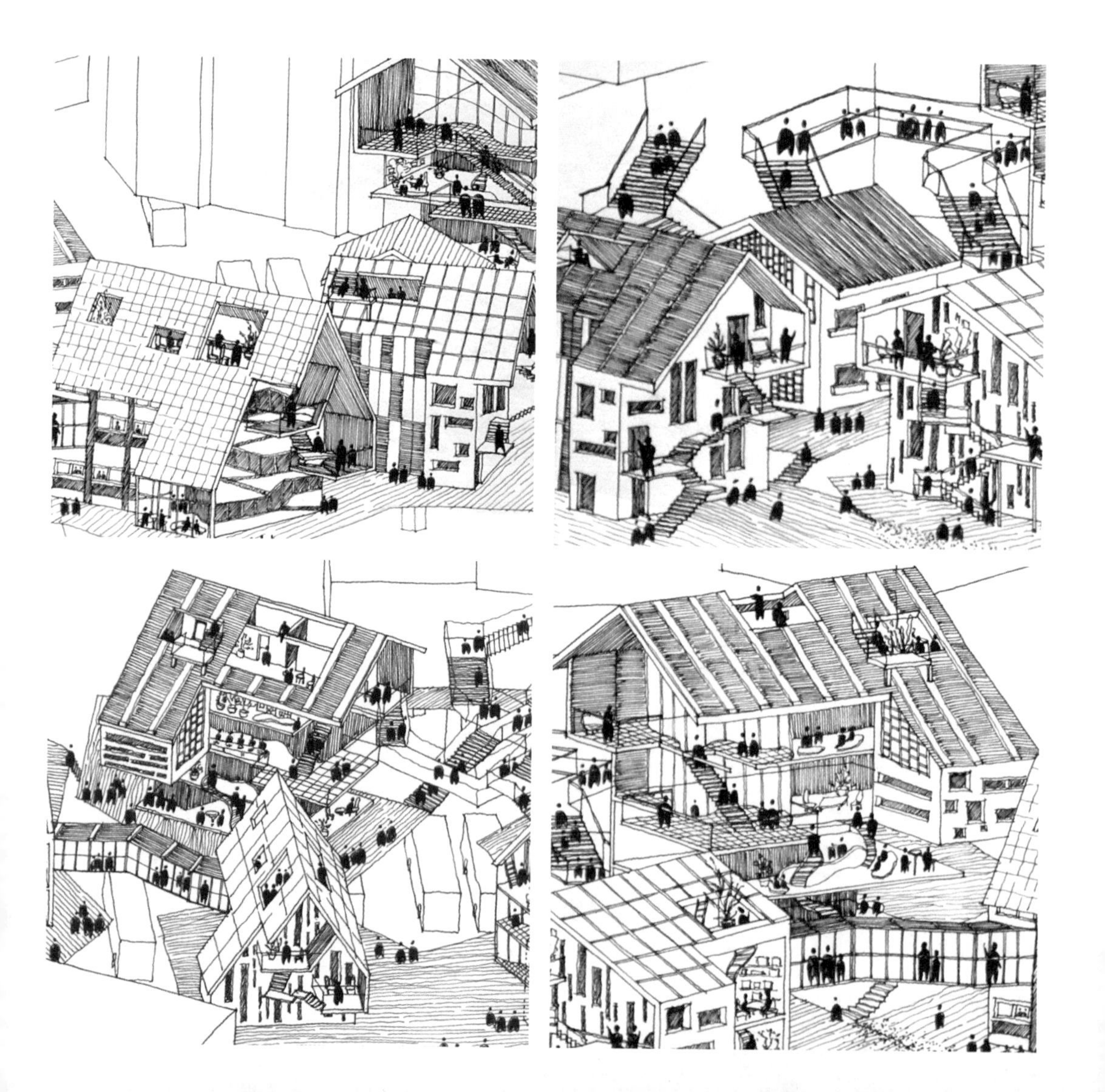

开篇：主方案图。**对页下图：**草图一微城市空间想象画。**本页上图：**总平面图。**本页下图：**山墙空间模型。

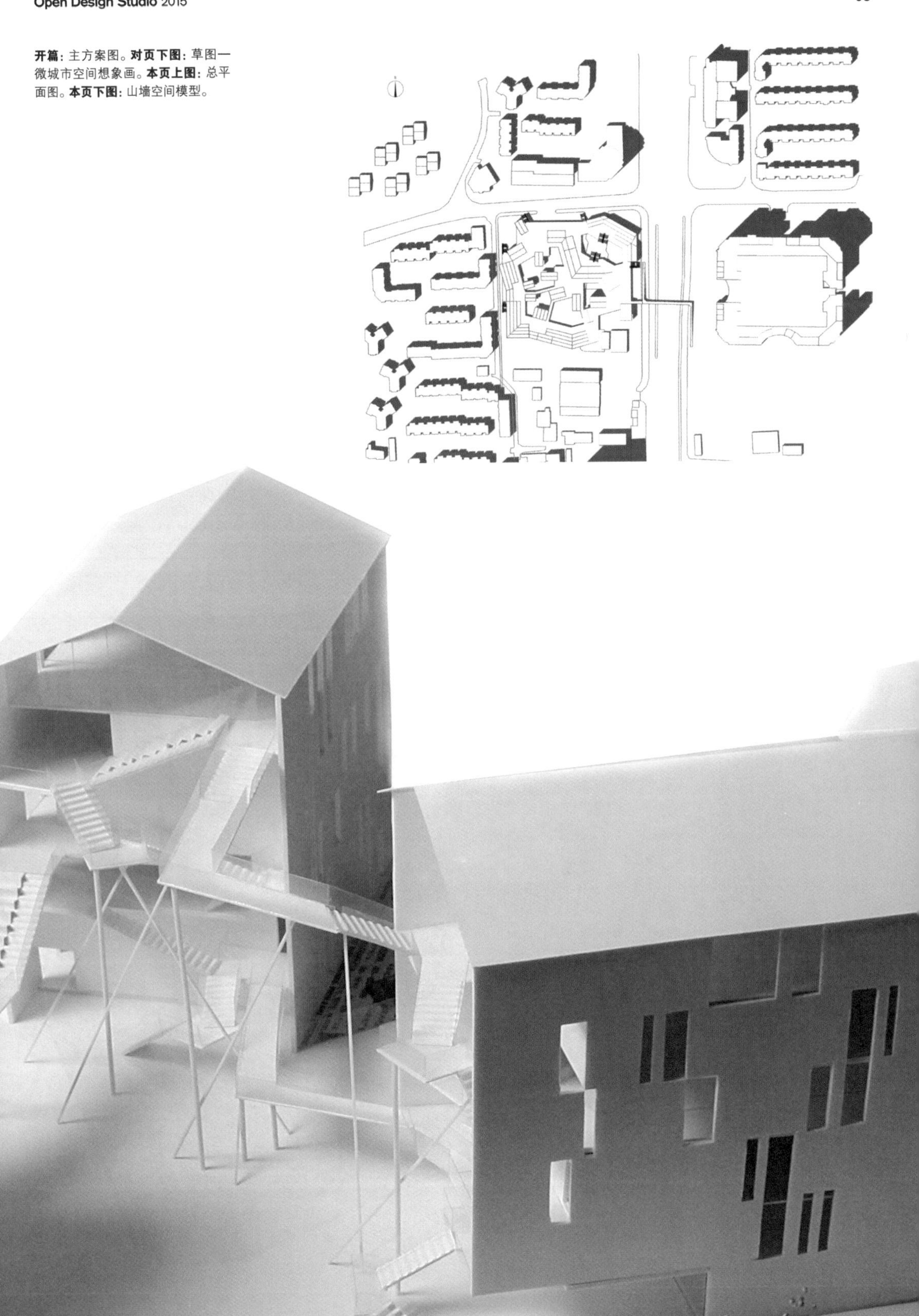

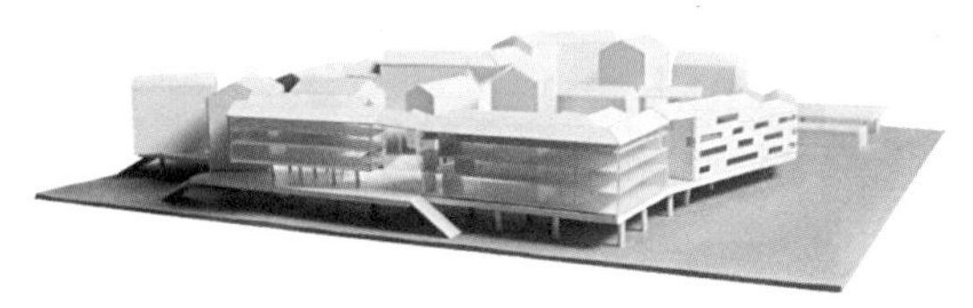

本页上图：效果图。本页下图：功能分析图。对页左图（从上至下）：微城市模型。对页右图：轴测分解图。对页下图（从左至右）：首层平面图。人流分析图。

The "ordinary" life state in imaginary painting is transferred into architectural language -- The simple stair element is amplified so as to connect the floor and construction of different functions, which becomes the carrier of the lives of young people. It is expected that the building users react in the transformation process of different scenarios.

The gable wall city is a solution of the "Micro-city of the Young People" with a straightforward and metaphysical means. The mining of usual life segments is a new posture of space innovation.

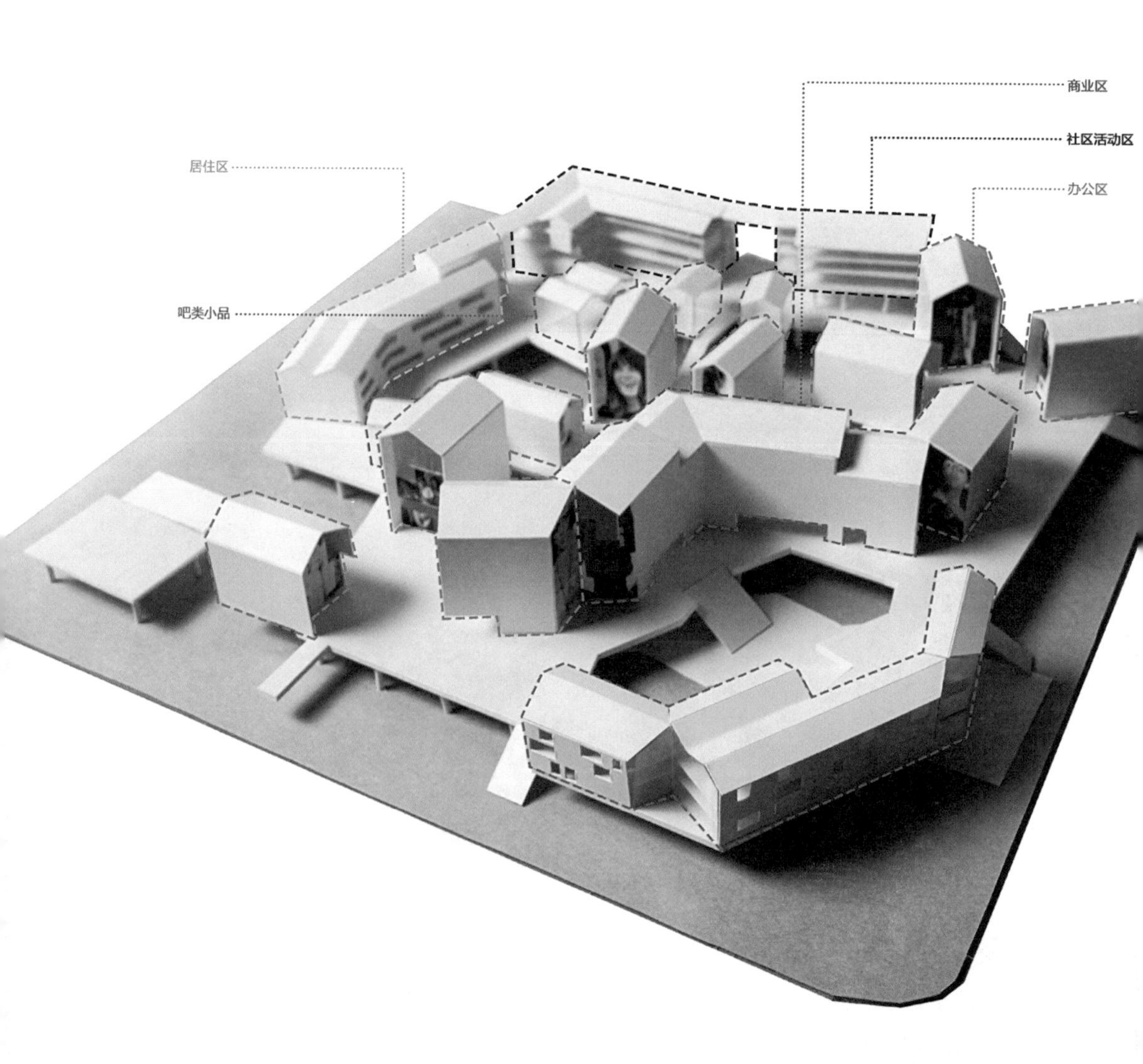

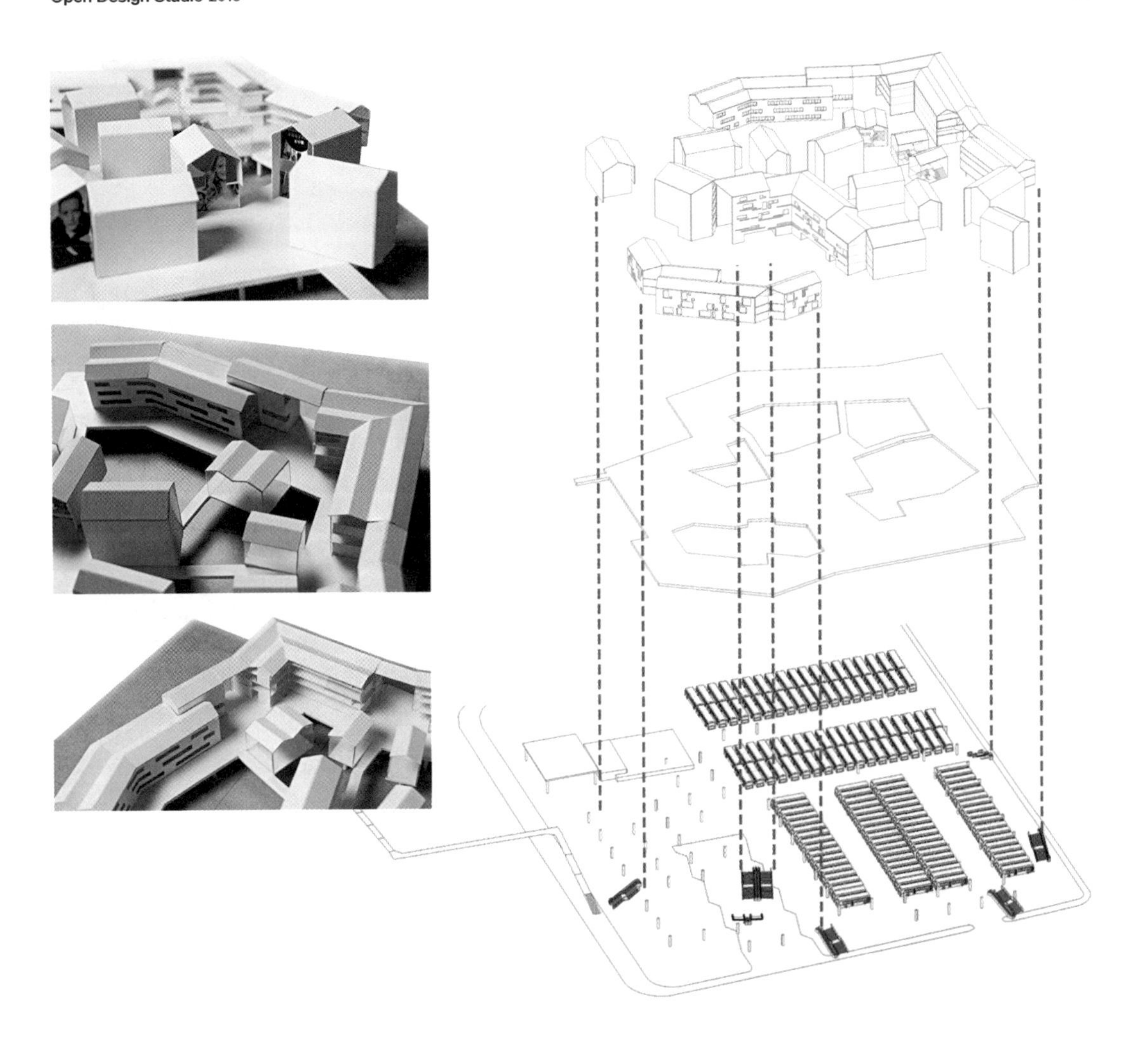

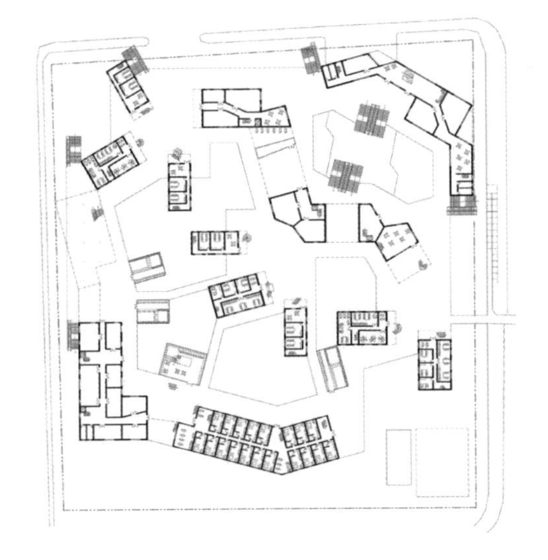

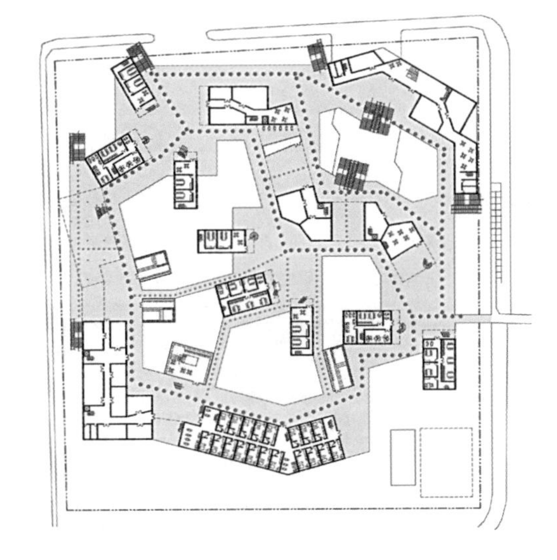

跨页图：建筑细部。

教师点评

本着对设计创作的热情和对城市生活的热爱，刘倩君同学将感性思考部分融入了年轻人的微城市设计构思中。基于对东京、上海等高密度街区城市的旅行和居住经验思考，她构思了一个充满邻里空间和活力生活的“山墙城市”概念。从概念阶段开始，刘倩君同学就通过绘画的方式构思了一幅栩栩如生的社区生活场景：在车场上方，一个由不同方向的坡顶建筑组成的居住和工作空间；年轻人通过打开的山墙和屋顶创造交流和共享空间；蜿蜒曲折的路径穿越社区平台。这些情景化的空间意象经过提炼，最终落实到“山墙生活”这个主题。设计深化阶段，建筑分布和组合方式也依据车场的实际场地建造而产生了调整，车场上方的建筑形态也相应地类型化，使微城市的营造更有序。在此基础上，把着力点放在“开放山墙”，使这一部分成为最有活力的因素。

最后的成果表达上，刘倩君同学制作了大比例的邻里空间模型，使故事性的叙述能够通过生活画面来表达，生动地诠释了“山墙城市”的细部生活场景。

Teacher's comments

With the passion for design and the love for city life, Liu Qianjun integrated the emotional thinking into design concept of the micro city of the young people. Based on the travel and living experiences in the cities composed of high-density blocks, including Tokyo and Shanghai, etc., she conceived a "gable city" concept full of neighborhood space and vitality of life.

Starting from the concept stage, Liu Qianjun conceived a vivid scene of community life through the way of painting: at the top of the car yard, there was a living and working space composed by buildings with sloping roofs in different directions, the young people created communication and sharing space by the open roof and gabled walls, the winding and zigzagging paths passed through the community platform. The scenic space image has been finally implemented into the theme of "gable life" after refinement.

In the development phase of the design, the distribution and combination of buildings were adjusted according to the actual construction of the car yard, the architectural forms at the top of yard were also correspondingly typed. Therefore, the micro city was created more orderly. On this basis, the focus was laid on "open gable", which became the most dynamic factor.

In the final review, Liu Qianjun produced a neighborhood space model in large scale, so that the story narration can be expressed through the pictures of life, which vividly interpreted the details of life scenes of "gable city".

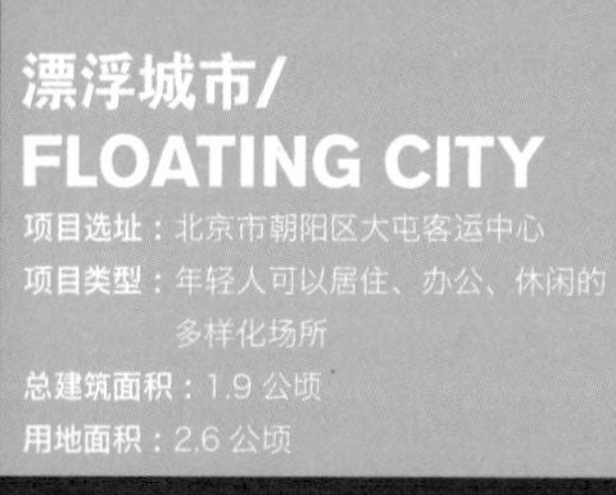

漂浮城市/ FLOATING CITY

项目选址：北京市朝阳区大屯客运中心

项目类型：年轻人可以居住、办公、休闲的多样化场所

总建筑面积：1.9 公顷

用地面积：2.6 公顷

方案设计：肖玉婷
指导教师：王辉
完成时间：2015

设计的选址在公交场站的上方，基地需要架起以保证车站的正常运作。我把设计分为场地和建筑两个部分来统筹。

场地部分，在保证了基本的乘车流线满足之后，通过布置一些地形起伏来丰富场地的运动场所。

建筑部分，在同样坡屋顶的原型控制下创造尽量多的空间可能，让年轻人自主去感受和选择，找到适合自己的去处，这是我所期望的多元化的城市。

此外为了增强场地的流通和连贯，所有的建筑都底层架空，设计中也采用了多种的架空方式来丰富场地的流线和视觉交互。

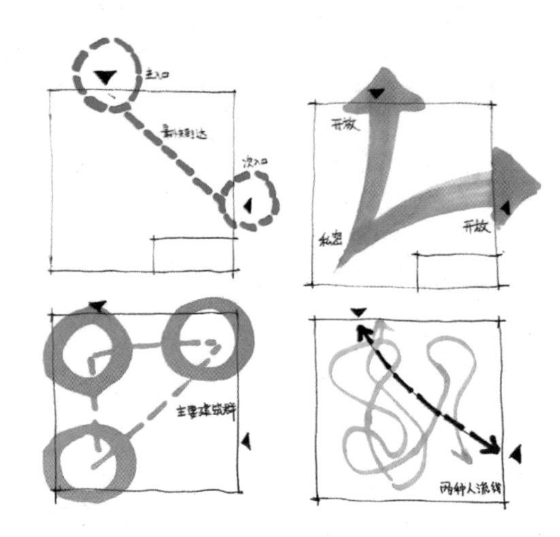

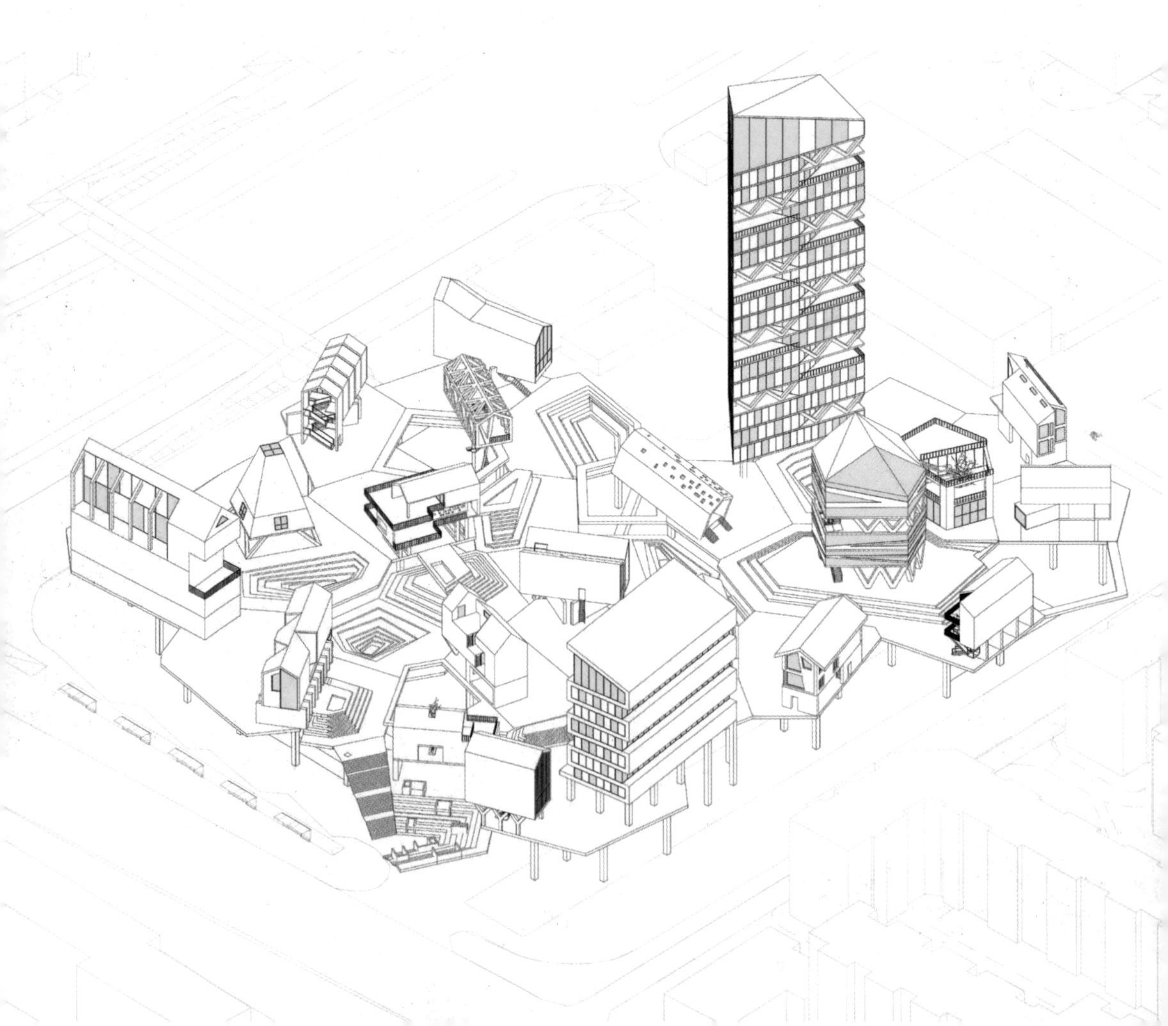

开篇：主要方案图。**对页上图**：概念草图。**对页下图**：主要方案图。**本页上图**：方案意向图。**本页下图（从上至下）**：生成过程。场地分析。

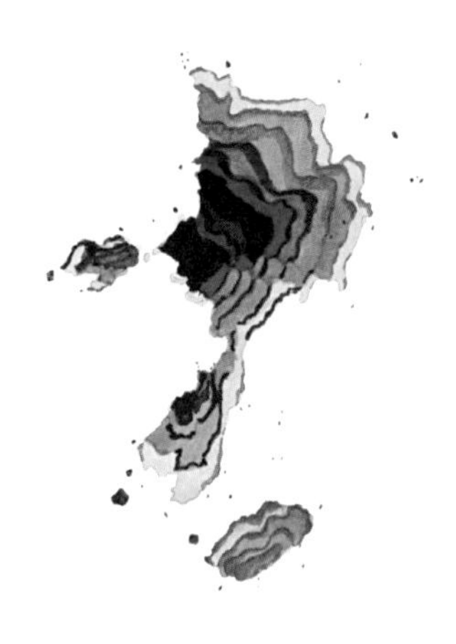

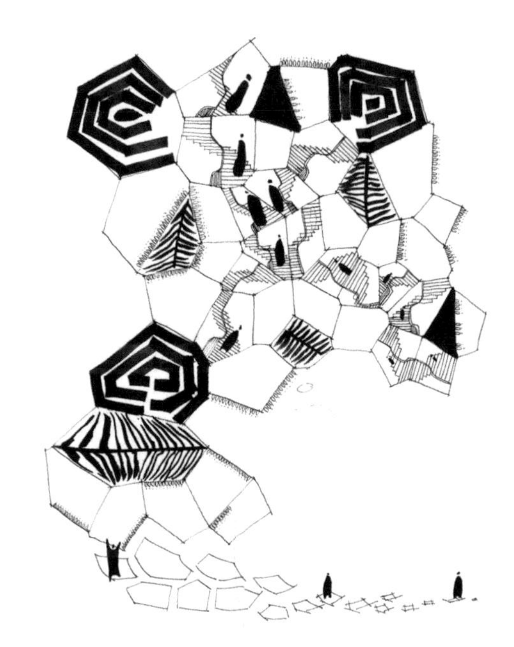

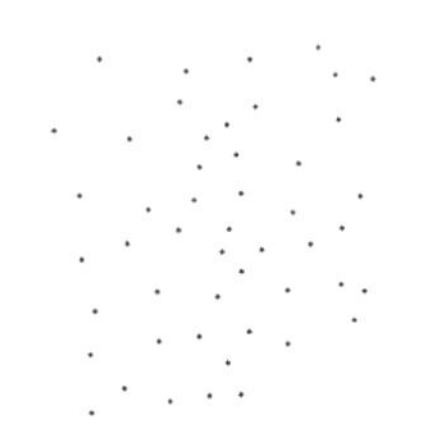

随机点生成

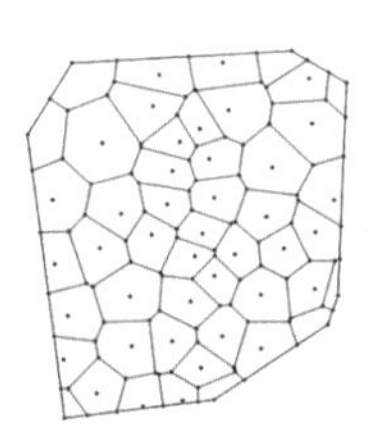

泰森多边形

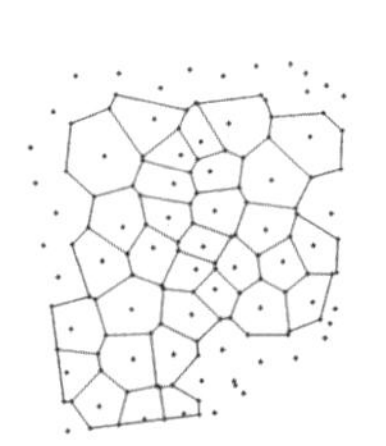

减法

功能分区

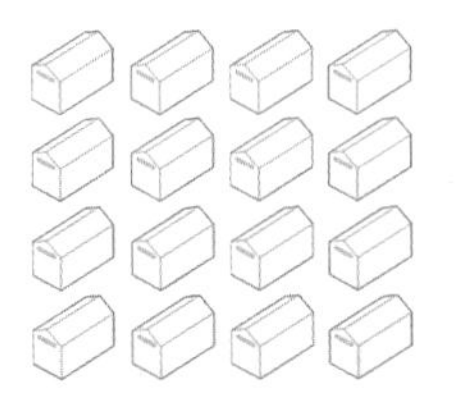

原型控制

底层架空

主要交通流线

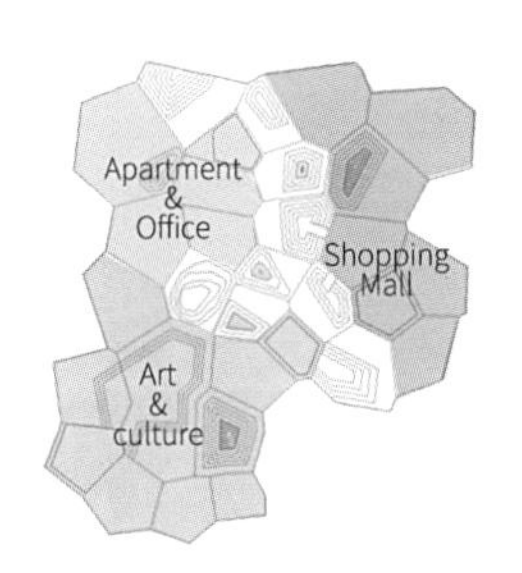

功能分区

绿化分布

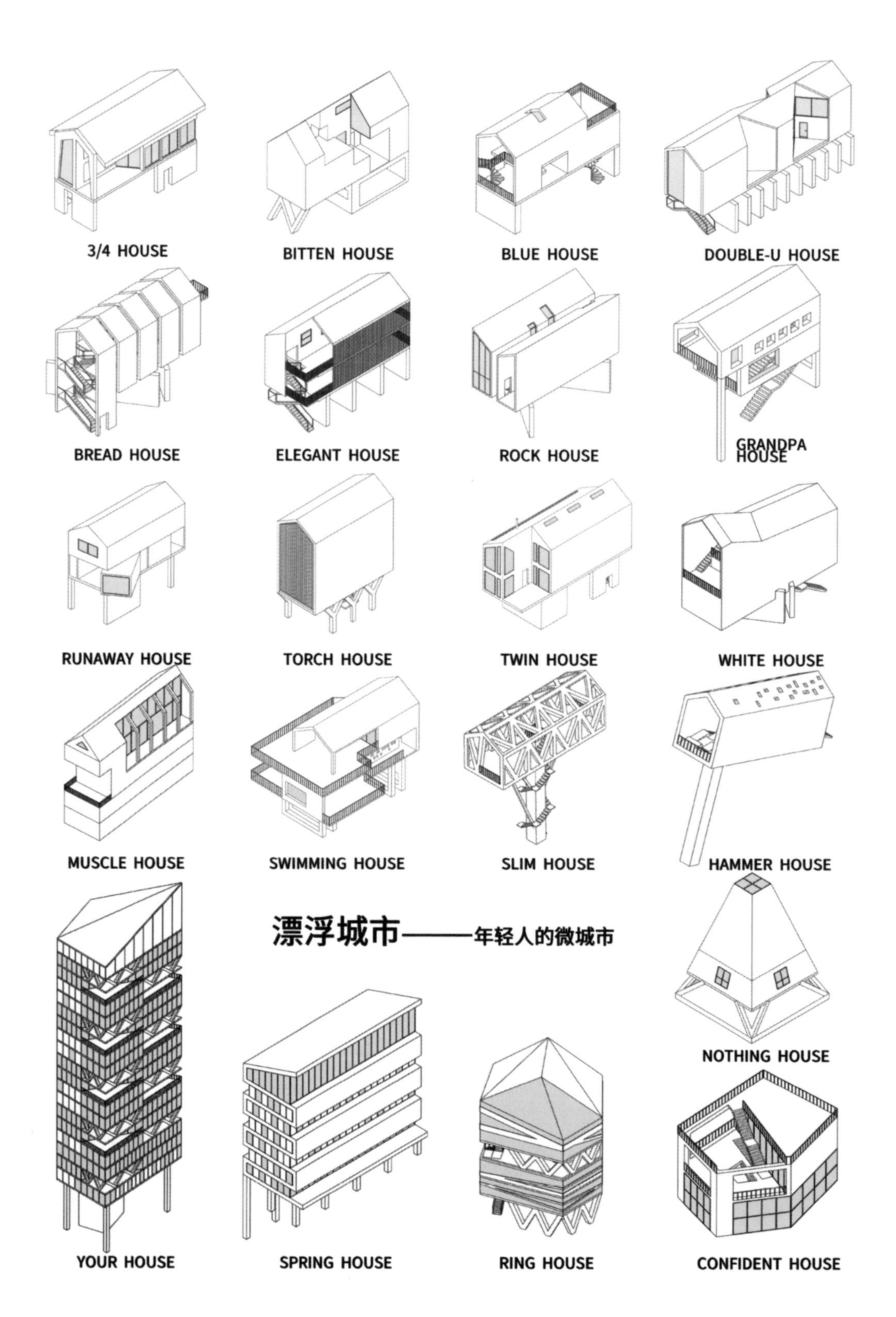
3/4 HOUSE
BITTEN HOUSE
BLUE HOUSE
DOUBLE-U HOUSE
BREAD HOUSE
ELEGANT HOUSE
ROCK HOUSE
GRANDPA HOUSE
RUNAWAY HOUSE
TORCH HOUSE
TWIN HOUSE
WHITE HOUSE
MUSCLE HOUSE
SWIMMING HOUSE
SLIM HOUSE
HAMMER HOUSE
漂浮城市——年轻人的微城市
NOTHING HOUSE
YOUR HOUSE
SPRING HOUSE
RING HOUSE
CONFIDENT HOUSE

对页图：21个房子整体展示。**本页上图：**总平面图。**本页下图（从左至右）：**一层平面图。二层平面图。

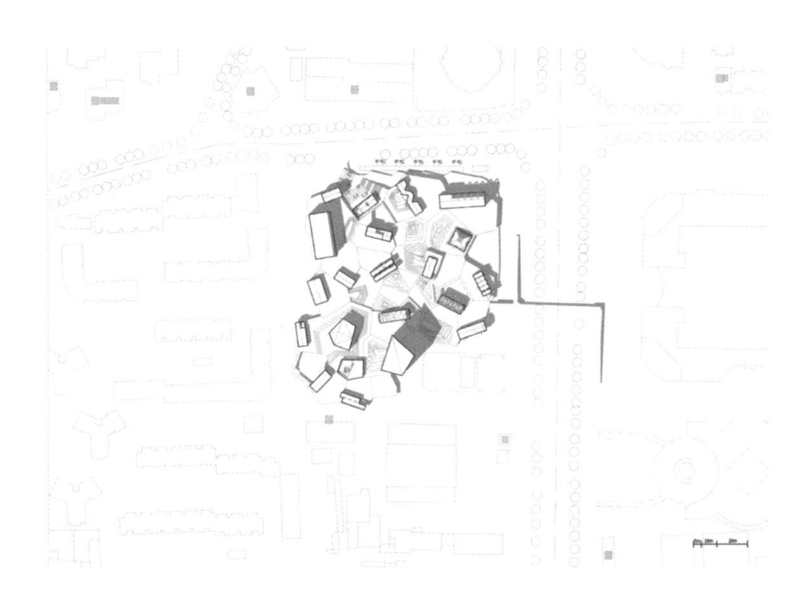

The site of the design is at the top of a bus station, and the base needs to be set up in order to ensure the normal operation of the station. I divided the design into two parts: the site and the building.

For the site, after the operation of basic bus lines is ensured, the sports venue of the site is enriched through the use of rugged topography.

For the construction part, create as much space as possible under the prototype control of sloping roof, so that young people can feel, choose and find their own place, this is the city diversification that I expect.

In addition, in order to enhance the circulation and continuity of the site, all the buildings are elevated on pilotis, a variety of elevating methods are used in the design to enrich the site's flow lines and visual interaction.

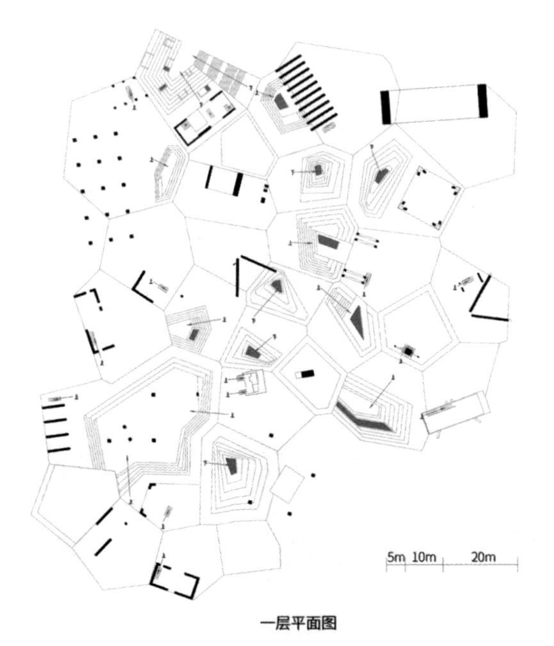

一层平面图

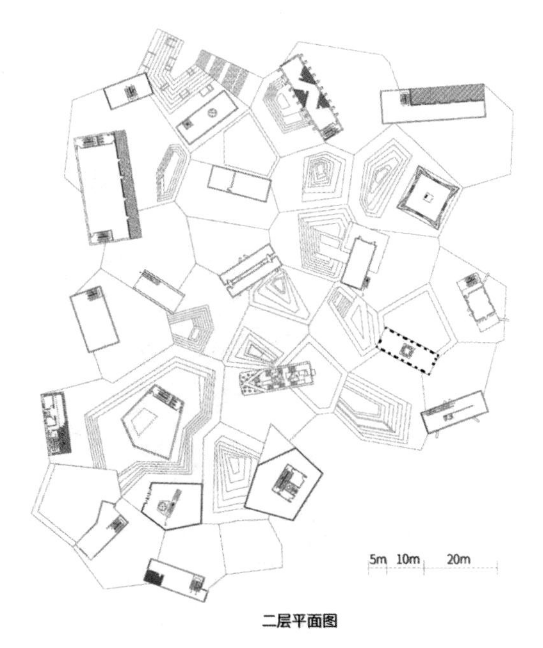

二层平面图

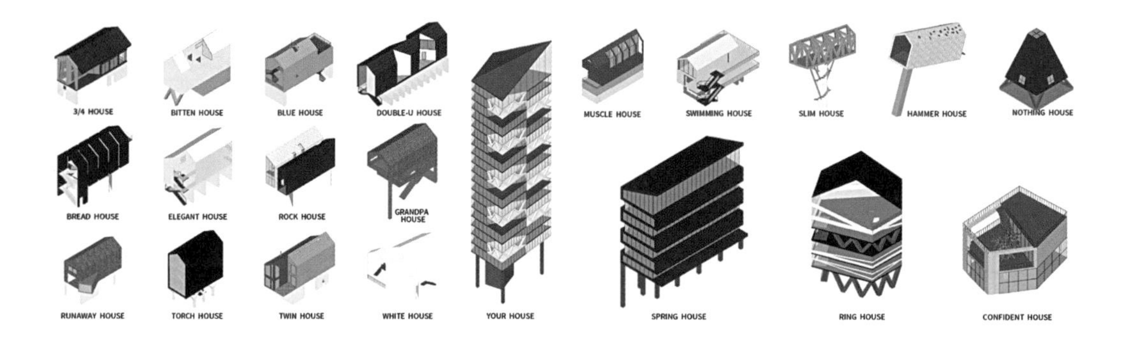
3/4 HOUSE
BITTEN HOUSE
BLUE HOUSE
DOUBLE-U HOUSE
MUSCLE HOUSE
SWIMMING HOUSE
SLIM HOUSE
HAMMER HOUSE
NOTHING HOUSE
BREAD HOUSE
ELEGANT HOUSE
ROCK HOUSE
GRANDPA HOUSE
RUNAWAY HOUSE
TORCH HOUSE
TWIN HOUSE
WHITE HOUSE
YOUR HOUSE
SPRING HOUSE
RING HOUSE
CONFIDENT HOUSE

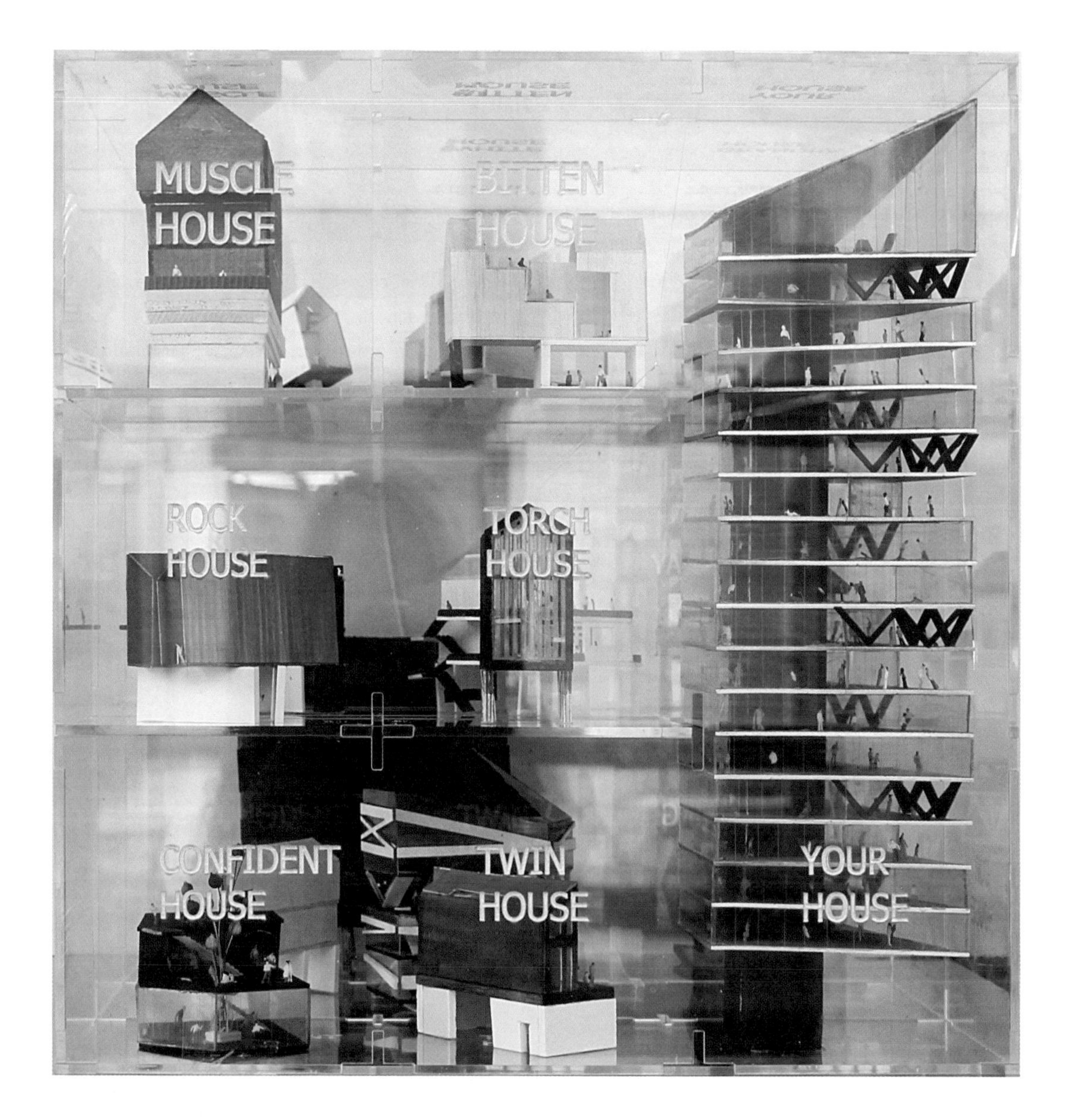
MUSCLE HOUSE
BITTEN HOUSE
ROCK HOUSE
TORCH HOUSE
CONFIDENT HOUSE
TWIN HOUSE
YOUR HOUSE

教师点评

肖玉婷同学的系统性思维能力和对设计的自明性表达能力很强。在对微城市构建的概念阶段，她就运用绘画的方式思考城市和聚落的空间意象，并随后迅速形成了由21个不同大小、形态的建筑组成的"漂浮城市"设计思路。

在完成图案式的场地建造后，肖玉婷同学采用了一种简单低调的形式建造她的房子。她认识到了物体—名字这一对偶的意义，命名了21种带坡顶建筑，并演绎出了每个房子的个性，以及这些房子的聚合所形成的社区的特色。

肖玉婷同学在最后评图时端出了一个有机玻璃箱，含了21 个小箱子，每一个小箱子里有个小房子，而这些箱子又是通过有机玻璃的壁厚所形成的“榫卯”搭接而成。大家对她的设计已很知晓，但这个透明的展示体还是让人眼前一亮。这让我真正感到一种欣慰：教学的目的不是传授知识和方法，而是引导学生能享受设计的快乐、并培养成对这个职业的热爱。

Teacher's comments

Xiao Yuting's systematic thinking ability and the ability of self - expression are very strong. In the concept stage of the micro city, she used the painting method to think about the space image of city and settlement, and then quickly formed the design idea of "floating city" with 21 buildings in different sizes and shapes.

After the pattern design of the site, Xiao Yuting used a simple form of low profile to build her house. She recognized the significance of the object-name relation and named the 21 kinds of building with sloping roof, and interpreted the personality of each house and the characteristics of community formed by the gather of these houses.

Xiao Yuting brought out a perspex box with 21 small boxes in the final review. In each small box there is a small house, and each box is made up by the structure with "mortise and tenon connections" formed by the thickness of perspex. Everyone has been familiar with her design, but this transparent display box still left a profound impression. It makes me really feel a sense of comfort: the purpose of teaching is not to impart knowledge and methods, but to guide students to enjoy the happiness of design, and to cultivate a love for the profession.

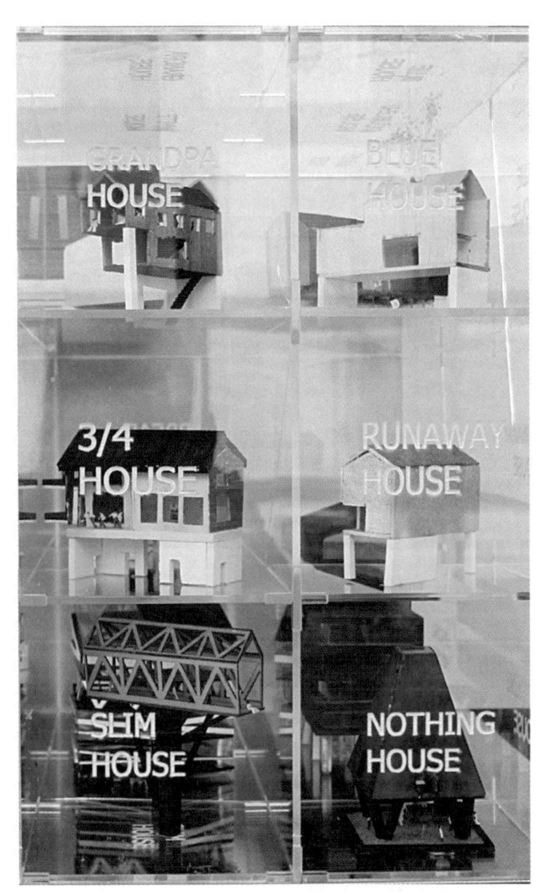

对页上图：21个房子整体展示。**对页下图**：模型正立面图；**本页上图**：模型正立面局部；**本页下图**：模型鸟瞰

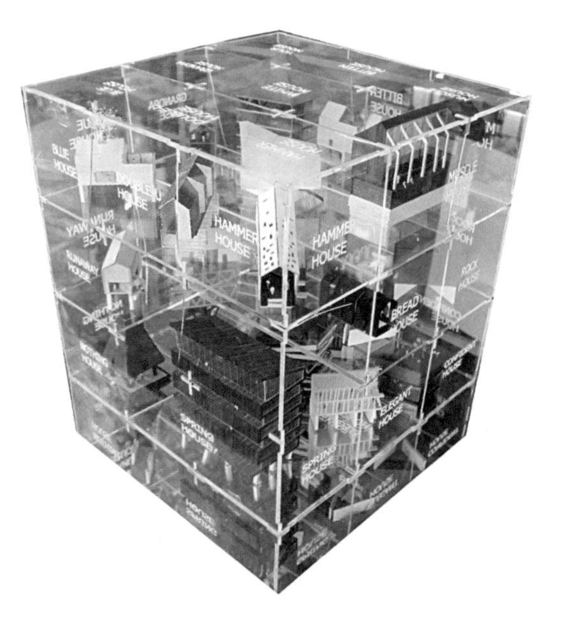

百宝箱/ TREASURE BOX

项目选址：朝阳区大屯公交场站

项目类型：城市设计

建筑面积：3 万 m^2

占地面积：2.6 万 m^2

上盖容积率：1.13m^2

SUN CAKES

方案设计：庞凌波
指导教师：王辉
完成时间：2015

由年轻人组成的群体具有独立、叛逆、创新等综合特质。为了在整个城市环境中，建立年轻人具有个性的社群，需要在城市现状下，展现异于周边肌理的独特秩序。为了与环境区别，分离的单元地块有着独特的肌理和密度；轴线与街道呈现角度关系，建立向心的秩序，使其产生由内向外的自生长性。

有了群体表达的基础，统一的单元形式赋予年轻人最大限度的个性彰显。异质的空间个性通过“橱窗”的概念向环境表达，使微城市中的活动因为丰富而不同。电路板意象与“橱窗”概念有异曲同工之处。线状趋势将线路分组，放大的端头与关键的电子元件相接，就类似于人在场地中的线性移动方式，以及向城市开敞的立面形式。

初具雏形的概念，因为被赋予功能而具有灵魂。形式对功能的几方面应对，分别反应在剖面、轴线、沿街立面等内容上。在概念发展的过程中，逐渐体现人们在微城市中的真实活动。

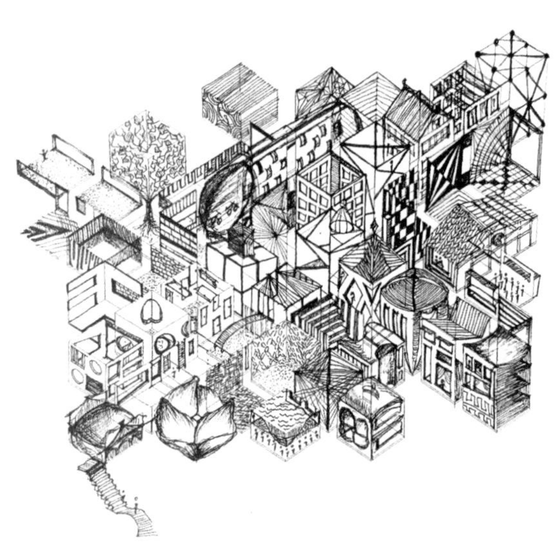

开篇：主要方案透视图。**本页上图：**概念效果图。**本页下图：**主要方案透视图。**对页：**首层平面图。

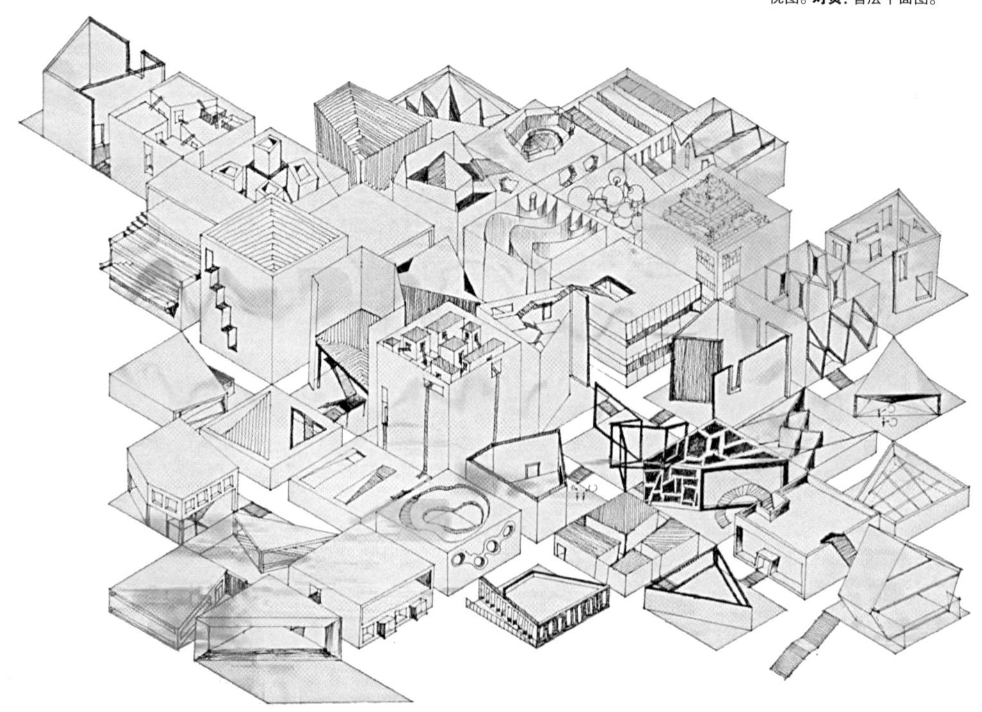

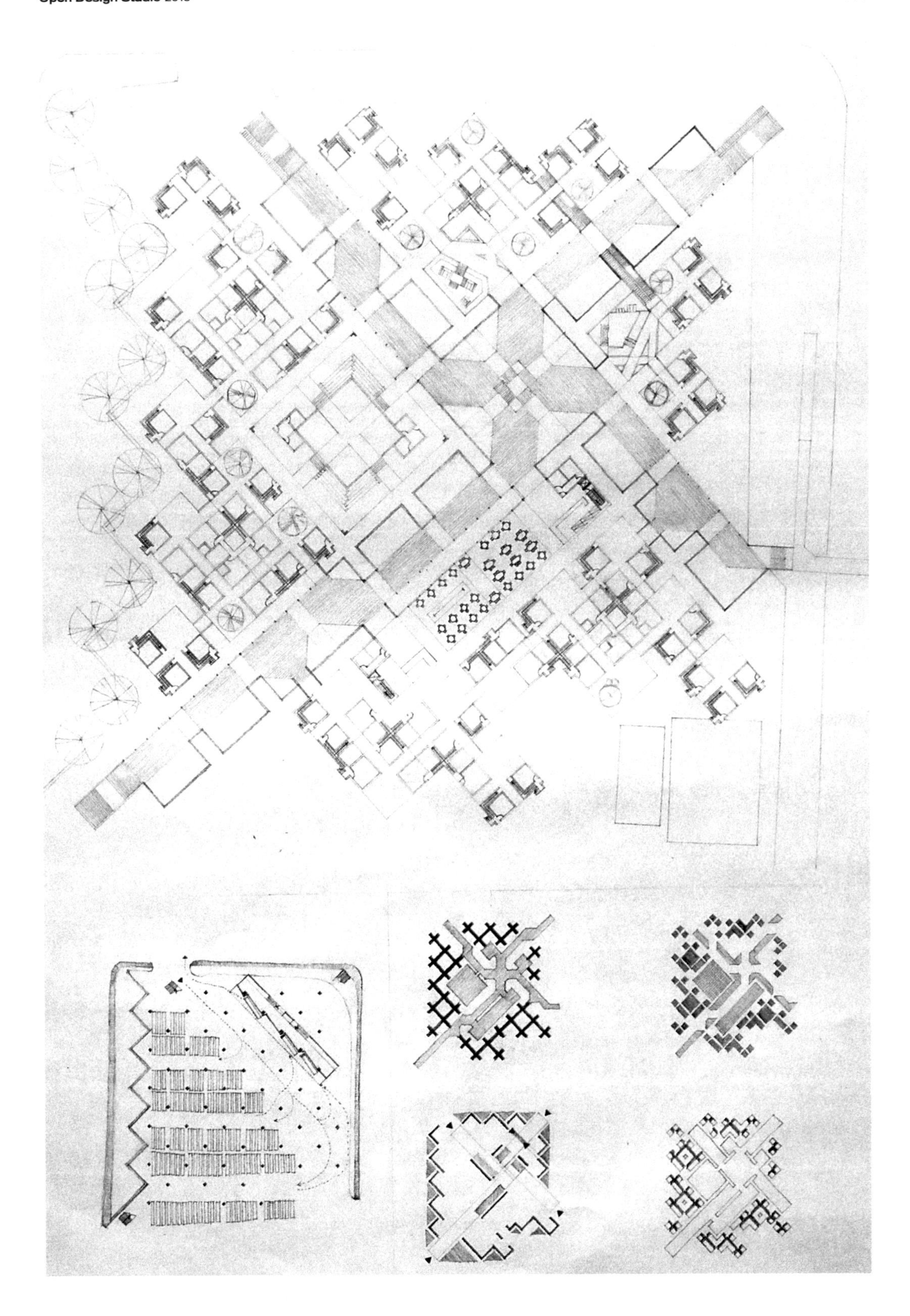

The groups composed of young people are featured by independence, rebellion, innovation and other characteristics. In order to build a community of personality for the young people in the whole urban environment, the unique order that is different from the surrounding texture should be presented in the current city. In order to be distinguished from the environment, the separated unit block has a unique texture and density; the axis and street presents the angular relationship, and self-growing nature from inside to outside is produced with the establishment of endocentric order.

With the basis of group expression, the unified unit form grants the young people maximum personality highlight. The heterogeneous space character makes expression to the environment through the concept of "window", which makes the activities of the micro city different for richness. The image of circuit board has different approaches but equally satisfactory results with the "window" concept. Linear trends group the lines, and the enlarged terminal connects with key electronic components, which is similar to the linear movement of people on the field, and the facade form opening to the city.

The original concept has soul for being endowed with the functions. The forms correspond to several aspects of the functions, which are respectively reflected in the sections, axes, facades along the street and other contents. In the development process of the concept, the real activities of people in the micro city are gradually reflected.

本页及对页：预设场景图。**下页：**建筑外观图及场景图。

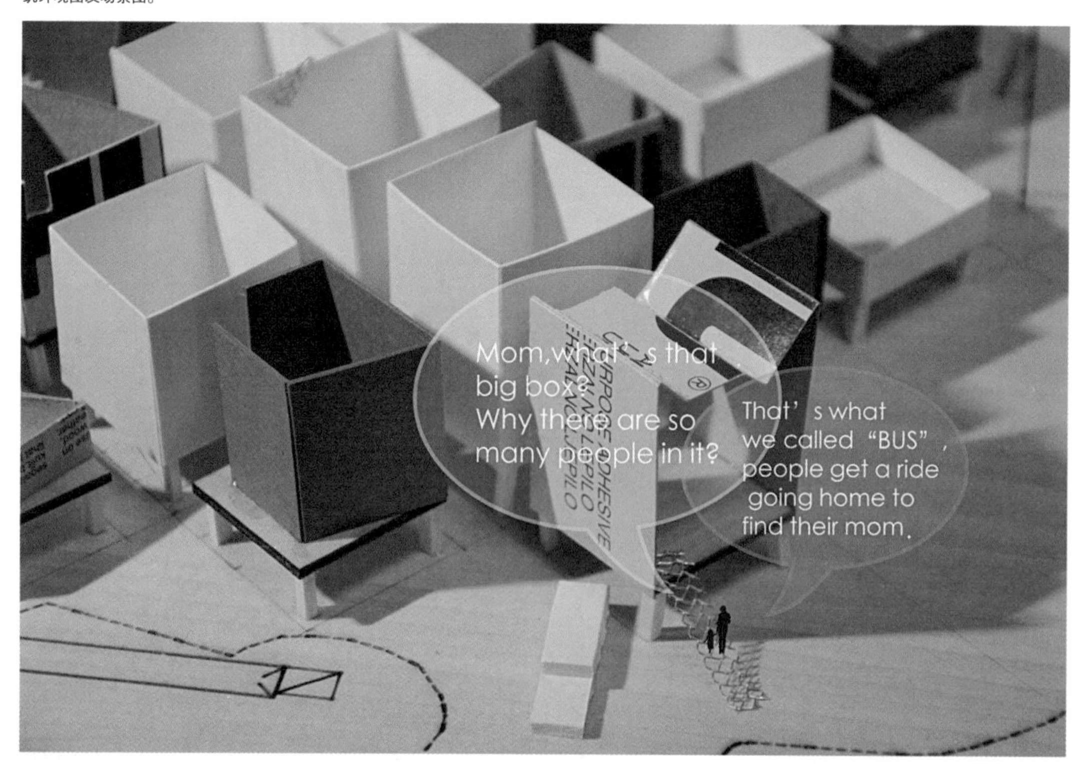

eave me
lone! I don’t
ant to see
ou any more!
Wait! I didn’t mean
it! Sandwich with milk
is just fit like you and
me! I’m sorry for the
words I said...

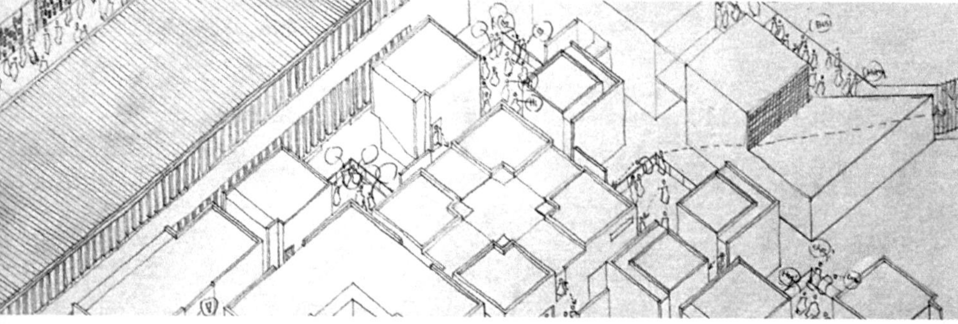

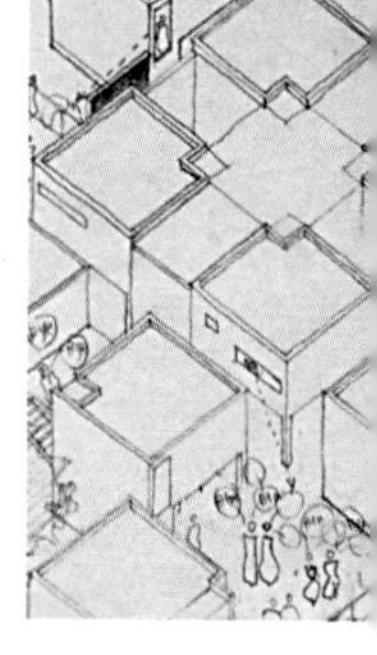

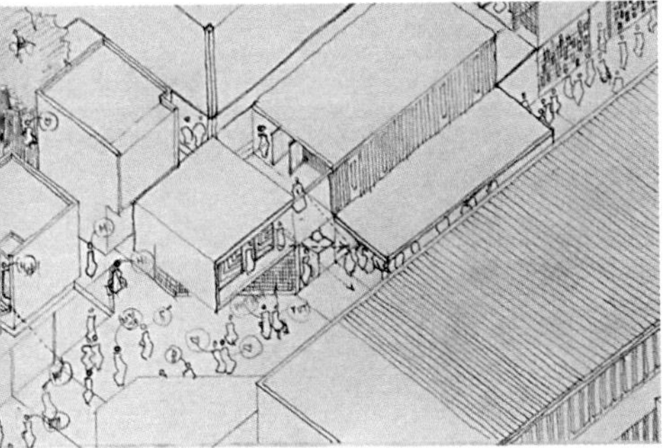

教师点评

庞凌波的这个设计可以说是个比较典型的"微城市",但并不等于说是"微缩城市"。二者的区别在于前者是功能性的,因为在这个尺度上城市的功能在运营,甚至运营得更好;而后者是形式性的,因为在这个尺度上城市只停留于模型式的视觉层面,而毫无功能可言。这个设计比较有趣的是在街区格网明确后,每个地块的更深入设计变成了类型学的操作。这种操作在浅层层面上是空间形态上的各种可能性的探索与取舍;而在深层层面上则是当每一个单体变成一个有自身逻辑的生命体后,如何形成一个有自主性的社会生态。因此,这个设计在选择了最普通的格网之后,设计者仿佛给自己画了个棋盘,不断地推演各种棋局。而在八周短短的时间内,随着自身用功度的加强,以及对棋局认知的加深,棋艺也不断长进。我相信通过这种模式,学生能从形式入手,最终却要摆脱形式,而能理解形而上的操作是种必要的语言培训,但其目的还是要来赞美建筑和城市中鲜活的生活。

Teacher's comments

Pang Lingbo's design can be viewed as a typical "micro city", but it is not meant to be a "miniature city". The difference between the two definitions is that the former is functional, because the function of the city still operates, or operates even better, in this scale; the latter is formal, because the city only stays on the model visual level but without function. The interesting point of the design is that after the grid of blocks is defined, the further design of each plot becomes the research of typology. In the shallow level it is about the exploration and choices of all possibilities of the space forms, while in the deep level it is about how to form social ecosystem with autonomy when each monomer becomes a life of its own logic. Therefore, after the most common grid was selected in the design, the designer seemed to paint a chessboard for herself and constantly deduce all solutions. Within a short period of eight weeks, with the intensified hardness and increased depth in the cognition of the chess game, her skill in chess playing has continued to progress. I believe that through this model, the student can start with the form and eventually get rid of the form to understand that the metaphysical operation is a necessary language training, but its purpose is to praise the fresh life in architecture and city.

城市碎片/ CITY FRAGMENTS

项目选址：北京市朝阳区大屯客运中心

项目类型：城市综合体

建筑面积：10 330m²

占地面积：26 802m²

方案设计：金爽
指导教师：王辉
完成时间：2015

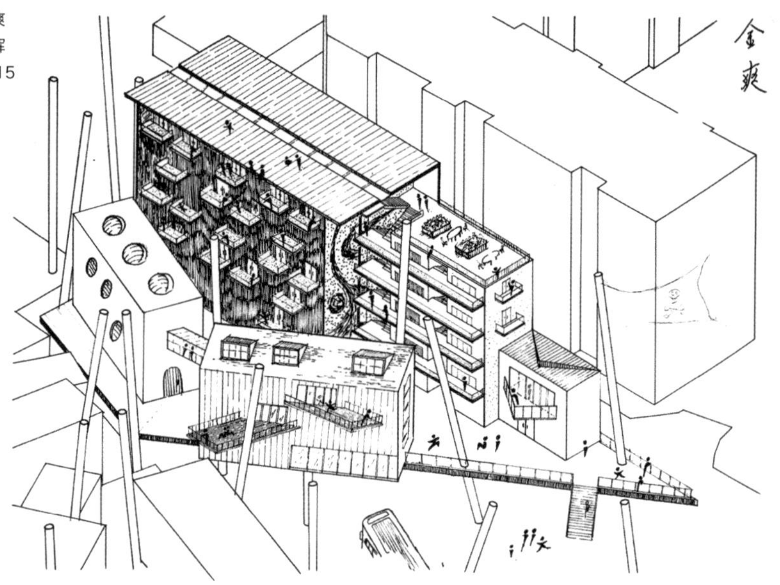

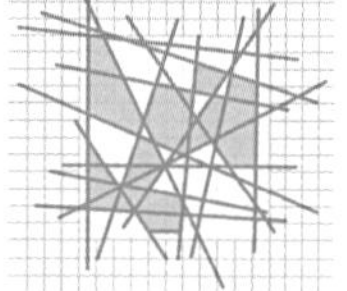

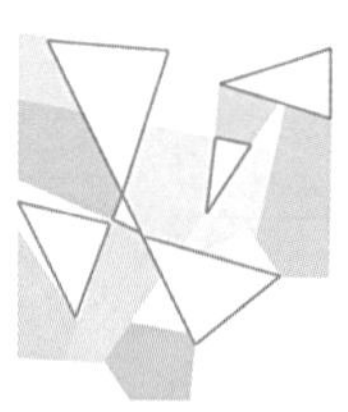

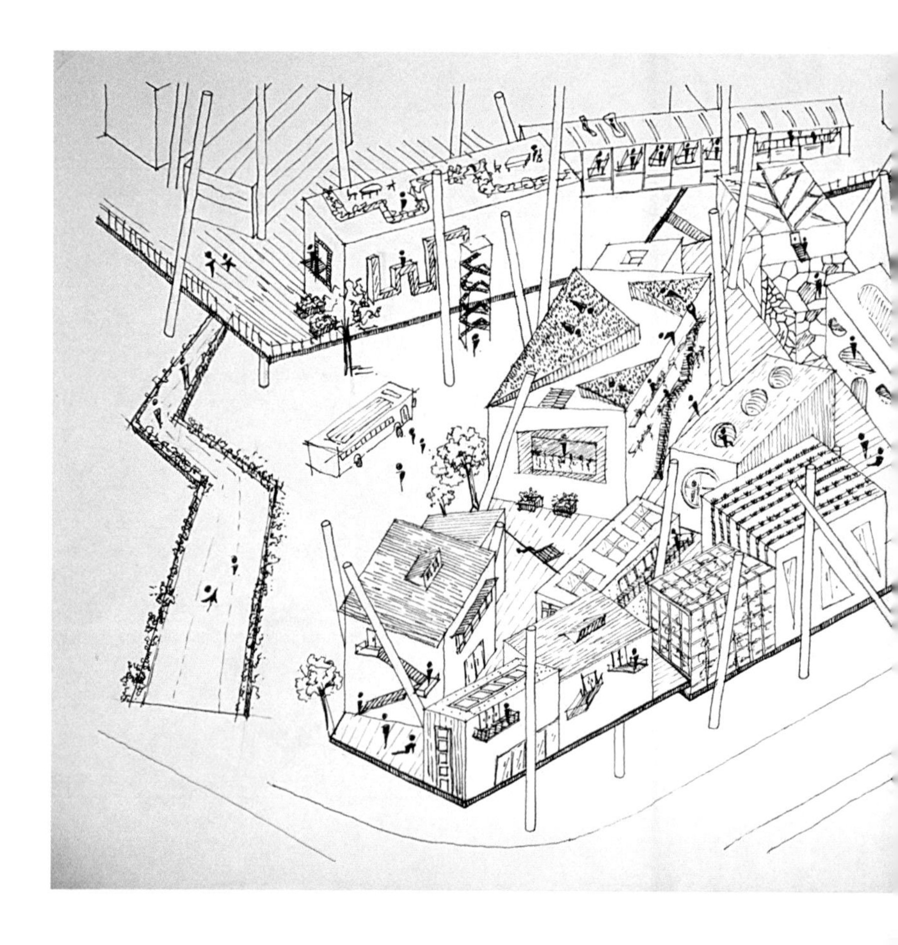

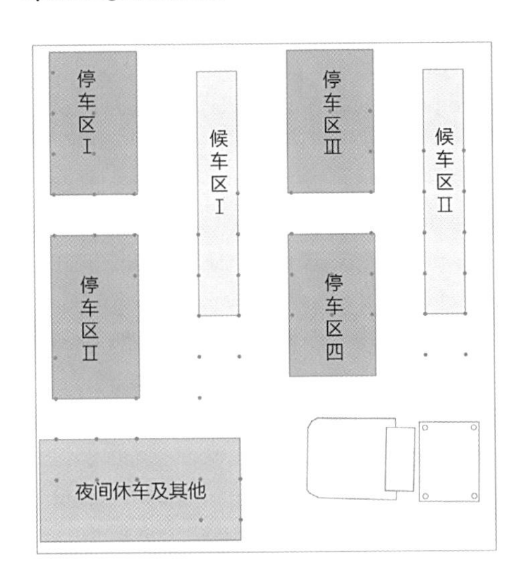

车场布置

在周边配套设施齐全的公交场站上设计一个年轻人的微城市，我的着眼点是探究社区内部活力的激发点。图书馆作为一种被普遍需要并热烈提倡的公共资源，一直以来是单体式或聚合态的，若将其按功能类型解构打散，以“碎片”方式填充进年轻人的微社区，成为细化区域的核心，不仅提供了平心静气阅读的场所，也沟通了周边为创客设计的办公、生活等空间，同时提供了一种新型的生活理念——即通过共享公共资源来展开社会行为，包括学习、工作、交友等。被打碎的图书馆作为社区配套设施，亦可辐射带动微城市外的区域，吸引不同年龄、阶层的人前来体验。夜晚来临时，围合在中间的玻璃图书馆将成为盏盏明灯，激活社区、点亮夜色。

开篇：主要方案图。**对页上图**：初期碎片意象。**对页下图**：上盖生成。**跨页**：初期碎片意象。**本页上图**：车场布置。**本页下图**：总平面图。

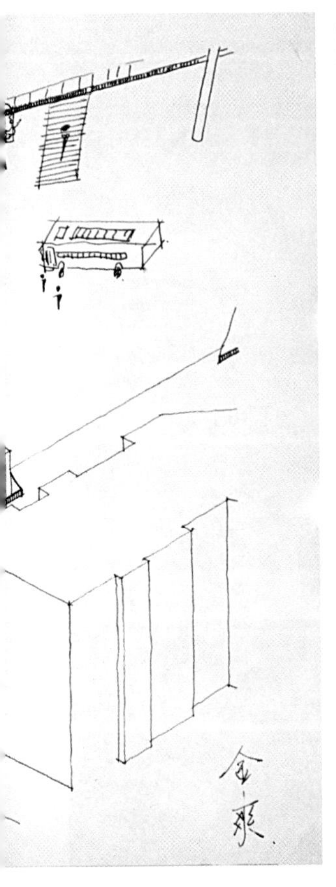

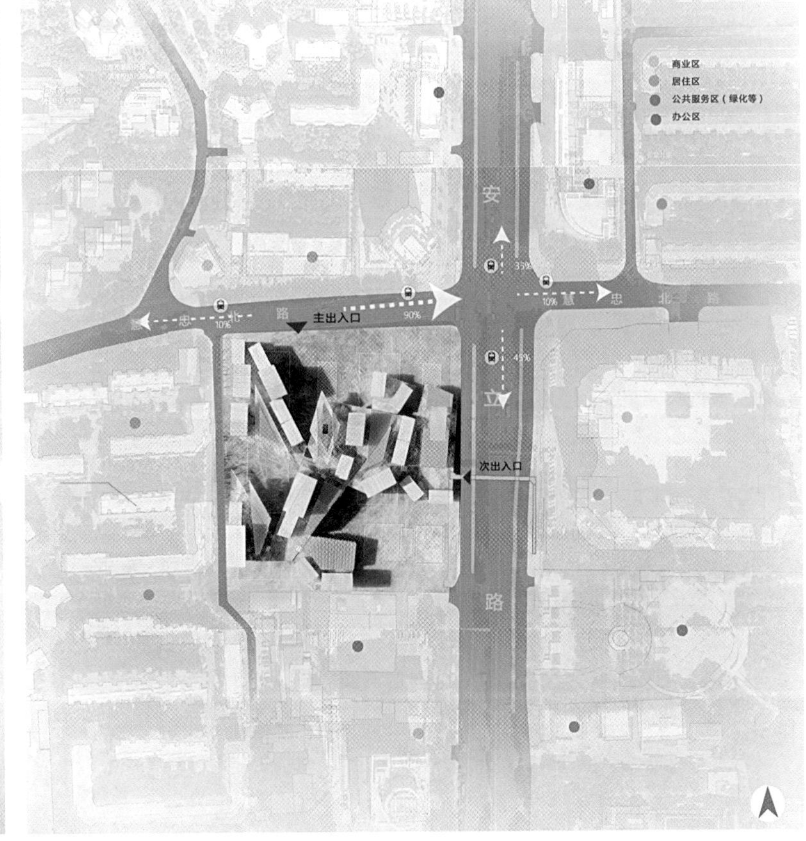

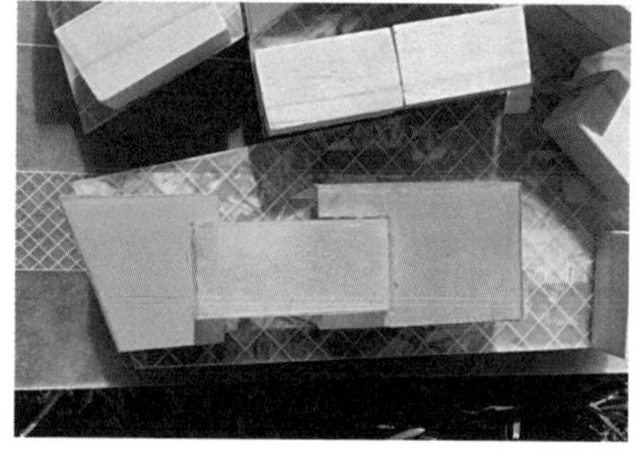

本页上图：模型细节。**本页下图**：夜景。
对页图：碎片辐射。

Form of original building group is adopted for this design, namely the rectangular box with layout vertical to the current pattern. Due to the requirements of traffic, I set a negative form of 13th apartment in the part with space next to the 13th apartment. The 13th apartment also extends to the both sides along the original direction, and connects with the east and west buildings. Half of the shape of the 13th apartment is left in both sides as the channels. This method breaks the original good sight accessibility, and people will not see the things behind the building. On this basis, I hope the flow line passing through the building can be separated with sight of people. So, negative form of 13th apartment is diagonal, which is contrasted with the straight system around. To make the front and rear spaces be more dramatic, I take the reflecting glass curtain wall as the building facade, so people seem to see the extension of the 13th apartment from one side, but will find another world after passing through the building.

教师点评

金爽同学的主题词是碎片。这个貌似不理性的词及其在规划上的表现，其实有非常理性的基础。作为一个几乎是在“孤岛”上的年轻人的社区，很可能成为平庸城市中更平庸的一个场所。制造它活力的第一推动力在哪里？金爽通过解构一个社区图书馆，解决了这个问题。社区图书馆按阅览功能被分解成若干部分，如儿童、社科、科技等等，成为微城市中几个核心，周围围绕着为创客设计的工坊、办公室、住宅、配套。这样，一个政府投入的社区配套，变成了一个激活一个社区的力量，同时自己也得到了新生。

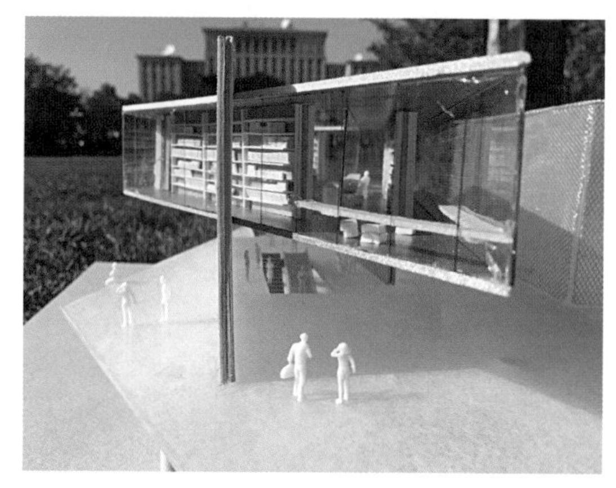

Teacher's comments

The subject term for Jin Shuang is "fragment", which is a seemingly irrational word but has a rational basis for its performance in planning. The community of the young on an "isolated island" is likely to become a mediocre place in the mediocre city. What is the primary impetus for its vigor? Jin Shuang solves this problem by deconstructing a community library, which is decomposed into many parts including children, social science, science & technology, etc. according to reading functions. These parts become the cores of a micro city, surrounded by the workshops, offices, residence and supporting facilities designed for makers. In this way, the community supporting facilities invested by a government become the power to activate a community, and the community will be reborn.

对页上图：建筑外观。**跨页图**：建筑外观。**本页下图（顺时针）**：书架围合之下。隐现的书架。底层停车场。

有故事的城市/ TALKING CITY

项目选址：慧忠里公交场站
项目类型：年轻人的微城市
建筑面积：1.8 万 m^2
用地面积：2.2 万 m^2

方案设计：唐诗童
指导教师：王辉
完成时间：2015

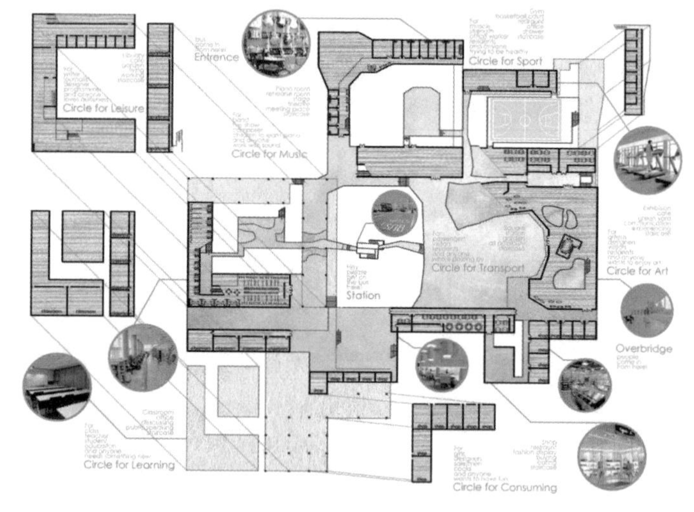

项目地段位于一个相对繁华的公交场站上方，题目是建起一个年轻人的微城市，因而，我期待这里能够成为一个有“故事”的城市。在功能上，“圈子”将微城市的受众更好地分类，提供面向公众开放的设施，汇聚人流，激活城市；在形式上，多变的楼梯让人们的速度放缓，目光放远，楼梯作为连接体在建筑中穿梭，成为了激发故事发生的场所。

年轻人的微城市/ YOUNG PEOPLE IN THE CITY

项目选址：北京市朝阳区大屯客运中心
项目类型：城市综合体、
建筑面积：$8000m^2$ **用地面积**：$12000m^2$

方案设计：孙仕轩
指导教师：王辉
完成时间：2015

方案考虑到年轻人流动性大、收入较低、生活方式多变的特点，选取集装箱作为基本单元组合出不同的功能模块，提供给年轻人自己动手改造空间的可能性。同时几个地块围合产生出两个核心绿化公园，吸引城市人流进入，激发相互间交流的契机。

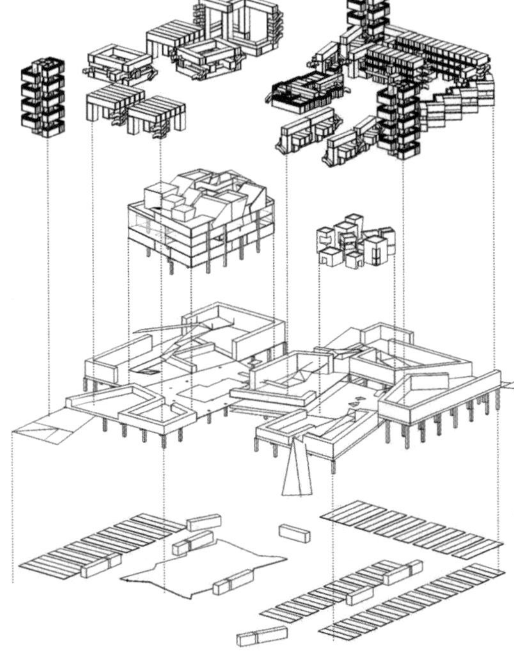

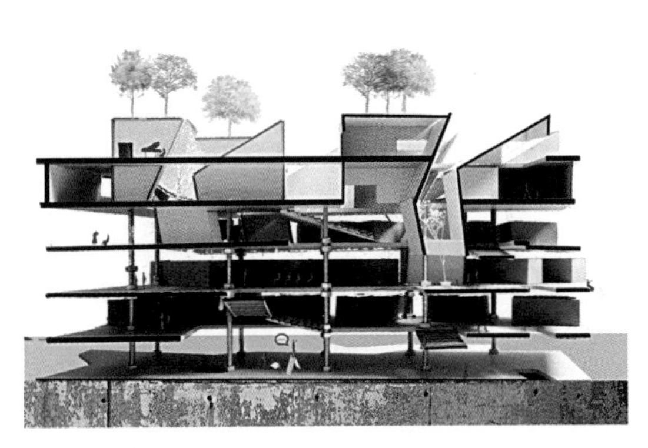

漫游城市/ROAMING CITY

项目选址：北京市朝阳区慧忠里公交场站
项目类型：年轻人综合体
建筑面积：27 000m²
用地面积：20 000m²

方案设计：唐雨霏
指导教师：王辉
完成时间：2015

设计从条带原型出发，尺寸相近的条形平台分别植入居住、交往、工作室等功能。平台之间利用绿色平台联系、刺激年轻人的交流，同时利用反光板使平台上下产生互动。各条带连续大空间中利用多样的剖断面让人的行走过程不断变化，使漫游其中的过程变得有趣、从而触发更多人与人交流的可能。

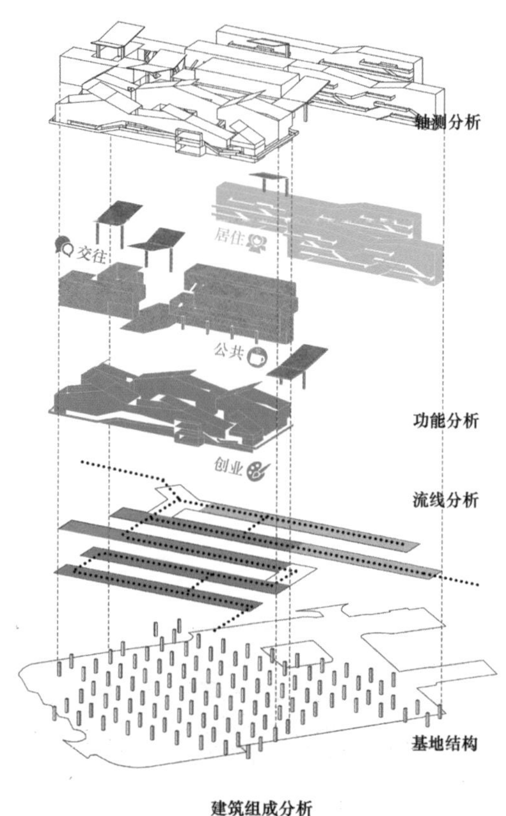

建筑组成分析

各条带平台分别植入居住、公共办公和艺术创业等功能。各平台之间利用开敞空间和顶部反光板构成年轻人的交往空间。底层基地结构为方格网柱网，整体方向斜向路口偏转。

城市工房/ STUDIO CITY

项目选址：北京市大屯公交场站
项目类型：商业、居住
建筑面积：15 000m²
用地面积：20 000m²

方案设计：毛宇帆
指导教师：王辉
完成时间：2015

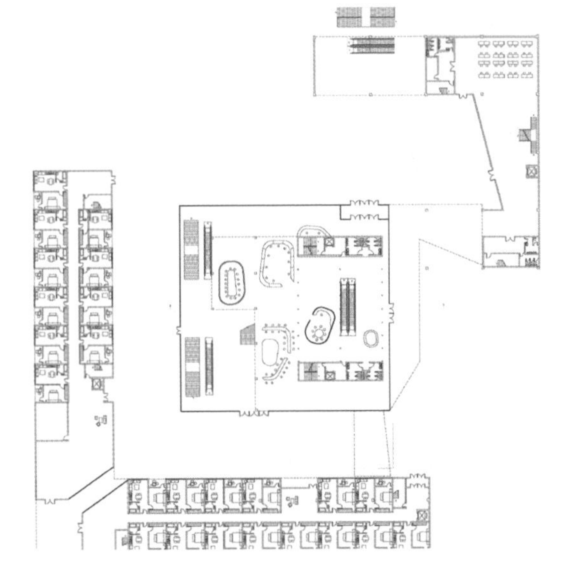

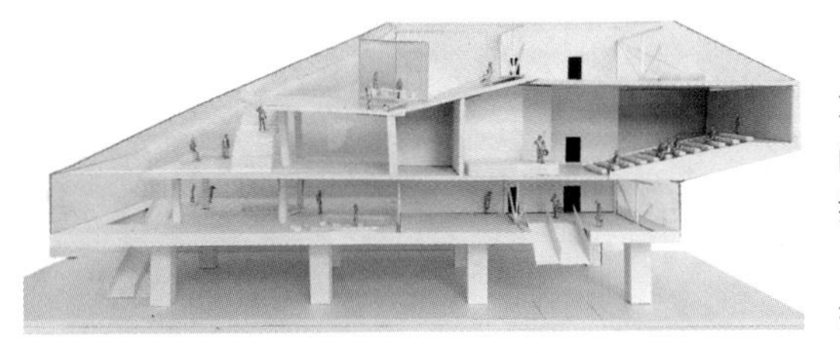

方案以“山”为概念，建筑群如同城市中的村落，意为年轻人的落脚点。通过四周起伏状的集合住宅，形成层峦，为年轻人提供低成本的住所，缓和的屋顶可进行休闲活动；中部布置“山石”状，向下直通公交场站的文化创意中心—城市工房，提供居住以外丰富的交流机会。

补丁城市/ PATCH CITY

项目选址：北京市朝阳区大屯公交客运中心
项目类型：城市综合体，包括社区、市场、文化中心与车站等核心功能
建筑面积：15500m² **用地面积：**22500m²

方案设计：王佳怡
指导教师：王辉
完成时间：2015

方案通过置入补丁元件，以空中院落将原本被公交场站割裂开的社区、商业服务设施和车站缝合串联，填补城市空隙，缝合空间业态。补丁院落呈开放的姿态，将邻里尺度还原于城市环境，将年轻人的社会形态实体化、网络化。多样的灰空间构筑起一连串的城市公共空间，进而激活城市，以小尺度的宜居环境作为年轻人与城市交换能量的场所。

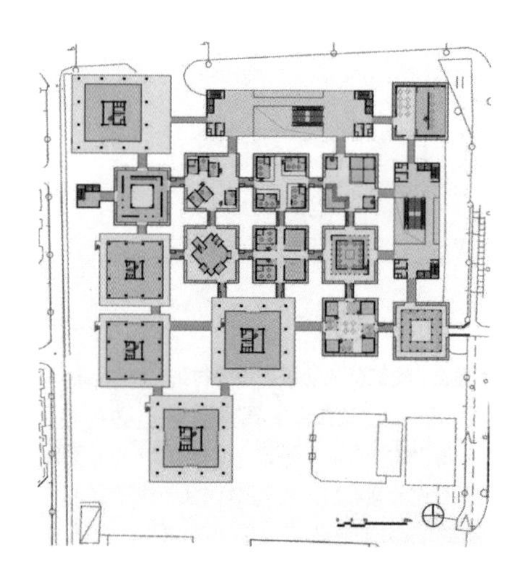

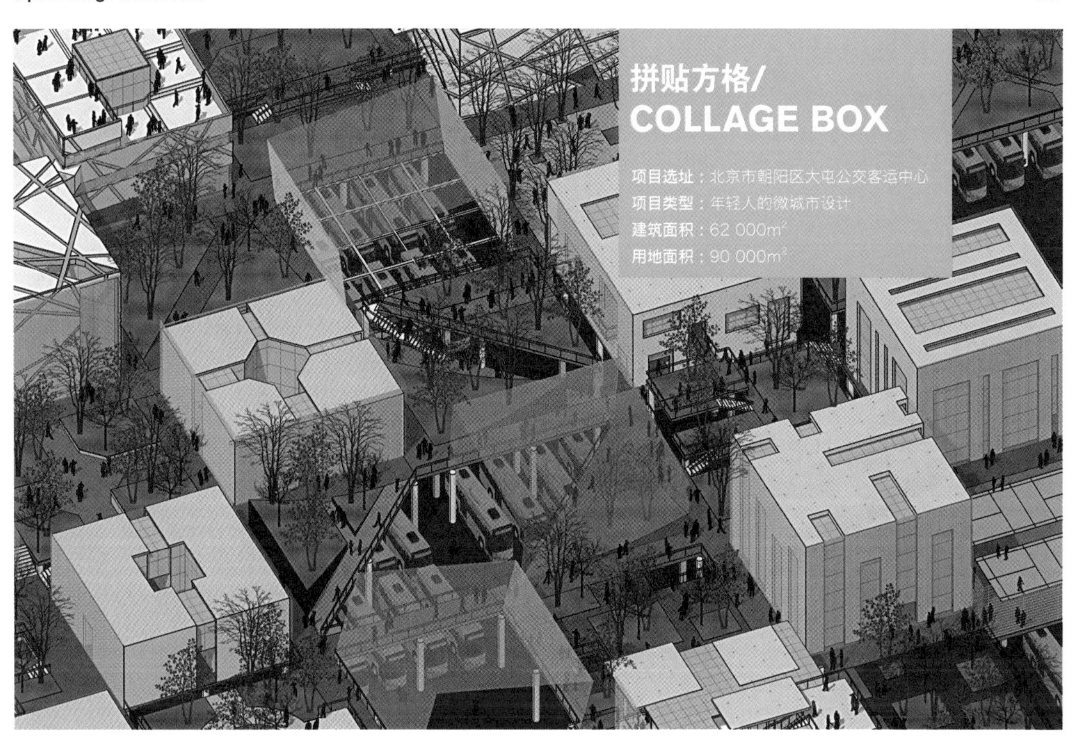

拼贴方格/ COLLAGE BOX

项目选址：北京市朝阳区大屯公交客运中心
项目类型：年轻人的微城市设计
建筑面积：62 000m²
用地面积：90 000m²

方案设计：吴承霖
指导教师：王辉
完成时间：2015

年轻人在这个城市里漂泊奋斗，却缺少一个任其施展个性的空间。而拼贴方格提供了一个供各种各样年轻人活动的场所，置放在一共36个网格里，靠方格状路网联系在一起。

16个建筑体为不同的功能，酒吧，网吧，书店，博物馆，展览馆，画廊，小剧场，KTV，艺术家工作室，园艺室，体育场，迷宫等，为年轻人提供了丰富而有趣的活动场所；而20个院落则围绕周围建筑形成院落和各有特色的绿地，成为良好的室外景观。利用公交场站原本的空闲上层空间，能够为支付能力有限的年轻人提供更优良的场所。36个方格36种特色，拼贴成为性格各异的年轻人活动的微城市。

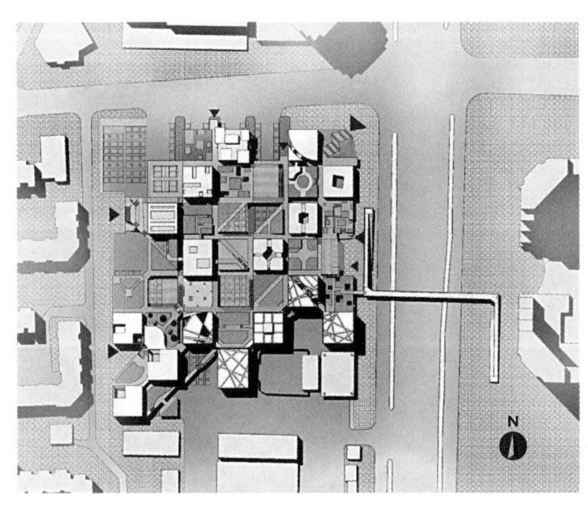

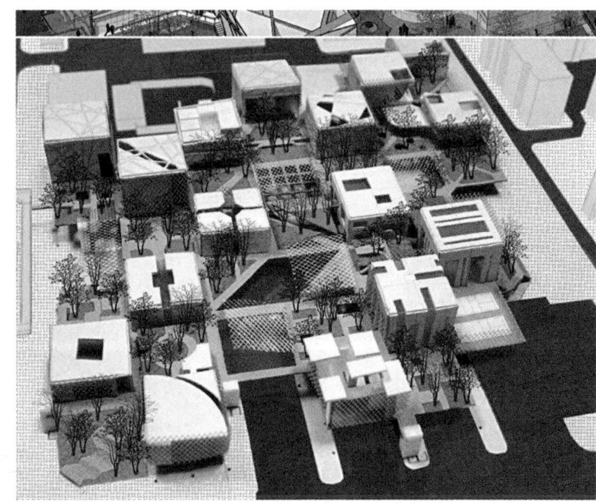

漫步田园/ WANDERING CITY

项目选址：大屯中心场站
项目类型：城市综合体
建筑面积：17 550m²
用地面积：12 000m²

方案设计：吴濯杭
指导教师：王辉
完成时间：2015

年轻一代留恋着繁华都市，又向往着田园风光，因而设计从都市农场切入，探讨建筑与城市的关系，用街道串联组团，用地标点亮聚落，就平台、灰空间、农场生活等不同专题进行了研究，希望为年轻人的城市生活增添一分活力。

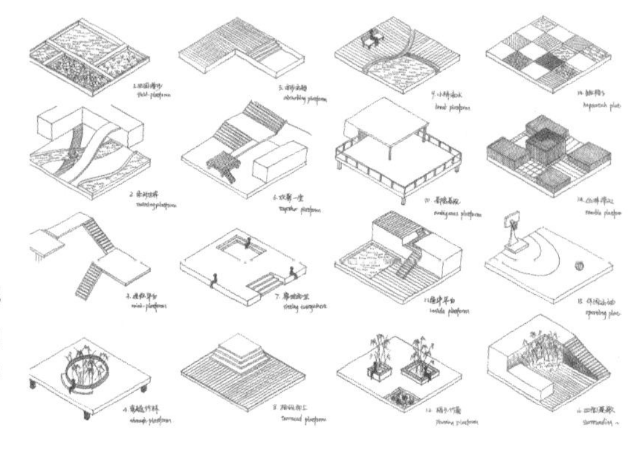

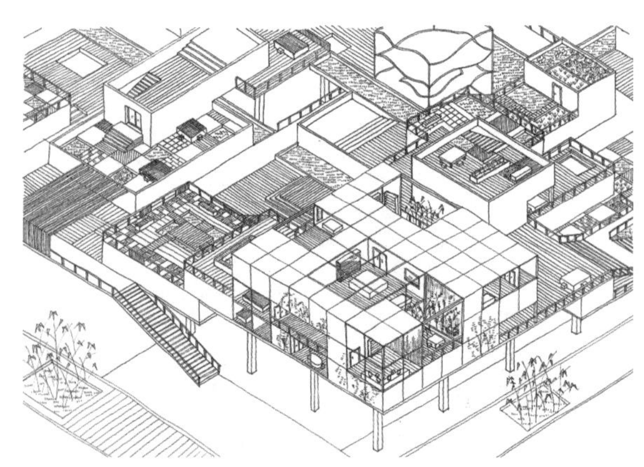

托邦城/ TOPIA-CITY

项目选址：北京市朝阳区大屯公交客运公司
项目类型：集商业、居住、娱乐于一体的城市综合体
建筑面积：18 000m²
用地面积：22 500m²

方案设计：白楠
指导教师：王辉
完成时间：2015

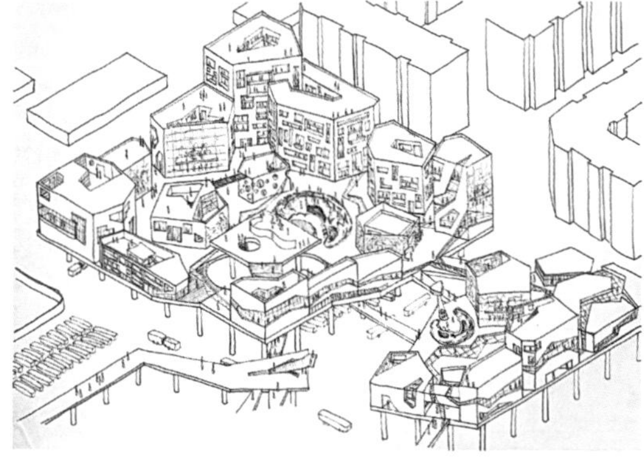

托邦城的设计，以年轻人反叛的性格与汇聚的活动为出发点，将涂鸦活动赋予建筑环境，使整个城市综合体成为建筑与涂鸦的集合体，年轻人在建筑中的活动始终会与遍布在建筑各处的涂鸦画面形成互动，这是一种看似Utopia的建筑理想，但用一种Heterotopia的方式展示出来，形成了年轻人活动模式的新鲜场所。

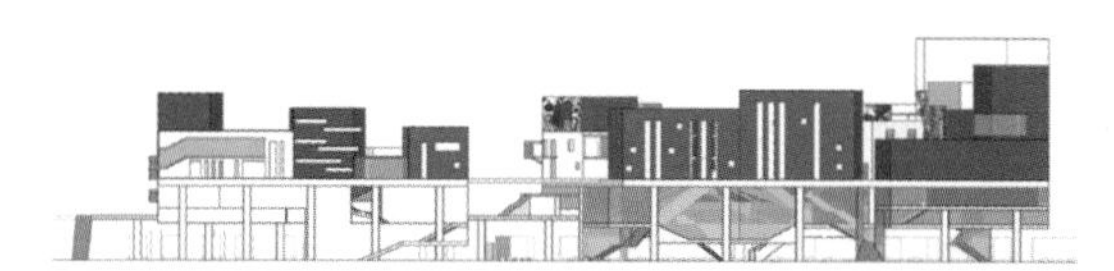

清华大学十

齐欣

齐欣建筑设计咨询有限公司
董事长，总建筑师

齐欣
齐欣建筑设计咨询有限公司
董事长，总建筑师

教育背景
1978年 – 1983年
清华大学建筑学院　学士
1985年 – 1988年
巴黎Belleville建筑学院　研究生
1984年 – 1991年
巴黎La Villette建筑学院　建筑师

工作经历
1990年 – 1993年
法国巴黎建筑与城市规划设计院　建筑师
1994年 – 1997年
英国福斯特亚洲建筑设计事务所　高级建筑师
1997年 – 2000年
清华大学 副教授
1997年至今
清华大学 特聘外国专家
1998年 – 2001年
京澳凯芬斯设计有限公司　总设计师
2001年 – 2002年
德国维思平建筑设计咨询有限公司　总设计师
2002年至今
北京齐欣建筑设计咨询有限公司　董事长兼总建筑师
2004年至今
清华大学 设计导师
2014年至今
中国科学大学 荣誉学衔教授

主要论著
齐欣.建筑的本土化和公共性 [J].时代建筑，2012(7): P74-79
齐欣.齐欣建筑—似合院 [J].a+u 建筑与都市，2009(8): P16-23
齐欣.合院谐趣—似合院 [J].世界建筑， 2008(6): P106-113
齐欣.北京奥林匹克中心区下沉广场 [J].建筑创作，2008(8): P44-45
齐欣.天创科技大厦 [J].建筑学报，2005(7): P68-71

设计获奖
2002年 – 获WA建筑奖
2003年 – 被评为中国房地产十佳建筑影响力青年设计师
2004年 – 获亚洲建筑推动奖
2004年 – 获法国文化部授予的艺术与文学骑士勋章
2010年 – 获全国优秀工程勘察设计行业一等奖
2011年 – 获北京国际设计三年展建筑设计奖

代表作品
北京国家会计学院（图1）、廊坊商业街坊（图2）、东莞松山湖管委会（图3）、杭州玉鸟流苏（图4）、南京江苏软件园（图5）、杭州西溪湿地三期（图6）、北京奥体公园下沉广场（图7）、天津于家堡写字楼（图8）、天津武清剧院（图9）、北京大栅栏北京坊（图10）

清华大学十三公寓改扩建

三年级建筑设计（6）设计任务书

指导教师：齐欣

扩张型的大规模建造已进入尾声，而地球接着转，城市到处有，人要继续活，房子还得盖。终于，我们离开了始发站，开始为现在经营过去，为未来准备曾经。

城市是一个在不断沿革的躯体，每一代人的设计都是一个逗号。如何在前人写下的句子后面填词，如何不在一张白纸上画出最新最美的图画，是这个题目的关切。

兵营式的多层住宅布局遍及大江南北，这一简单的格式为来者预留了无限的可能。清华园里的十三公寓就处在这一格局中，也曾经历过局部的改造，但仍在族群中最小最矮，从而奠定了厚积薄发的基础。

在这里，行列式布局并非是我们要嘲讽的落伍者，更不是我们要唾弃的对象。相反，它成了继往开来的基石，引导我们寻求地域本身的特质，并通过有针对性的设计，造就出多姿多彩的局面，让生活更精彩，让城市更美好。

十三公寓介绍

清华园的西南家属区始建于1930年。1958年建成十三公寓时，这里已有几十栋版式公寓和点式别墅。经过四十年的建设，点式别墅逐渐被板楼替代，形成明显的行列式肌理。

十三公寓由五个一梯两户的单元组成，砖混结构，进深10m，总长60m，共三层，四坡顶，檐口高10m，初始外观呈清水红砖墙。1984年对楼体进行过一次加固，并扩建出南北阳台。新添部分的外墙为水刷石。总建筑面积1567m^2。

地段周边情况

十三公寓的南侧立着同时期建造的三层高的二公寓，为平顶的灰砖建筑。两者间现有一排临建板房，已基本被废弃。十三公寓的北侧是一片空场，树木茂盛，用于停车和居民的户外活动。空场的北侧立着五层高的西10号楼。再向北，则是两岸成荫的校河，以及一条由西向东的道路。这条路，也是从清华正面（西门）进入校园后，通往学校中心区的必经之路。

总体上，该区域内的建筑多为五层高的砖混住宅，功能相对单一，楼间距较大，为后期加建留有充足的机会。为了避免新建筑对现有住宅构成日照遮挡，设计组还运用日照软件反求建筑的极致形体，得出了在不同用地范围内建筑的极限设计高度。

从居委会提供的资料显示：西南家属区共有住户1006户，常住人口3181人，以清华大学离退休教师为主，文化程度较高，老年人居多。

课程要求

通过一个单体建筑的加改建，激活一个社区，使其整体风貌和生活质量有所提升。加建出的建筑面积不小于现状，主要为范居住功能，可配置不大于30%新添建筑面积的社区公共服务功能。扩建后的十三公寓不可影响周边建筑的日照和交通，加出的建筑部分在有依据的前提下可规避日照规范。用地范围自定。

教学安排

第1周，集体分工合作

1.追寻社区历史

2.调研社区需求

3.整理现状建筑图纸

4.从事日照分析

5.设想范居住功能

第2周，制定、确定个人设计任务书

第3~4周，概念设计

第5~8周，深化设计

十三公寓改扩建设计/ RECONSTRUCTION AND EXPANSION OF NO.13 APARTMENT BLOCK INTSINGHUA UNIVERSITY

项目选址：清华大学照澜院十三公寓及周边

项目类型：公寓改扩建，仍用于居住

建筑面积：11000m^2

占地面积：4700m^2

方案设计：何文轩
指导教师：齐欣
完成时间：2015

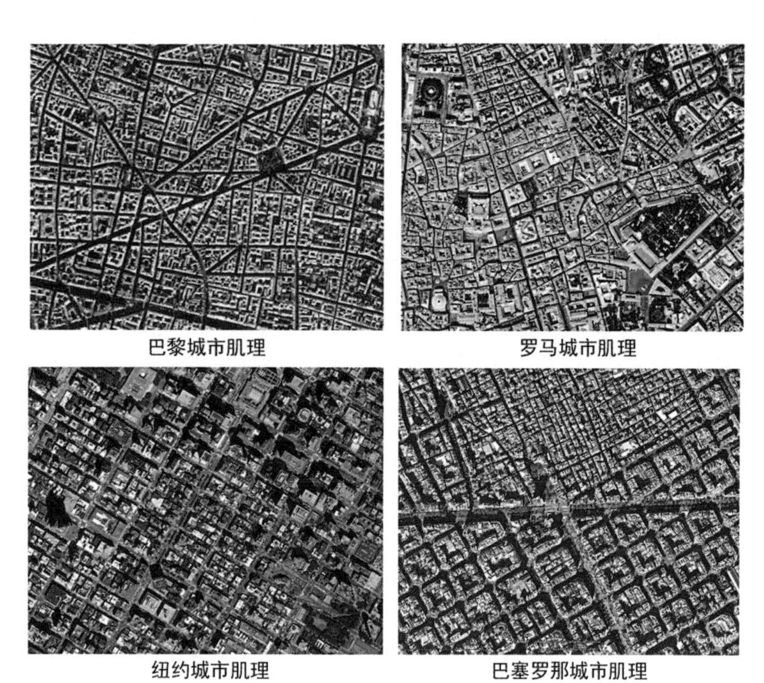

巴黎城市肌理　罗马城市肌理

纽约城市肌理　巴塞罗那城市肌理

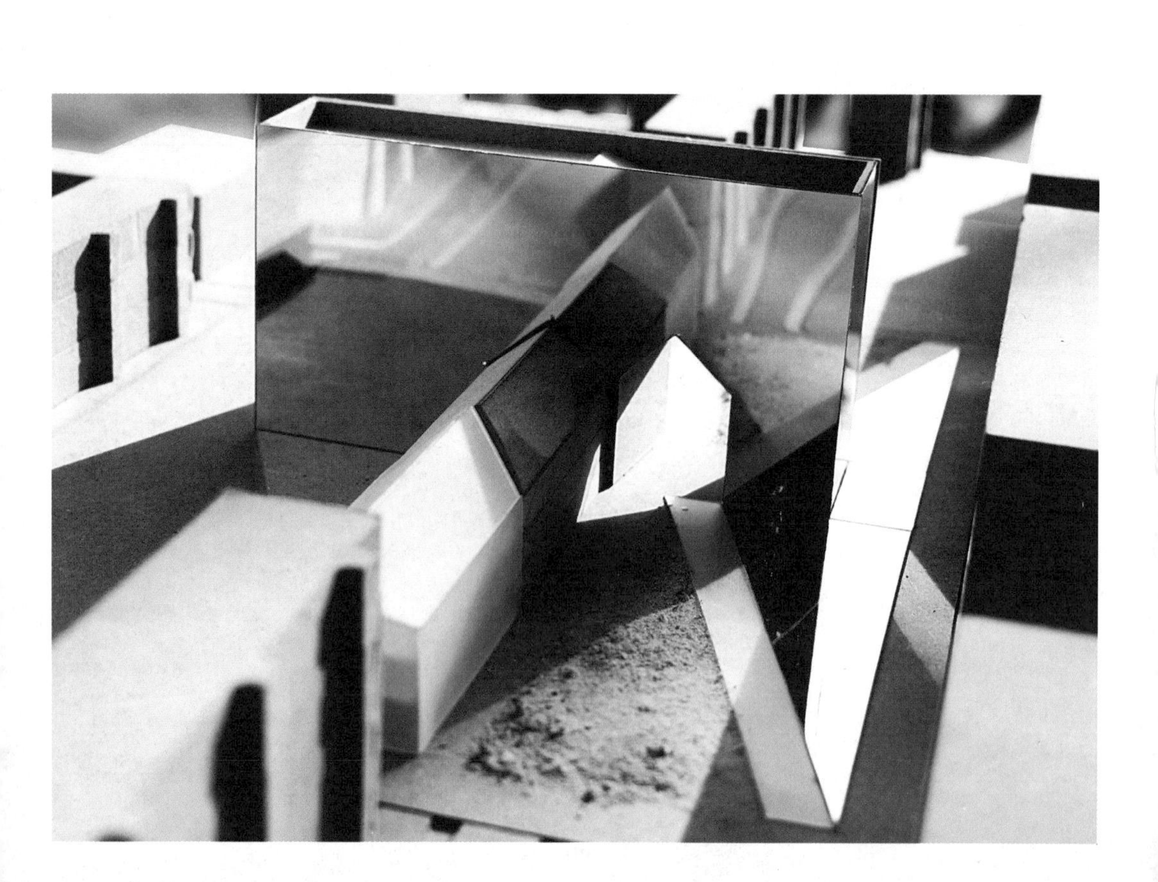

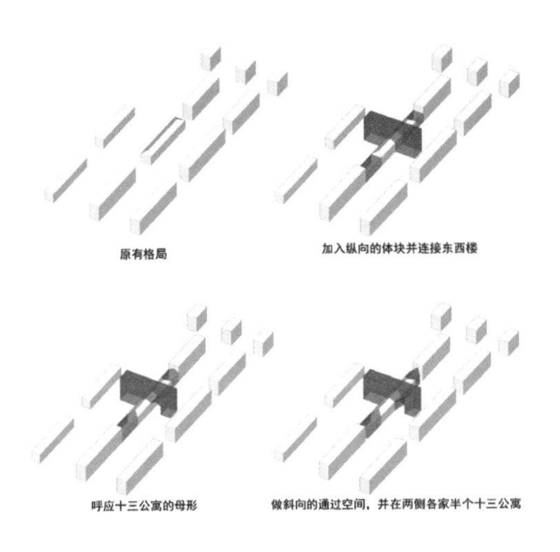

原有格局
加入纵向的体块并连接东西楼
呼应十三公寓的母形
做斜向的通过空间，并在两侧各家半个十三公寓

开篇：主要方案图。**对页上图**：概念来源。**对页下图**：整体鸟瞰。**本页上图（从左至右）**：生成过程。西侧道路。**本页下图**：主要方案图。

本设计采用了原有建筑群的形式，即长方形的方盒子，却用了与现有格局垂直的布局。由于交通要求，在紧邻十三公寓的部分设置了一个十三公寓的负形。而十三公寓本身也沿原有方向向两侧延伸，与东西楼相连接，两侧各留出半个十三公寓的形作为通道。这种做法带来的效果是原先良好的视线通达性被打破了，人们看不到我的建筑后面是什么。在这样的基础上，我希望人们穿过建筑的流线能和视线相分离，所以十三公寓的负形是斜的，与四周横平竖直的体系形成对比。为了前后空间更加戏剧化，用反射玻璃幕墙作为建筑立面，人们在一边看到的仿佛是十三公寓在延伸，而穿过建筑后发现又是另外一片洞天。

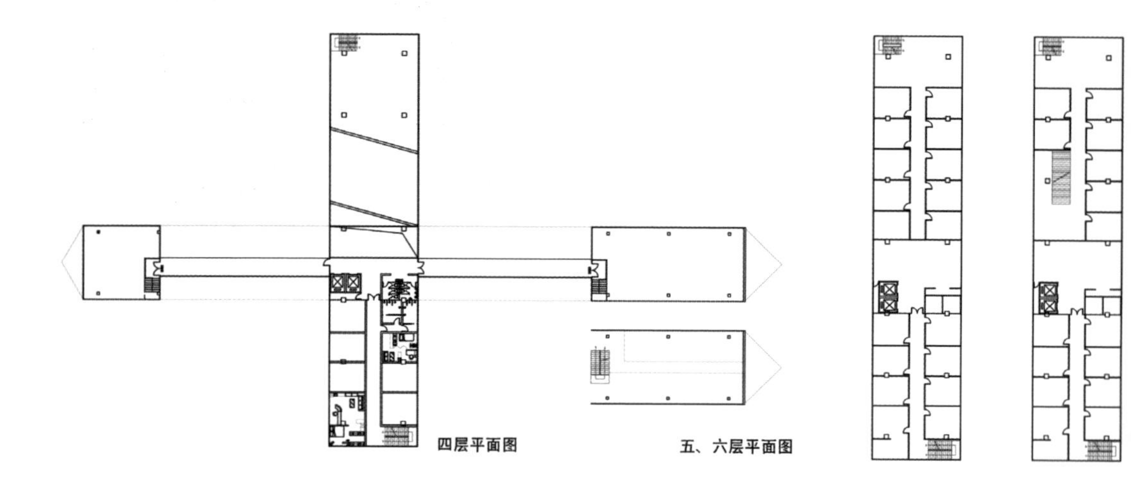

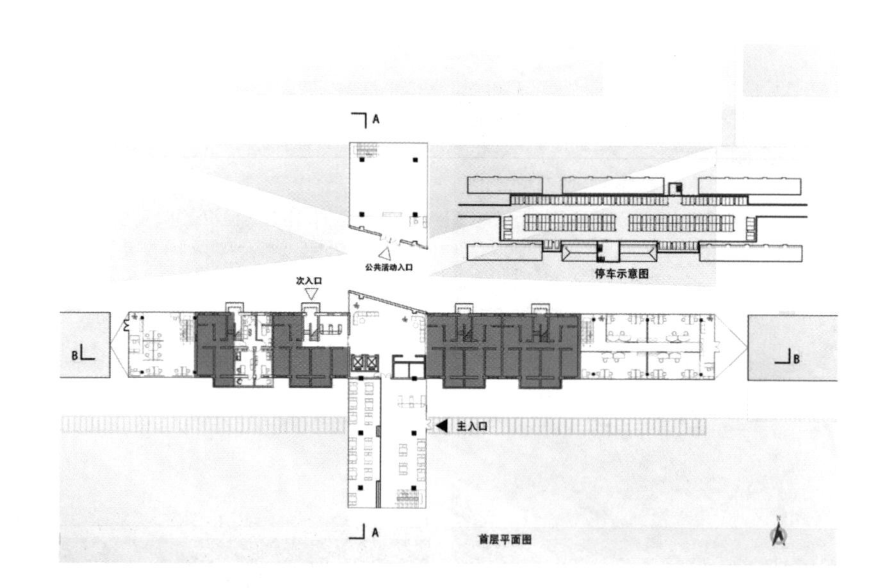

Form of the original building group is adopted for this design, namely the rectangular box, but with the layout vertical to the current one. Due to the requirements of traffic, I set a negative form of the NO.13 apartment at the part next to it. The NO.13 apartment also extends to both sides along the original direction, and connects with the east and west buildings. Half of the shape of the NO.13 apartment is left in both sides as passage. This method breaks the original good sight accessibility, and people will not see the things behind the building. On this basis, I hope the flow line passing through the building can be separated with sight of people. So, negative form of NO.13 apartment

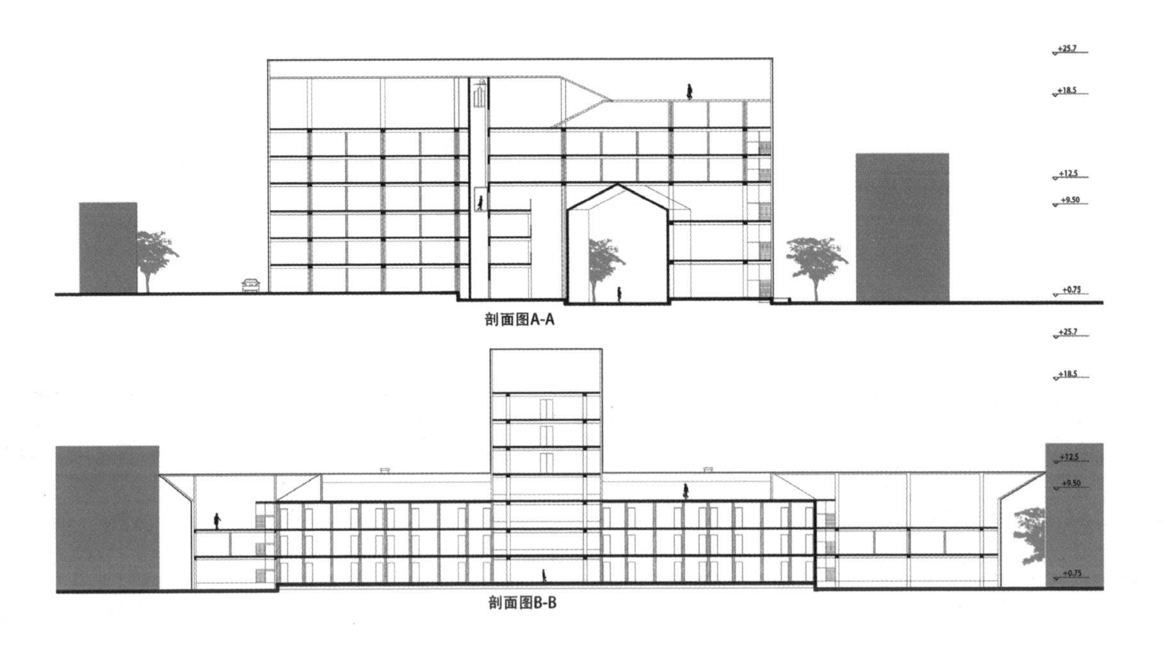

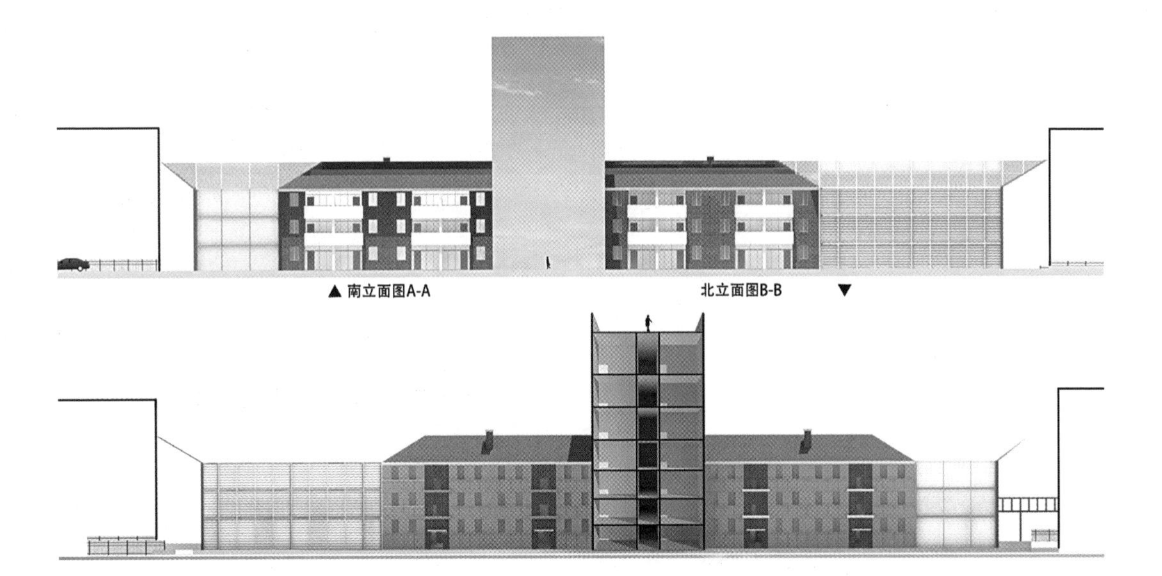

对页上图：标准层平面。**对页下图：**一层平面图。**本页上图：**剖面图。**本页下图：**立面图。

is diagonal, which is contrasted with the straight system around. To make the front and rear spaces more dramatic, I take the reflecting glass curtain wall as the building facade, so people seem to see the extension of the NO.13 apartment from one side, but will find another world after passing through the building.

本页上图：入口。**跨页图：**夜景鸟瞰。**对页上图：**屋顶连廊。**对页下图：**西侧透视。

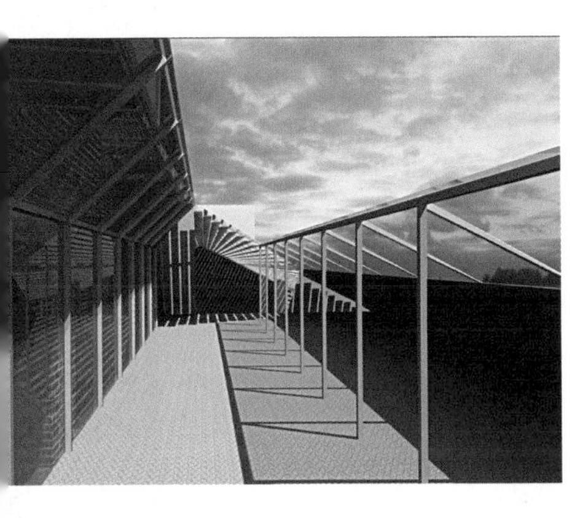

教师点评

基于对行列式布局变异的基本研究,摸索出顺势发力的途径:将间断的横向物体合并,并在楔入的纵向体上敷设镜面反射材质。不经意间,成行而松散的建筑搭着臂膀连成极致的整体;不经意间,区域中唯一的坡屋顶建筑轮廓由正到负,由合到分;不经意间,被阻隔了的带状空间仍在延展;不经意间,透视的原点飘忽不定;不经意间,地面浮现了楼宇,墙面飘出了云朵,空中生长着树木。土地、建筑、天空,在未知的向量中蔓延。

Teacher's comments

Based on the basic research on the variation of parallel layouts, the homeopathic measures were explored: merge the discontinuous horizontal objects, and lay specular reflection materials on the wedged vertical objects. Inadvertently, the rows of loose building draped over the arm to connect into an integral whole body; inadvertently, the only construction with slope roof in the area changes from positive to negative and from combination to separation in the contour; inadvertently, the separated belt-shaped space is still extended; inadvertently, the origin of perspective is uncertain; inadvertently, the building emerges from the ground, the cloud comes out from the metope, the trees grow in the sky. The land, the buildings, and the sky are spreading in unknown vectors.

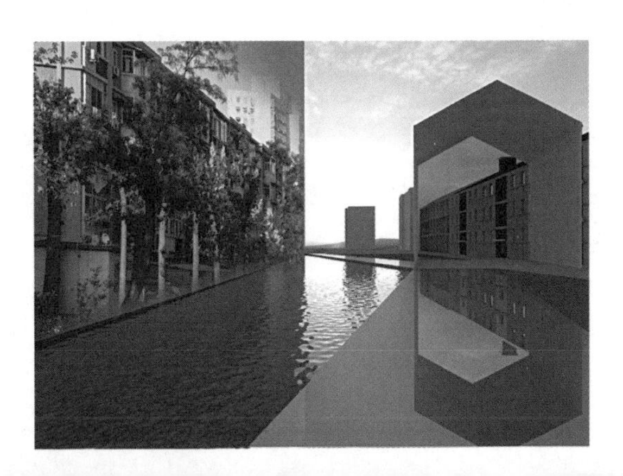

十三公寓改扩建设计/RECONSTRUCTION AND EXPANSION OF NO.13 APARTMENT BLOCK INTSINGHUA UNIVERSITY

项目选址：清华大学西南家属区十三公寓

项目类型：公寓与社区活动（住宅改扩建）

原建筑面积：1800m^2

现建筑面积：5751m^2

方案设计：周川源
指导教师：齐欣
完成时间：2015

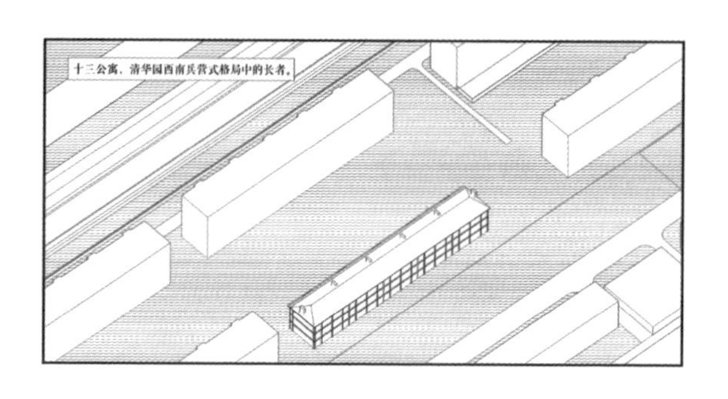

十三公寓改扩建是一个关于平凡房子的平凡课题：我们可以抛开所有的预期，用自己的本心去观察和探索，找到一个转变的契机。我希望能够引导人们重新发现十三公寓，让它从毫无表情的机械体转化为社区生活中鲜活的角色，从历史中向死而生。

“舞台”，成为一个可行的构想。我试图用一种简单的方式彻底改变十三公寓的空间结构：剖面上，加建体量呈两个倒L形，将十三公寓框在其间。加建体量在南北方向上错动，形成空间的张弛与围合。每个体量挖出灰空间，彼此连绵成连续、顿挫的公共生活舞台，并最终通向十三公寓屋顶部分拆除后形成的平台。

但这些都是设计逻辑方面的事情，事实上，当我决定在剖面上用一个框与十三公寓发生关系时，很多结构性的转变就自然而然发生了。

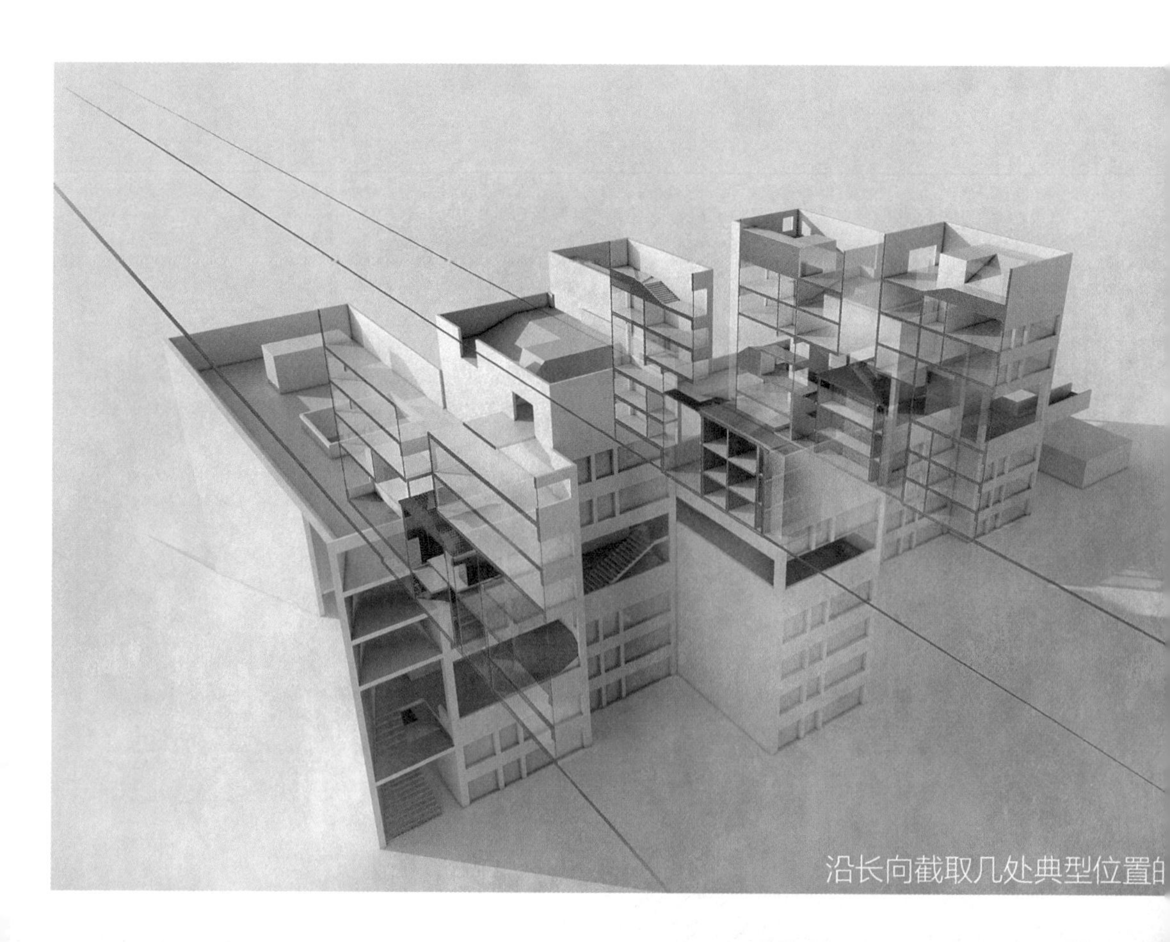

开篇：主要方案图。**对页及本页图：**十三公寓改造故事。**跨页图：**剖面演进分析。

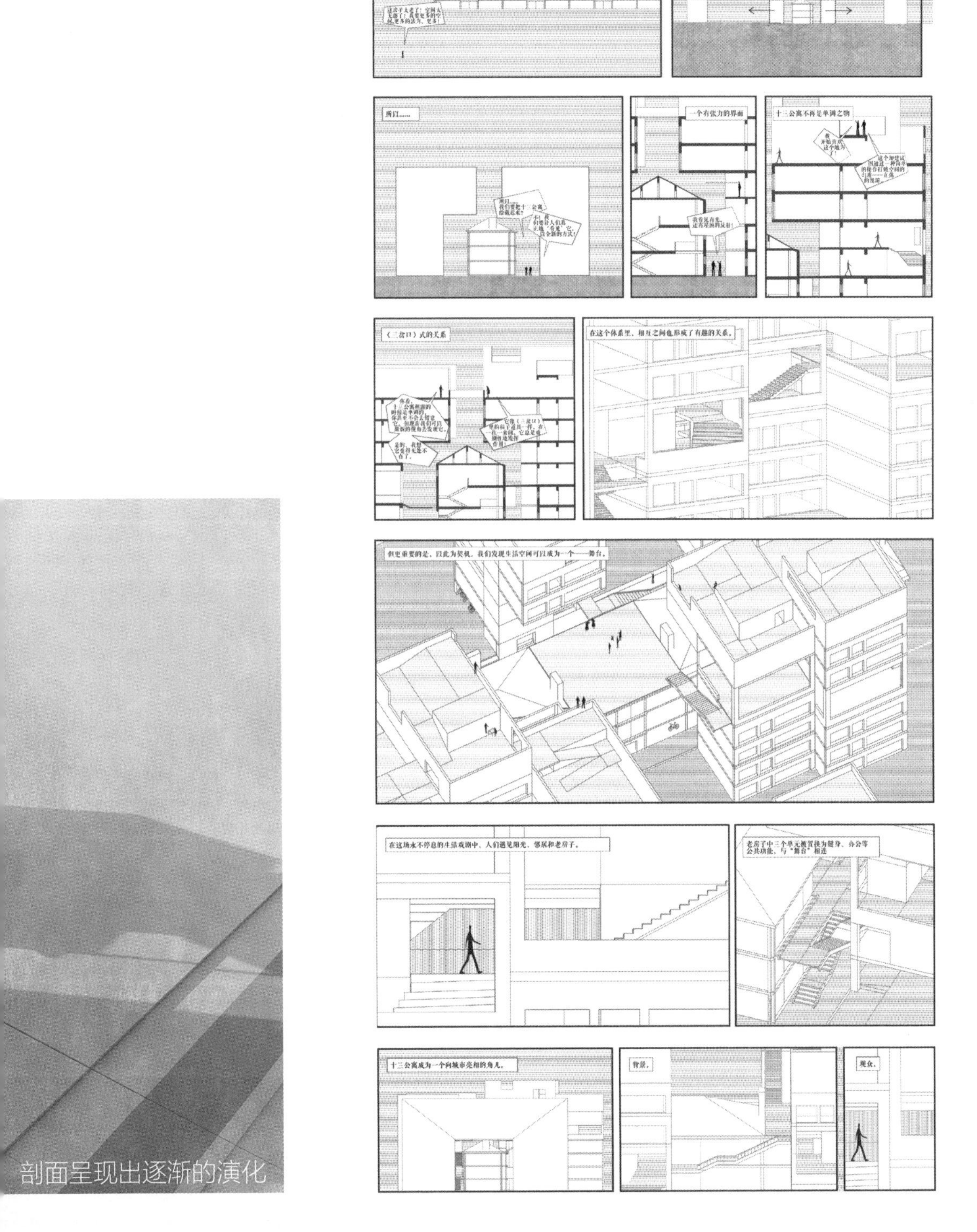

Reconstruction and expansion of the NO.13 apartment is an ordinary topic about ordinary houses; we could ignore all assumptions, conduct observation and exploration with our original heart, and find out an opportunity for change. I hope guide people to rediscover the NO.13 apartment, to make the apartment change from an expressionless mechanical object to a vivacious role in the community life, and reborn from the past.

"Stage" becomes a feasible idea. I attempt to thoroughly change spatial structure of the NO.13 apartment in a simple way. On the section, the added space presents two inverted L shapes, and frames the apartment. The add space changes from south to north, forming various open and enclosed space. The gray spaces are connected as a

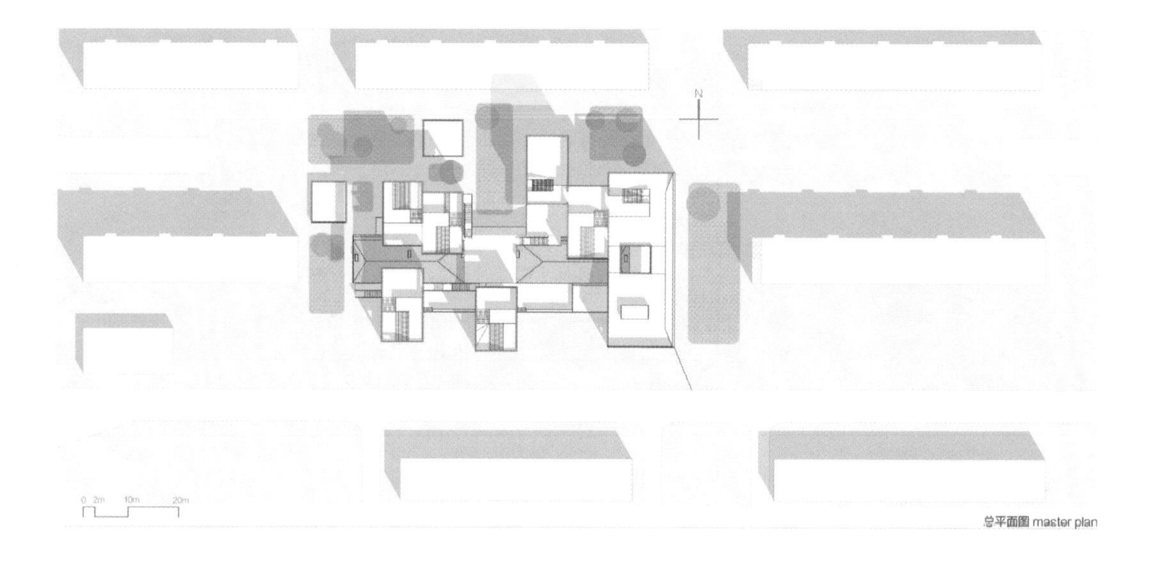

总平面图 master plan

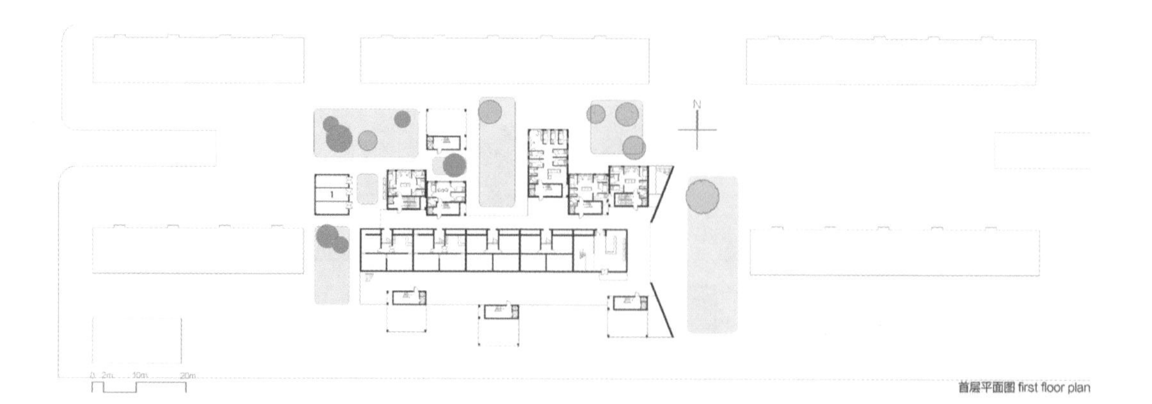

首层平面图 first floor plan

continuous and transitional public life stage, which finally lead to the platform formed by the removal of the roof of the NO.13 apartment.

But these are issues in the aspect of design logic. In fact, at the time when I decided to use a frame on the section to connect the NO.13 apartment, many structural changes happen naturally.

本页上图：总平面。**本页下图：**首层平面。**对页上图：**剖面概念图。**对页下图：**公寓单元平面。

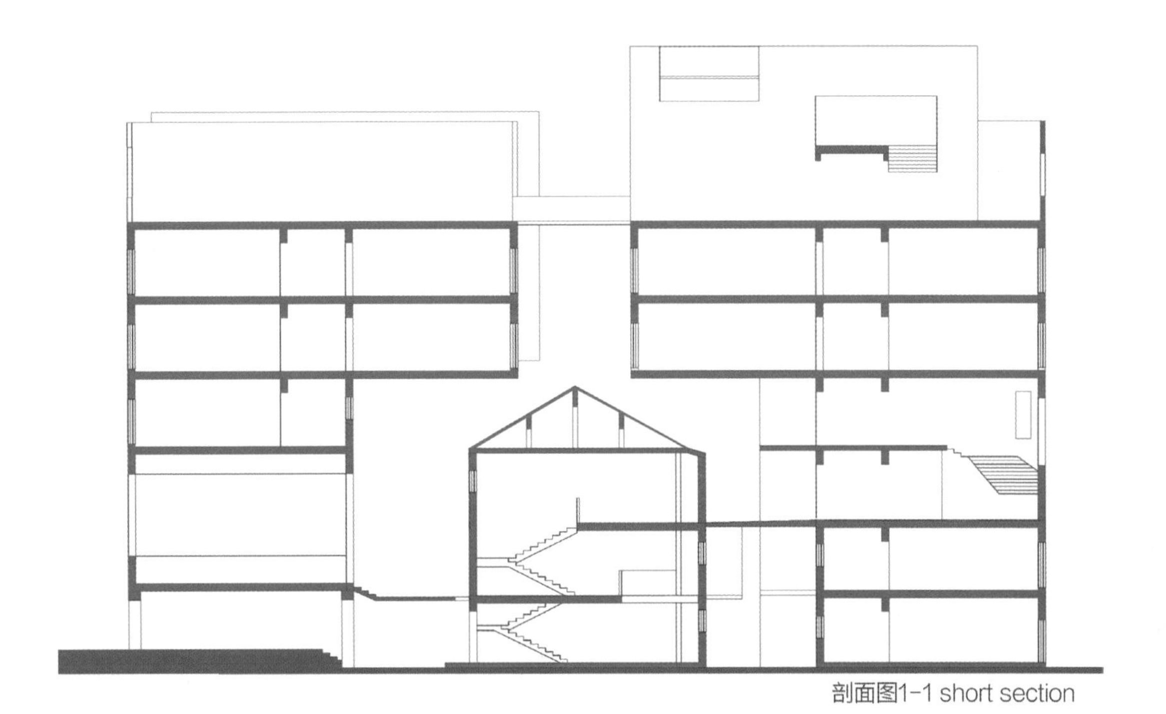

剖面图1-1 short section

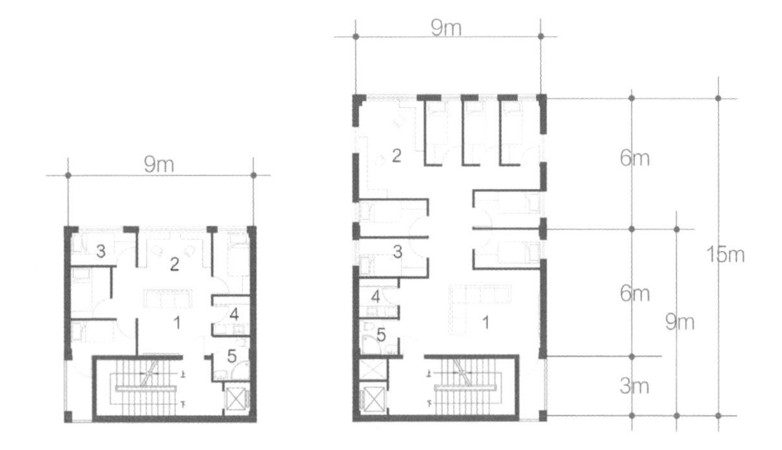

本页上图：人视效果图。**跨页下图**：交通空间效果图。**对页上图**：庭院空间效果图。

教师点评

当一个看似平庸的物体被放在镜框里，物体的价值会徒然倍增；当场景变成舞台时，人物就变成了角色；角色，也在欣赏着场景。不断更换的场景，给人们带来连绵不断的惊喜。在这里，老建筑焕发青春，在新建筑群中扮演角色；新建筑朝气蓬勃，在老建筑搭出的舞台上精彩亮相。新老建筑中的居民及过客都进入了戏场，同时充当着观众和演员。生活，是这里的主角，社区就是舞台。社区也是角色，登上了城市的舞台。

Teacher's comments

When a seemingly mediocre object is placed in a picture frame, the value of the object will be doubled; when the scene turns into a stage, the characters will become roles who are also enjoying the scene. The constantly changing scenes bring people continuous surprises. Here, the old building blooms and plays a role in the new buildings; the new buildings are full of vigor and vitality with amazing performance on the stage built by the old building. The residents and travelers in the new and old buildings have entered the scene, while serving as the audience and actors. Life is the protagonist here, while the community is the stage. Community is also the role that boards the stage of the city.

十三公寓改扩建设计/
RECONSTRUCTION AND EXPANSION OF NO.13 APARTMENT BLOCK INTSINGHUA UNIVERSITY

项目选址：清华大学照澜院家属区十三公寓
项目类型：旧建筑加改建
功能定位：清华 X-Lab 活动场地（办公 / 居住 / 社区中心）
建筑面积：20246m^2
用地面积：4000m^2

方案设计：林雨铭
指导教师：齐欣
完成时间：2015

开篇：主要方案图。**本页左图（从上至下）：**结构分析。功能分析。加建分析。**跨页图：**主要方案图。

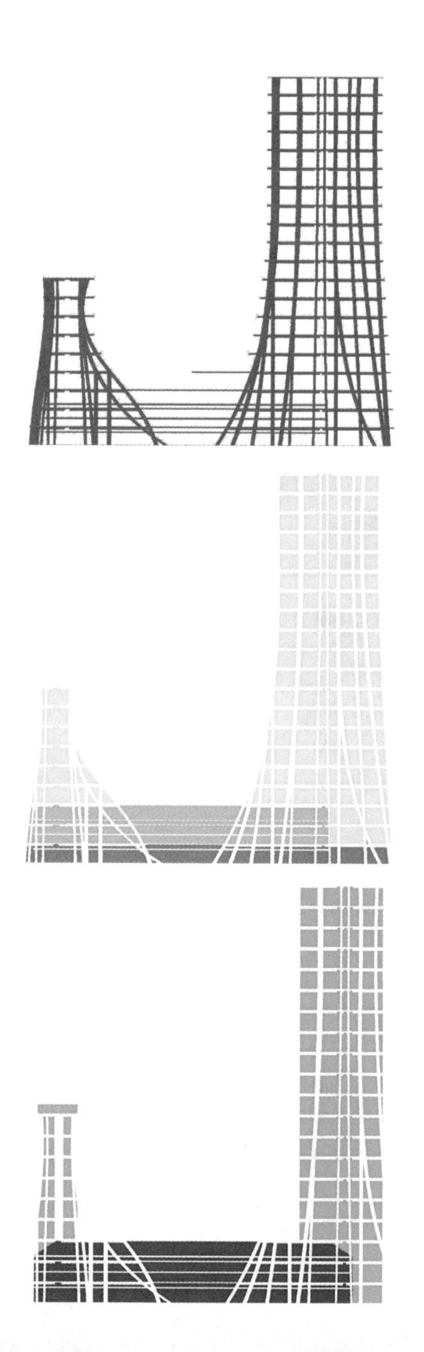

在最为普通的建筑格局中，隐藏着未曾被人发觉的巨大体量可能。方案充分利用了日照遮挡的盲点，因势利导，在遵循规范的前提下将土地资源利用到极致，在均质社区的一个点上做出密度和高度的突破，以巨大的体量达成对兵营式格局的破解。富于现代性的曲线与横平竖直的格局相互辉映，现有住区的低密度建筑和高密度路网被充分利用，超高层建筑对市政、交通造成的压力得到了稀释、缓解。方案使得均质的空间呈现出强烈的中心性，吸引的大量人流为社区带来了生机，硕大的建筑体量使社区从一个沉寂的历史中迅速醒来，走向未来。

In the most ordinary architectural pattern, there is hidden possibility for undiscovered huge space. The scheme takes full use of the blind spot of sun shade, makes the best use of the circumstances, takes ultimate use of the land resources on the premise of following the specifications, and breaks through the density and height on one point of the homogeneous community, so as to achieve the solution for barrack-like buildings with huge space. The modern curves echo with the straight pattern. The low-density buildings and high-density road network are fully used. The pressure for municipal administration and traffic

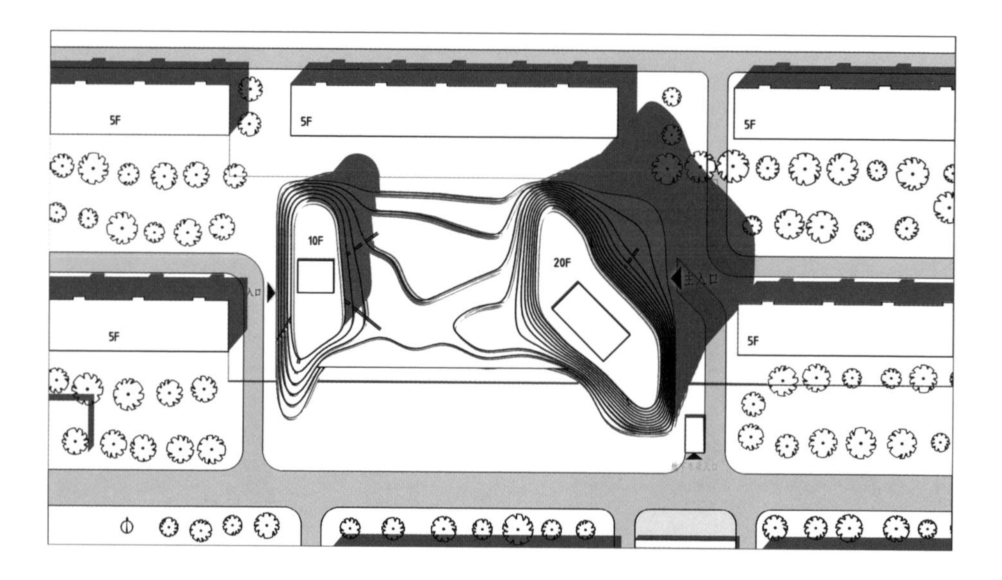

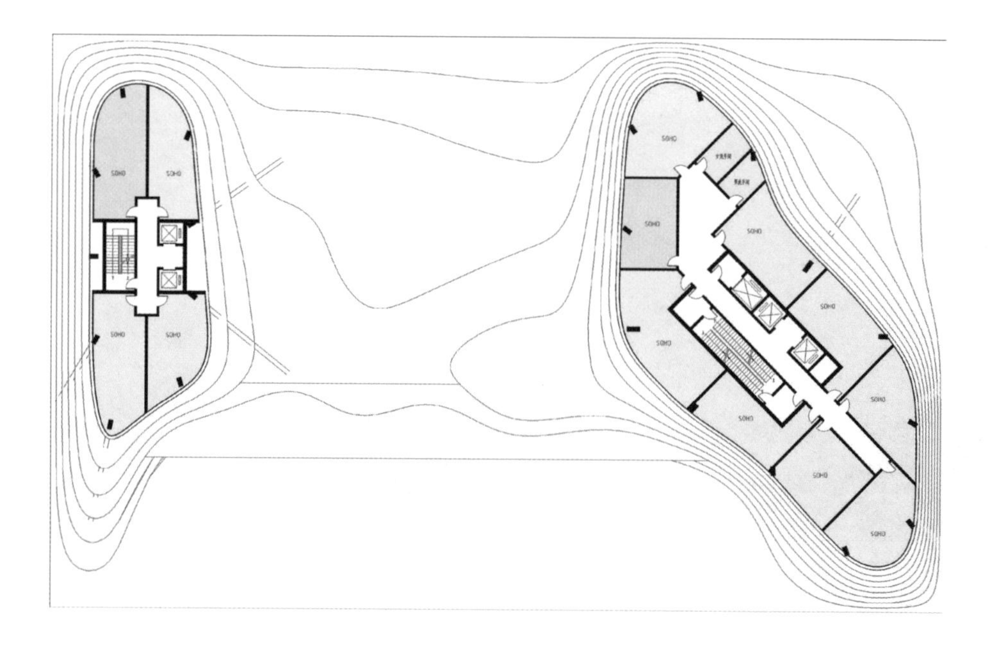

caused by super high-rise building is diluted and relieved. The scheme gives the homogeneous space strong centrality, and attracts heavy crowds to bring vitality to the community. The huge building volume rapidly wakes up the community from a quiet history to step into the future.

对页上图: 总平面图。**对页下图:** 标准层面图。**本页上图:** 房间平面图。**本页下图:** 剖面图。

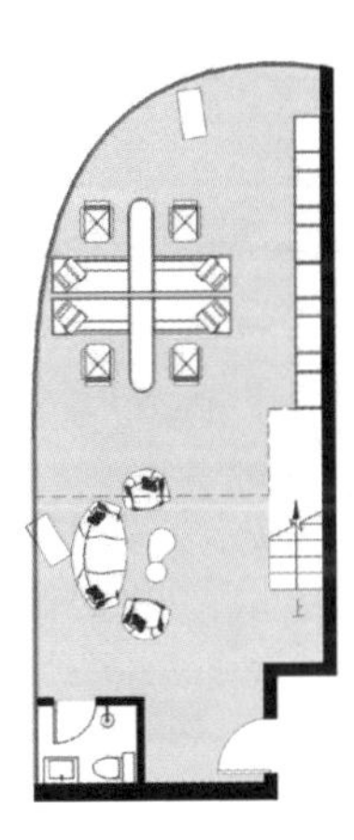

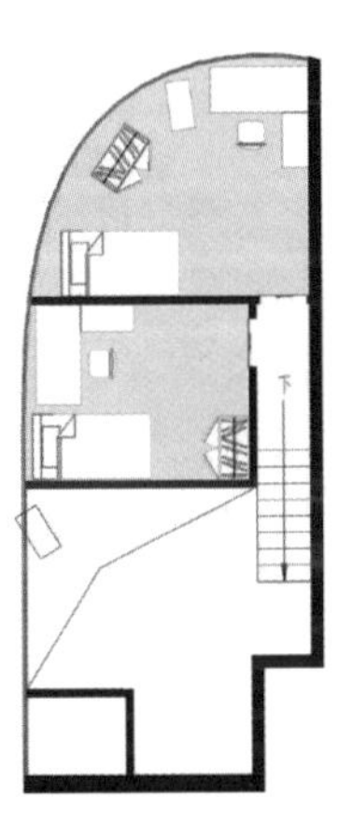

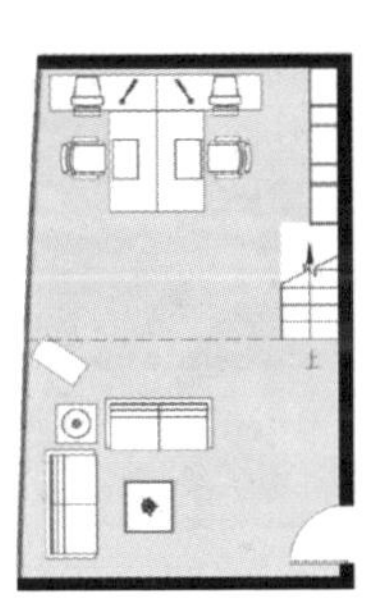

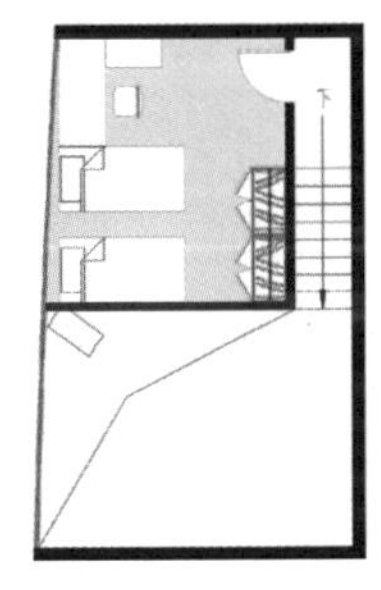

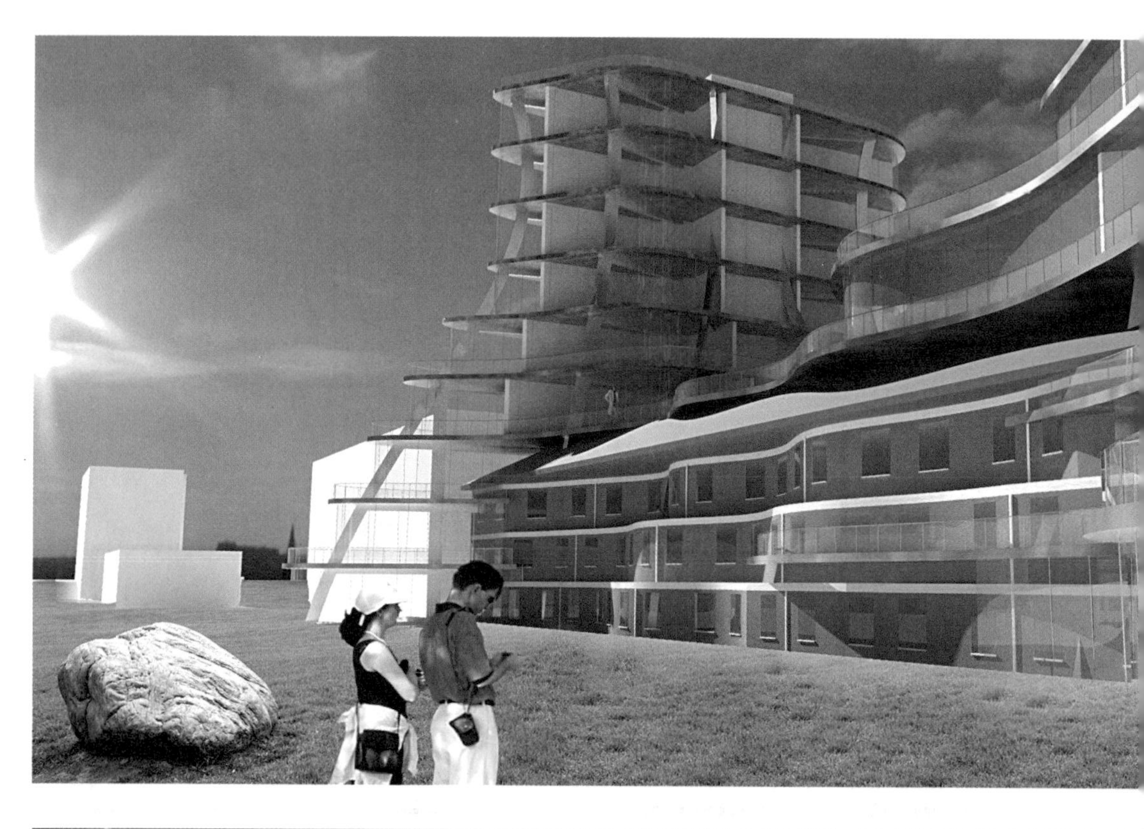

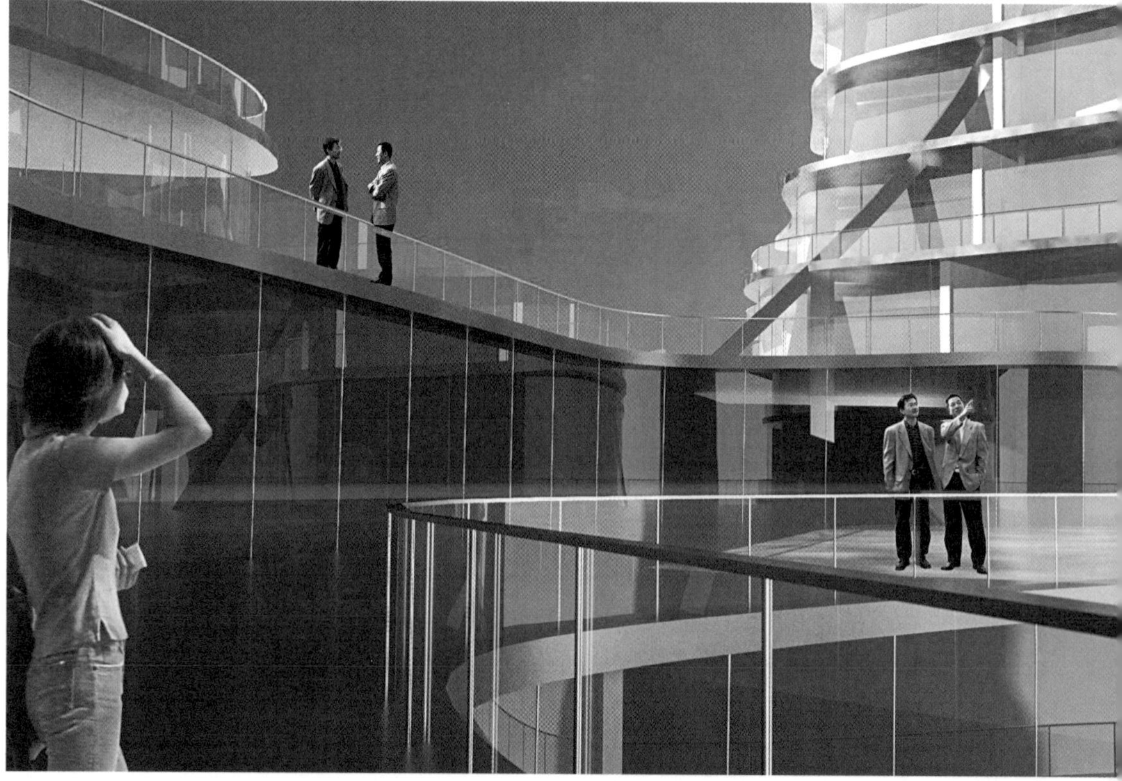

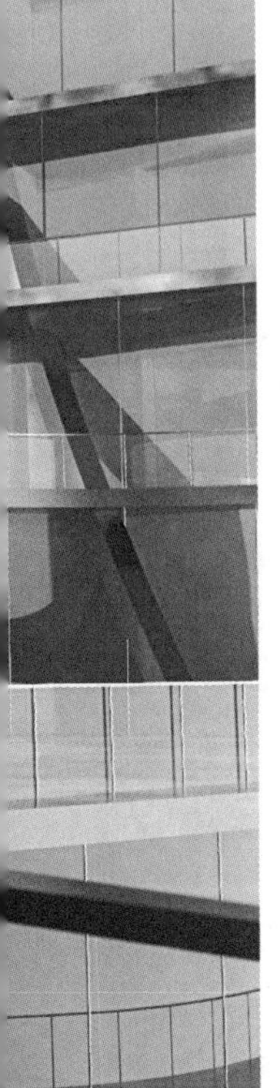

教师点评

土地是资源。土地资源的节省还能相应节省人的精力和能源的消耗。方案充分利用了日照遮挡的盲点，因势利导，在均质社区的一个点上做出密度和高度的突破，并利用现有住区的低密度和密布的路网，将超高层建筑对市政、交通造成的压力稀释、缓解。人流为社区带来了生机，均质空间中呈现出中心，硕大的建筑体量使社区面貌从一个沉寂的历史中转世，迅速走向未来。二者相辅相成，交相辉映。

Teacher's comments

Land is a resource. The saving of land resources may also save the consumption of human and natural energy. The scheme makes full use of the blind spots of sunshine shading and takes advantage of the new situation, making a breakthrough in the density and height at a point in a homogeneous community; relieves the pressure on the municipality and traffic caused by high rise buildings with the use of the low dense road network of existing residence. The stream of people brings vitality for the community. The central and huge building volume presented in the homogeneous space makes the community change from a quiet history towards the future rapidly. The two of them are interaction.

跨页上图：主立面效果图。**跨页下图：**室外平台效果图。**本页上图：**东北角透视。**本页下图：**西南角透视。

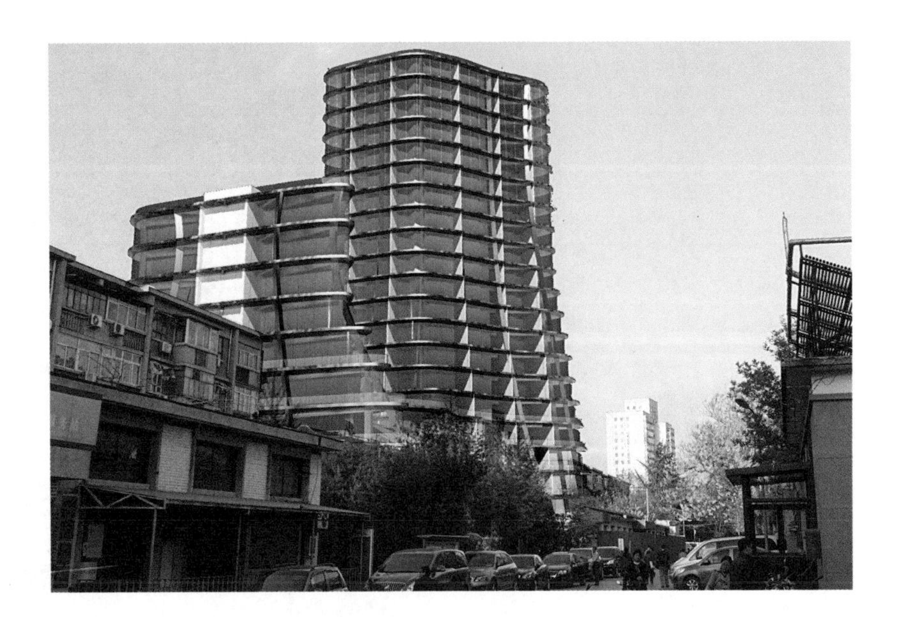

十三公寓改扩建设计/ RECONSTRUCTION AND EXPANSION OF NO.13 APARTMENT BLOCK INTSINGHUA UNIVERSITY

项目选址：清华大学十三公寓
项目类型：泛住宅加改建
建筑面积：6150m²
（保留 2000m²+ 加减 4940m²）
用地面积：4800m²

方案设计：陈达
指导教师：齐欣
完成时间：2015

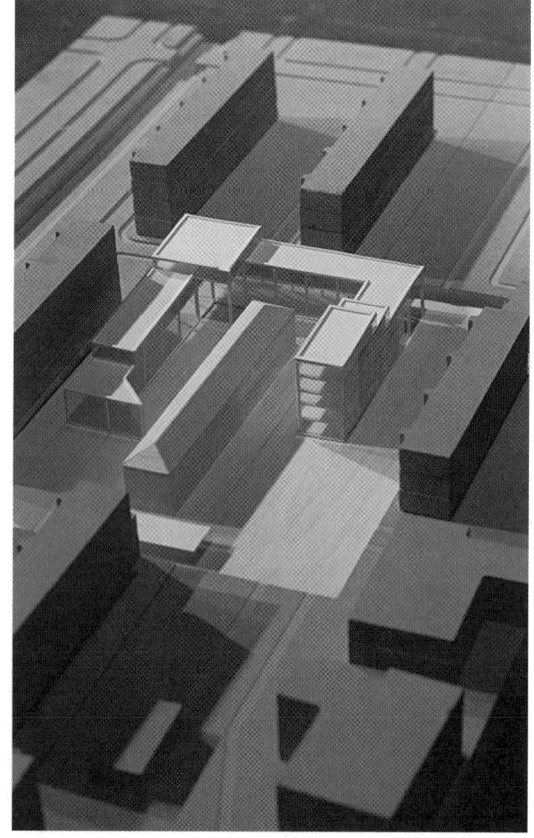

规整、线性的兵营式住区布局之中是否蕴藏着跳动的活力？将一条曲线置于规整之中，通过高低错动的操作赋予置入物以方向，进而牵动户外空间的开合与疏密。横看成岭侧成峰，这个"闯入者"以不同视角的丰富形态扰动了平静，带来了生机。

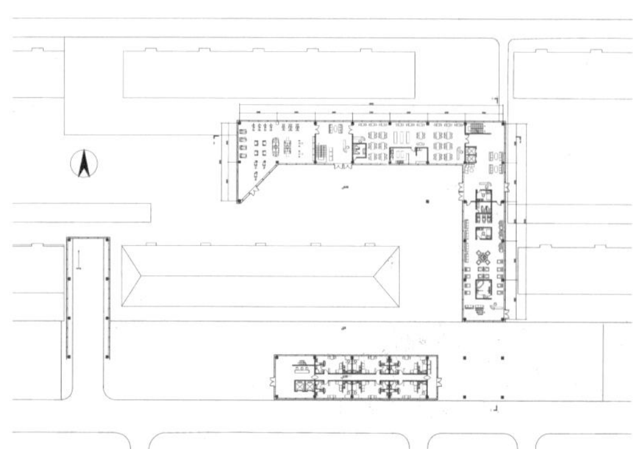

私搭加建/PRIVATE BUILD UP

项目选址：清华大学照澜院家属区十三公寓
项目类型：住宅楼加建，公寓式酒店
原建筑面积：2000m² 加建面积：3850m²
用地面积：12000m² 绿化率：20%

方案设计：胡德民
指导教师：齐欣
完成时间：2015

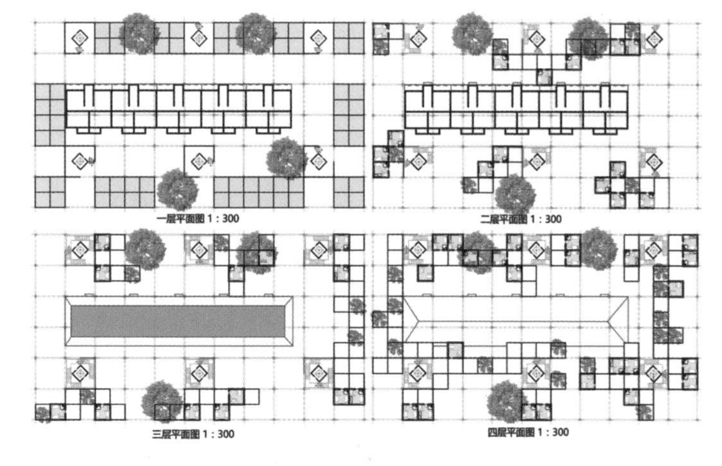

清华大学照澜院内存在大量私搭乱建的房屋“私搭乱建”。这些自发的行为使用的材料是多样的：木头、红砖、瓦楞钢、塑料等等。

同时在现代社区中，打印店/理发店/照相馆等小商业的生存空间被不断挤压，这些小商业在社区中扮演不可替代的重要角色，却只能委身于各种私搭乱建的构筑物之中，他们应当成为社区重要的公共空间。

还有清华聚集了大量以学习为目的的人员，这些或考研或参加各种培训的青年人的居住是以学习为主要目的。本设计出发点之一也就是为这些年轻人提供小面积的公寓式酒店。

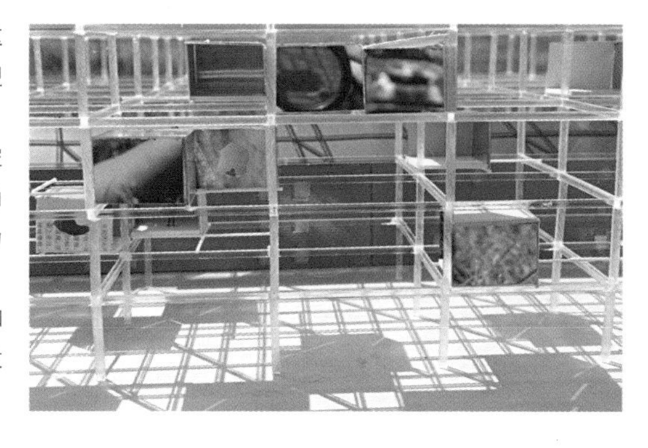

十三公寓改扩建设计/ RECONSTRUCTION AND EXPANSION OF NO.13 APARTMENT BLOCK INTSINGHUA UNIVERSITY

项目选址：清华大学 十三公寓
项目定位：公寓
建筑面积：3240m²
用地面积：5000m²
绿化率：36%

方案设计：金日哲
指导教师：齐欣
完成时间：2015

十三公寓住宅区内的老人很多，但是区域内又缺少针对老年人的公共活动空间以及配套设施。 因此设计者构想了一个居住和公共空间结合在一起的建筑结合体。概念出发于将十三公寓与周边环境中的校河与树林结合起来，并尝试用最为简洁的手法去突出十三公寓的“L”型的空间特点，与活动场所更好地融合。

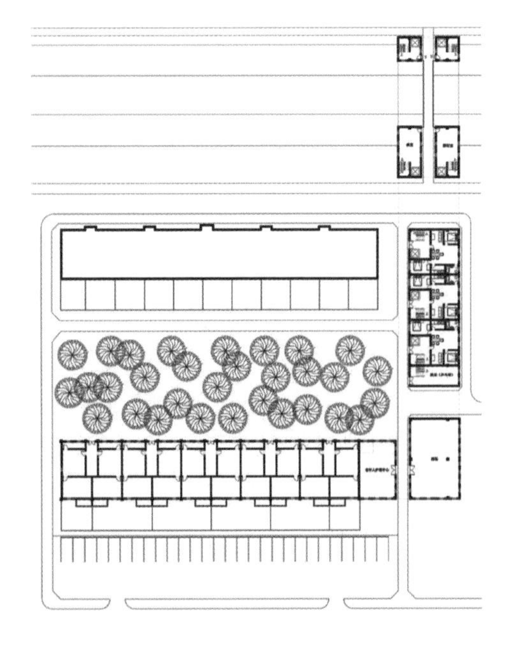

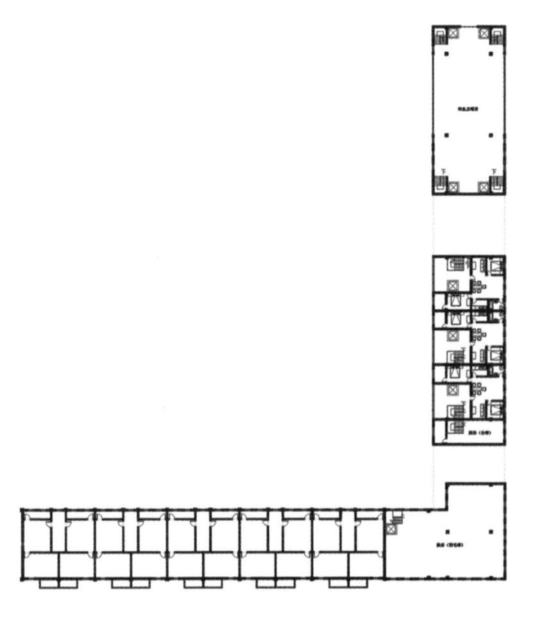

塔与沟壑/ TOWER OVER THE RAVINE

项目选址：清华大学十三公寓
项目类型：泛住宅加改建
建筑面积：6940m² **用地面积：**4596m²
容积率：1.51 **绿化率：**45%

方案设计：刘田
指导教师：齐欣
完成时间：2015

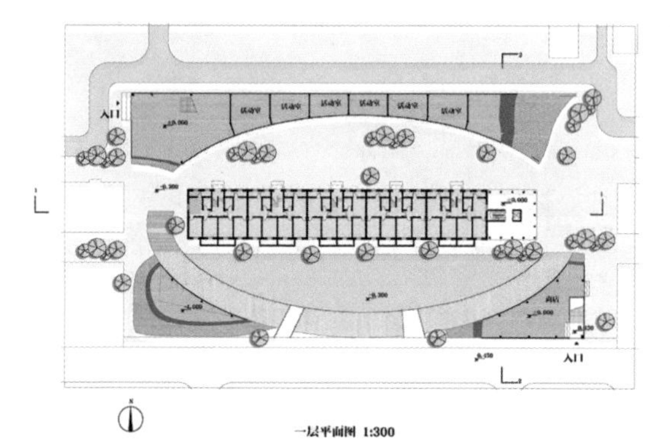

一层平面图 1:300

本设计从十三公寓所处的兵营式格局与公寓单体两个层次对现有问题进行了积极的应对。住区层面上，本设计在横平竖直的兵营式网格中植入了椭圆的沟壑，以形破形。沟壑作为地景，在地面建筑不断加密的趋势下，承担起公共活动和商业的功能。在单体层面上，采用超高容积率的塔楼来最大限度满足人口的居住需求，在设计中以十三公寓为母题，将其竖起，最终形成塔与沟壑的完整形象。

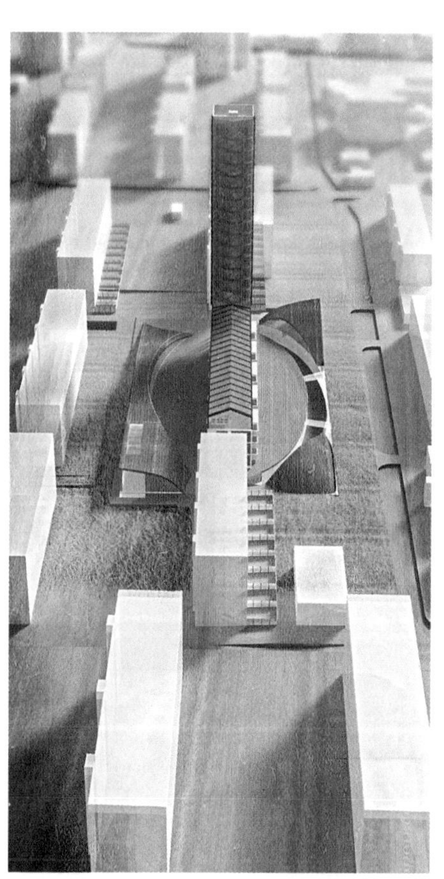

青年公寓设计/ APARTMENT FOR YOUTH STUDENTS

项目选址：清华大学南老公寓区内十三公寓

项目类型：青年 soho

建筑面积：4500m^2

用地面积：5530m^2

方案设计：刘炫育

指导教师：齐欣

完成时间：2015

20世纪，国家经历着大规模的建设浪潮，兵营式住宅作为一种高效均好的住宅形式广泛出。时至今日，扩张型的大规模建设已接近尾声城市却将继续变革，遍及大江南北的兵营式住宅以其简单格局为来者预留着无限的可能。

清华大学十三公寓就身处这一格局之中，并居于区域中心，虽经局部改造，仍在族群中最低最矮，奠定了厚积薄发的基础。本设计针对十三公寓社区老化严重的状况，力图发掘周边大量租住青年的人群潜力，将十三公寓改造为集居住，工作等功能为一体的青年公寓，激活社区活力，同时对兵营式住宅演进的未来进行有益的探索。

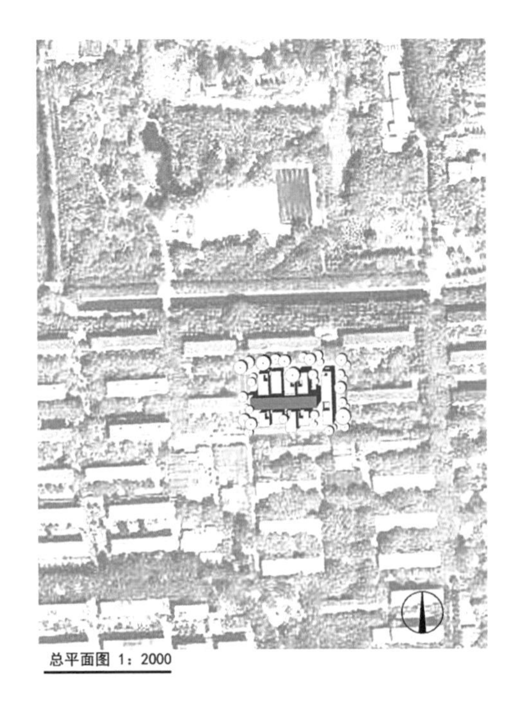

总平面图 1：2000

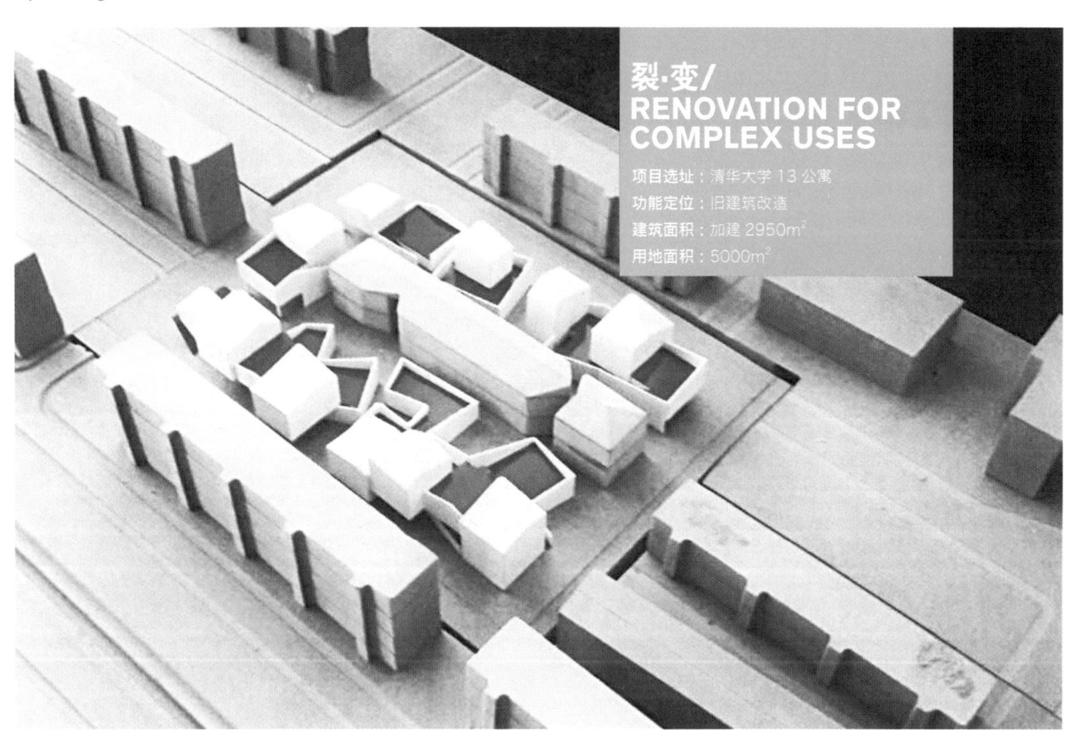

裂·变/ RENOVATION FOR COMPLEX USES

项目选址： 清华大学 13 公寓
功能定位： 旧建筑改造
建筑面积： 加建 2950m²
用地面积： 5000m²

方案设计： 马宏涛
指导教师： 齐欣
完成时间： 2015

兵营式住宅在其南北两侧形成了狭长、消极的场地，成为了人们生活中忽视的内容。改造首先将原有建筑进行“剪切”，获得了新的天际线和用于线上销售、线下展示的橱窗断面，并沟通整个场地。在场地底层引入多样的复合功能，对片区形成补充，并限定丰富的院落；上层错落的青年公寓和屋顶平台则提供了更多样的视线关系和交流机会。变异的坡顶和起伏的山墙，亦无不在强化十三公寓作为片区唯一坡顶建筑的气场。

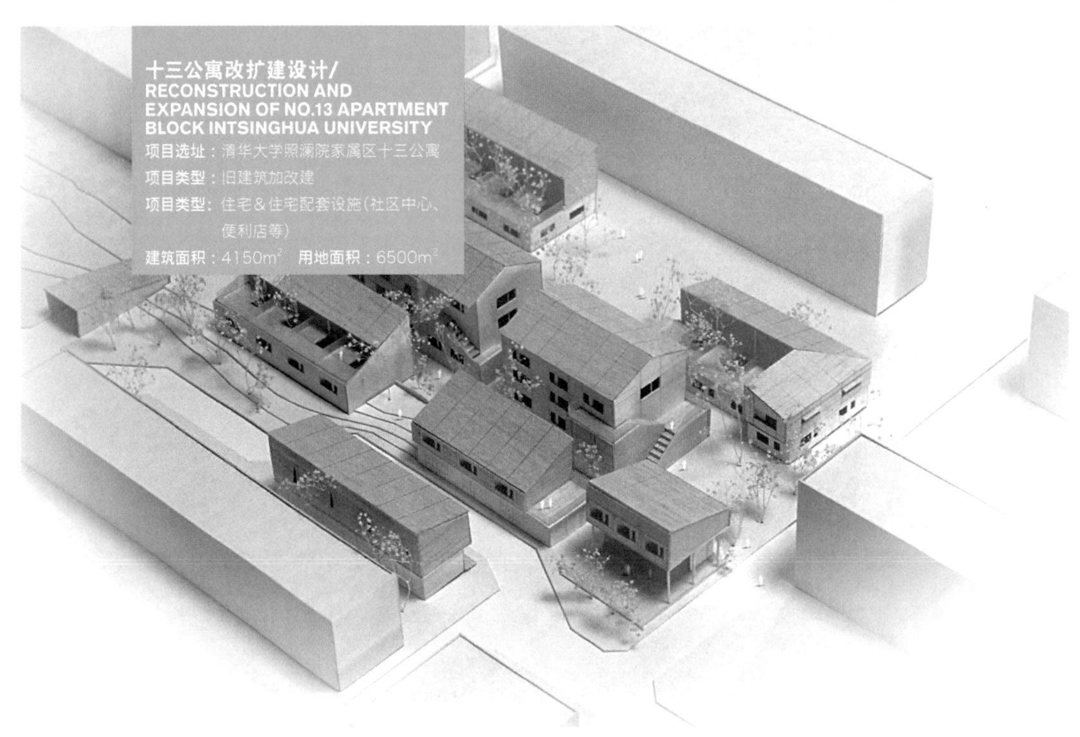

十三公寓改扩建设计/ RECONSTRUCTION AND EXPANSION OF NO.13 APARTMENT BLOCK INTSINGHUA UNIVERSITY

项目选址： 清华大学照澜院家属区十三公寓

项目类型： 旧建筑加改建

项目类型： 住宅＆住宅配套设施（社区中心、便利店等）

建筑面积： 4150m² **用地面积：** 6500m²

方案设计： 钱漪远

指导教师： 齐欣

完成时间： 2015

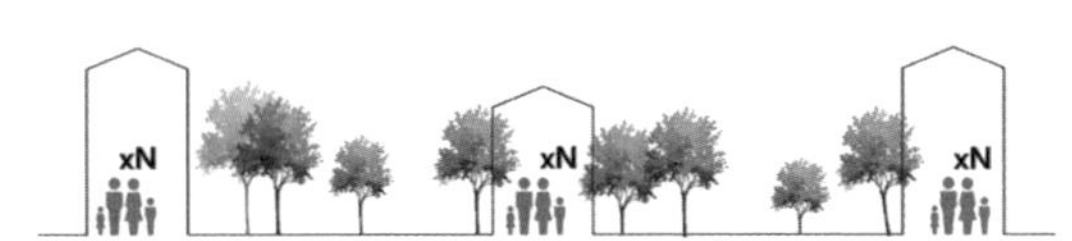

设计从场地自身的特色入手，抓住其两个特点：第一，场地中有茂盛的树木；第二，兵营式的格局看似死板，但蕴含着极大的空间潜力。

本方案以十三公寓为中心，围绕其“生长”出尺度宜人的群落。这样将闲置在空旷场地上的树木纳入私家的院落，新的体量通过围合与引导，激活了原本的城市负空间，兵营式格局得到了化解。

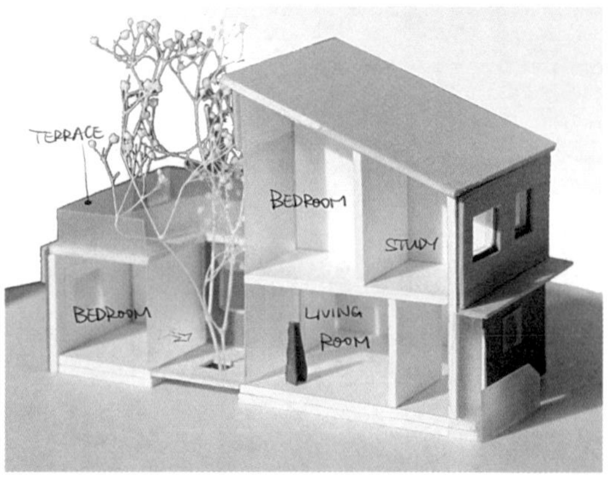

十三公寓改扩建设计/ RECONSTRUCTION AND EXPANSION OF NO.13 APARTMENT BLOCK INTSINGHUA UNIVERSITY

项目选址：清华大学照澜院家属区十三公寓
建筑功能：居住、商业服务
建筑面积：6560m^2
用地面积：4500m^2

方案设计：王文
指导教师：齐欣
完成时间：2015

兵营式住宅是本设计的问题引发点。本案即以问题导向，探讨在旧有格局下如何切实有效提高小区容积率、引入更多现代的城市功能，使旧建筑于改扩建得到的新的部分，共同容纳更丰富的现代城市居民生活。本案采取“一”生“三”的策略，其扩展的空间，目的不止于使用空间的扩展，而在于住宅模式更新的探索。

自然启

朱锫

朱锫建筑设计事务所创始人
主持建筑师

发设计

朱锫

朱锫建筑设计事务所创始人

主持建筑师

教育背景

1988年 – 1991年

清华大学建筑学院建筑学　硕士

1999年 – 2000年

美国加州伯克利大学　硕士

工作经历

1991年 – 1994年

清华大学建筑学院助教、讲师

2005年至今

朱锫建筑设计事务所 创建合伙人、主持设计师

2014年 – 2016年

美国哥伦比亚大学 客座教授

主要论著

《Discussion: Regional Development of Architecture》[J]. A+U (日本), 2016/03

《Minsheng Contemporary Art Museum-Full Metal Jacket》[J]. Architectural Record (美国), 2016/02

《Minsheng Contemporary Art Museum》[J]. Arketipo (意大利), 2016/02

《Letter from China-Listening to Nature》[J]. The Plan (意大利), 2015/11

《民生现代美术馆》[J]. 建筑学报, 2015/09

《"无用之用"—民生现代美术馆的态度》[J]. 世界建筑, 2015/06

《自然启发设计 建筑语言: 光/结构/材料》[J]. 建筑创作, 2015/04

《Minsheng Museum of Contemporary Art》、《OCT Design Museum》收录于《New Museum in China》[M]. Princeton University Press (美国), 2014

《从有形到无限-OCT设计博物馆》[J]. domus, 2012/04

《CaiGuo-Qiang Courtyard House Renovation》收录于《Architecture Now! Houses》[M]. TASCHEN (德国), 2011

《Atlas Architectures of the 21st Century, Asia and Pacific》[M]. Fundación BBVA (西班牙), 2010

《Digital Beijing》收录于《Architecture Now》[M]. TASCHEN (德国), 2008

《朱锫—北京新一代建筑家所追求的传统与现代的关系》[J]. 建筑NOTE (日本), 2008/08

《跨界思维与批判性探索-朱锫事务所特辑》[J]. A+U建筑与都市, 2008/06

《Made in China》[M]. DVA (德国), 2005

设计获奖

2011 –"当今世界最具影响力的5位(50岁以下)建筑师之一", 美国赫芬顿邮报

2007 –被美国Architectural Record杂志评为"全球设计先锋"

2009 –被英国Wallpaper杂志授予"库瓦西耶设计奖"

2015 –北京世园会设计方案获得美国建筑师协会(AIA)纽约分会荣誉奖

2009 –"年度创作奖", 美国Surface

2009 –"1949-2009 中国建筑学会建筑创作大奖"

2008 –"亚洲最高荣誉设计大奖", "亚洲文化优异设计大奖", 中国香港

2008 –"最优秀建筑", 中国香港

2005 –"中国建筑奖", 美国

2004 –WA"中国建筑奖", 中国

2004 –获奥运建筑数字北京国际设计竞赛一等奖

2009 –"1949-2009 中国建筑学会建筑创作大奖"

1989 –"设计特别奖", 国际建筑协会、联合国教科文组织

2014 –OCT设计博物馆被ArchDaily评为21世纪20座经典博物馆之一

2011 –OCT设计博物馆被《Frame》评为全球10个充满启发性顶级博物馆之一

代表作品

大理杨丽萍表演艺术中心(图1)、大理美术馆(图2、图3)、寿县文化艺术中心(图4、图5)、蔡国强四合院改造(图6、图7)、民生现代美术馆(图8)、阿布扎比古根海姆艺术馆(图9)

自然启发设计

三年级建筑设计（6）设计任务书

指导教师: 朱锫
助理教师: 杨圣晨　吴禛和

第一部分

作为实验性设计，本设计课程将侧重建筑语言：光，结构和材料

第一部分需要学生分析学习特定运用明确建筑语言创作的经典作品。

线性 / 混凝土

The Convent of La Tourette, Le · Corbusier, 1953 拉图维特修道院，勒 • 柯布西耶,1953

Kimbell art museam, Louis Kahn,1972 金贝尔美术馆，路易斯 • 康 , 1972

平面 / 玻璃

Country House，Mies Van Der Rohe，1923 乡村住宅，密斯 • 凡德罗,1923

体 / 木质

Vuoksenniska Church, Alver Aalto, 1959 沃奥克森尼斯卡教堂，阿尔瓦 • 阿尔托, 1959

体 / 混凝土

The Pilgrimage Chapel of Notre Dame du Haut at Ron-champ，Le Corbusier，1955 朗香教堂，勒 • 柯布西耶，1955

此练习将基于分析图纸和研究模型表达的建筑语言。

第二部分

学生需要结合第一部分的研究，选择针对任意一个经典案例，用真实材料创造一个综合性结构。物理模型的尺寸为 50cmx50cmx50cm。第一部分和第二部分一共用时 1.5 周。

第三部分

艺术 / 诗歌 / 声音空间

中国，北京，清华大学

场地位于清华校园中心，毗邻标志性建筑之一清华学堂。1000~2000 m^2以艺术 / 诗歌 / 声音为主 题的学生活动中心 / 博物馆，目标成为连接历史校园和现在校园的催化剂和社交场所。相对灵活开放的功能定义给学生更多可能的想象，从而创作出新建筑是研究课题的核心问题。

最后汇报主要将以 1:50 的物理模型(可为局部剖面小模型)，一段两分钟的视频呈现特定材料与光的空间感和质量，辅以图纸。用时 6.5 周。

建筑设计是团队性的，学生需要两人一组进行创作。

为对话而生的建筑/ARCHITECTURE FOR DIALOGUE

项目选址：清华学堂

项目类型：以艺术 / 诗歌 / 声音为主题的学生活动中心 / 博物馆，目标成为连接历史校园和现在校园的催化剂和社交场所

建筑面积：700m^2　**用地面积**：1500m^2

方案设计：陈嘉禾
指导教师：朱锫
完成时间：2015

方案设计：丛菡
指导教师：朱锫
完成时间：2015

方案设计：付之航
指导教师：朱锫
完成时间：2015

开篇：主要方案图。**本页上图**：总平面图。**本页下图**：剖面图。**对页**：光与人类情感的联系。

谈话是大学的核心，同时也是发生场所最草率的活动，这个矛盾激发了这次设计。我们从金贝尔美术馆的学习中探索光和空间的关系，从对传统绘画中寻找光对气氛的影响，最终针对不同谈话的气氛需要，塑造了不同尺度、结构的建筑空间。这些体量被半埋入地下，以求和历史场地契合，并减少干扰以突出光的影响。地上部分形成了排布带有禅意的朴拙雕塑群，激活院落空间。

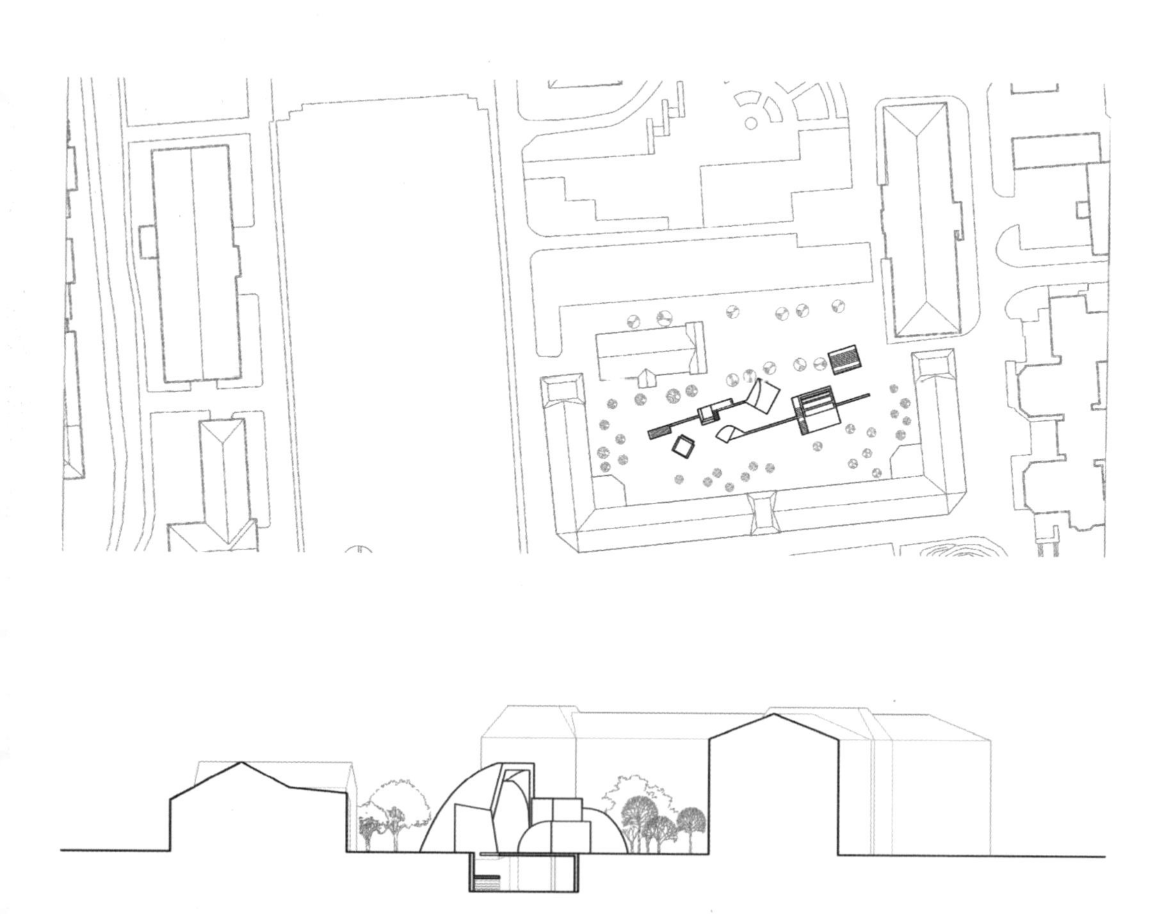

来自下方的光
——焦躁不安、压抑

蒙克《青春期》

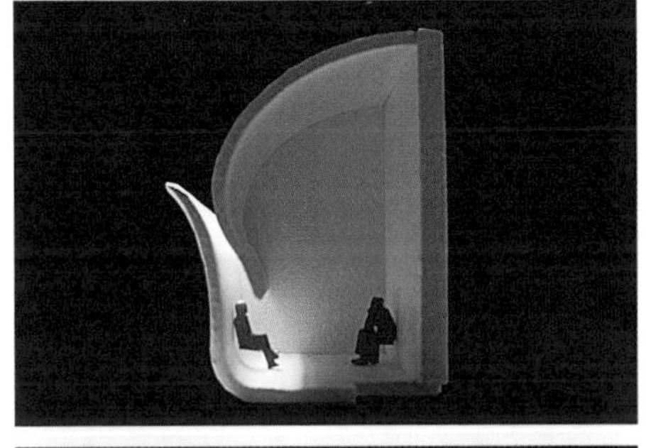

特定物体被照亮
——荒诞

柏克林《死亡之岛》

烛光、神秘、小范围的光明
——传递、柔弱的新生

伦勃朗《婴儿》

指向性的光
——启示、等级

卡拉瓦乔《召唤圣马太》

柔和而均匀的光
(Kimbell Museum)
——祥和、专注

维米尔《倒牛奶的女仆》

Conversation is the core of the university, as well as the activity which can take part in anywhere. This contradiction inspires the design. We explore the relationship between light and space in the study of Museum Kimbell, and find out the influence of light on atmosphere in traditional painting. Finally, according to the different needs of the atmosphere of the conversation, the building spaces of different scales and structures are shaped. These volumes are half-buried underground with an attempt to fit the historical site, and reduce the interference to highlight the impact of light. The simple sculpture groups with deep meditation purpose in arrangement are formed on the ground, which activate the courtyard space.

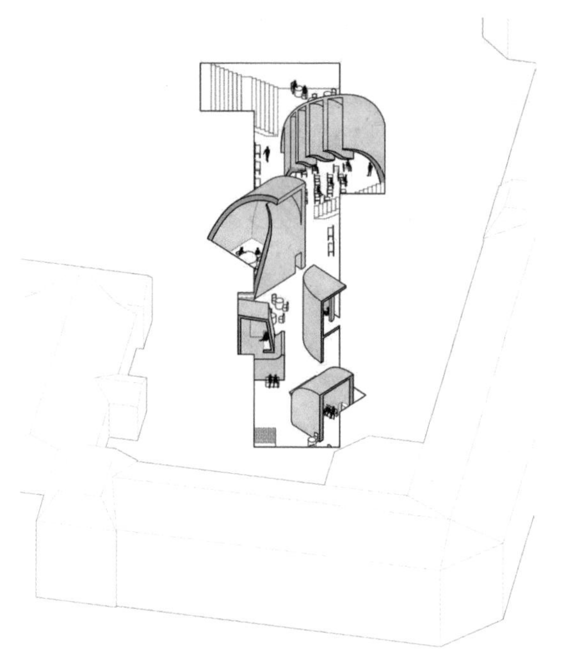

本页上图：轴测图。**本页下图**：鸟瞰。**对页图**：模型材料与尺度展示。

本页图：模型局部效果。**对页图**：模型局部效果。

教师点评

"为对话而生的建筑"找到了交流在大学生中的重要性，特别是在清华偏向工科的趋势中显得尤为重要。他们先从传统绘画的研究开始，寻找光对于人类交谈气氛的影响，然后根据不同的谈话需要，塑造不同尺度、结构和材料的建筑体量。为了和原有的历史场地契合，一半体量埋入地下，同时隔绝了外部环境的影响突出了光的氛围。在地上形成了富有禅意的类似雕塑似的建筑群，激活了院落的空间。

Teacher's comments

"Building for the Conversation" found the importance of communication to college students, especially in the tendency that the engineering courses in Tsinghua University. They started with the research of traditional painting to look for the influence of light on the atmosphere of conversation, and then shaped the building massing of different scales, structures and materials according to the different conversation needs. In order to fit the original historical sites, half of the building mass was buried underground, while isolating the impact of the external environment and highlighting the atmosphere of light. The building group similar to the sculpture was formed on the ground to activate the courtyard space.

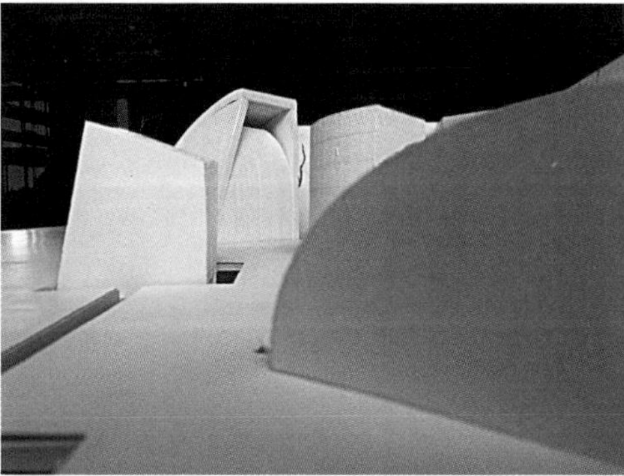

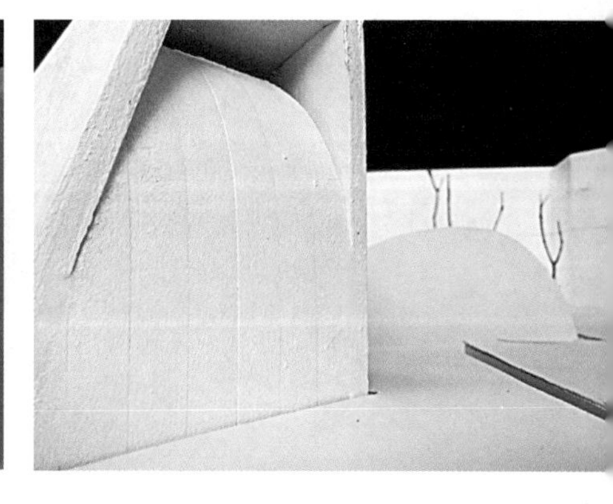

隔与不隔/ PARTITION AND INTEGRATION

项目选址：清华大学同方部南侧
项目类型：公共建筑
建筑面积：1200m²
占地面积：1700m²

方案设计：邓慧姝
指导教师：朱锫
完成时间：2015

方案设计：王章宇
指导教师：朱锫
完成时间：2015

摒除急躁的学术心态，回归文人面对自然、内视本心地读书、学习和思考的宁静状态。

通过光线的漫反射和自然光影的变化、交融，激起深思和灵感。

隔——半地下昏暗空间，柔和扩散的光线，隔开外界的喧扰，进入纯粹自然的世界。

不隔——人从变化万千的自然中得到灵感。

用投影和漫反射手法，向空间中引入自然因素。

树影、水光和水波在墙上的漫反射投影、黄昏时刻，光色弥漫整个空间。

用与无用——提供场所氛围，但不界定功能。可坐可躺的弧形空间，自由散漫的空间形态。

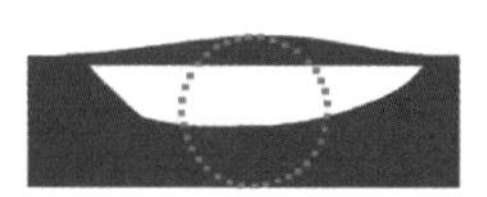

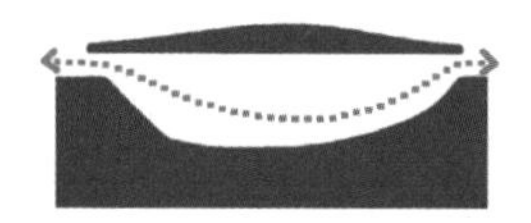

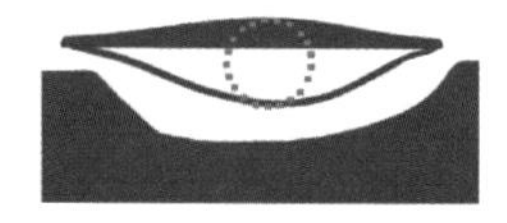

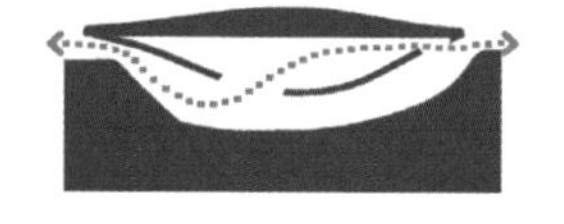

开篇：主要方案图。**对页上图**：生成分析。**跨页图**：主要方案图。**本页上图（从左至右）**：平面图及总平面图。

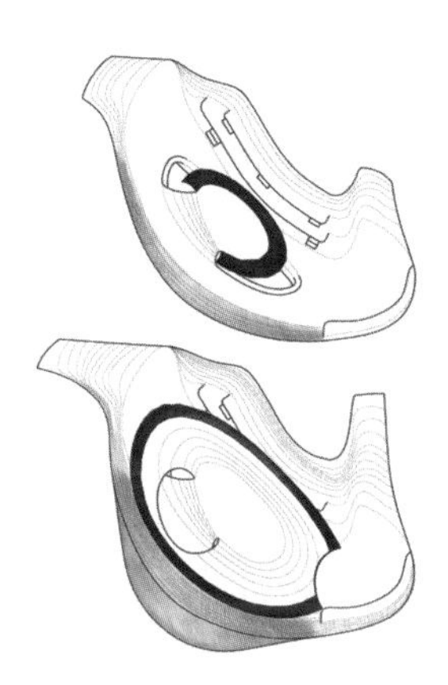

From the research on the light and atmosphere of Monastery of La Tourette, this design helps to find that light of different types coexists in the same space. Interactions of lights provide the trait of space spirit. Therefore, it puts forward the research direction of extracting the construction of introduced lights.

In the design, light experiment is conducted on the basis of the research:

In different time, change trait of the light Integration and collision of different kinds of light Spiritual space experience.

Separated - It is intended to calm people down with gloomy environment and deep and serene light. The specific application on design is gloomy cave and softly diffused light.

Not Separated - In order to make people obtain inspiration from the change and integration of natural lighting and shadow specifically applies to design. As time goes by, diffusely reflect the natural light, flickering shadow of the trees, leaped light and colorful dusk.

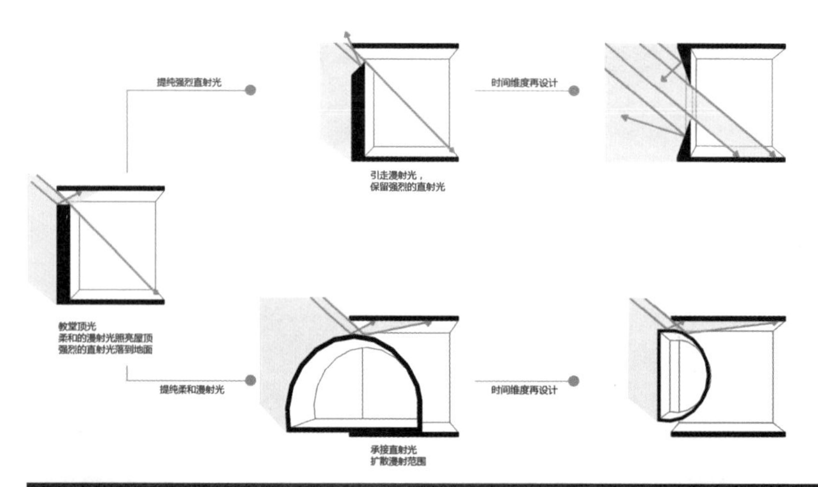

本页上图：光线研究分析。**本页下图：**壳体东部。**对页上图：**地段入口人视。**对页下图：**壳体西部。

对页上图（从左至右）：北向交流空间—黄昏。一天的光线来源。**跨页图**：南侧出口。**本页上图（从左至右）**：早上。下午。**本页中图**：光线分析。**本页下图**：壳体内—黄昏。

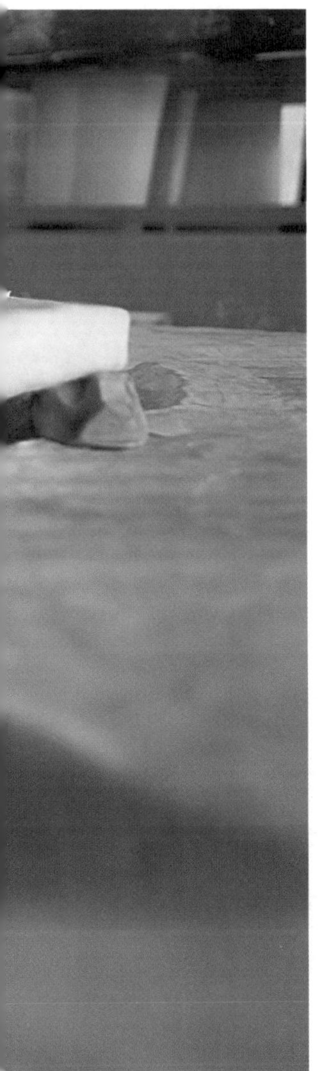

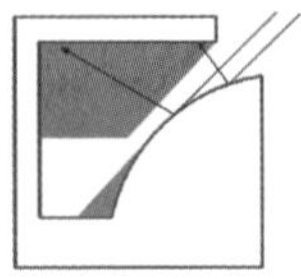

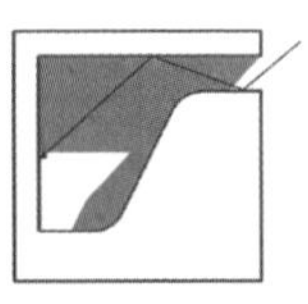

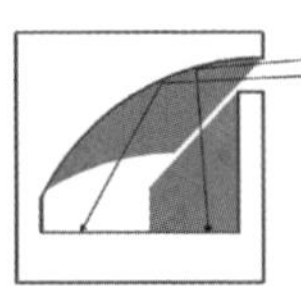

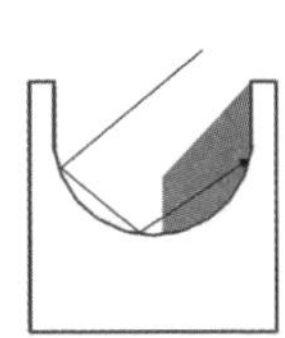

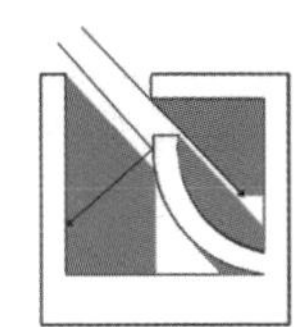

教师点评

作品的灵感则来源于富有诗意和清华文脉的“荷塘月色”,在场地上创造了一个既与外界相隔,但又通过光,流水等元素能感受的外界的“混沌空间”,在建筑内也分为两个空间,外围空间较明亮宽敞,提供给同学阅览交流,内部被包围空间较幽静且光线富于变化,更适合学生反思冥想。

Teacher's comments

The inspiration of the work is derived from the Lotus Pool by Moonlight that is rich in poetry and Tsinghua context, the "chaotic space" which is separated from outside but can be felt through the light, water and other elements is created on the site. In the building, it's also divided into two spaces, the outer space is spacious and bright, which is available for the students to read books and communicate with each other, and the enveloped interior space is more secluded and full of changes in light, which is more suitable for the students to reflect and meditate.

美术馆设计/ART MUSEUM DESIGN

项目选址：清华学堂北
项目类型：美术馆
建筑面积：1000m²
用地面积：1500m²

方案设计：李天颖
指导教师：朱锫
完成时间：2015

方案设计：张昊天
指导教师：朱锫
完成时间：2015

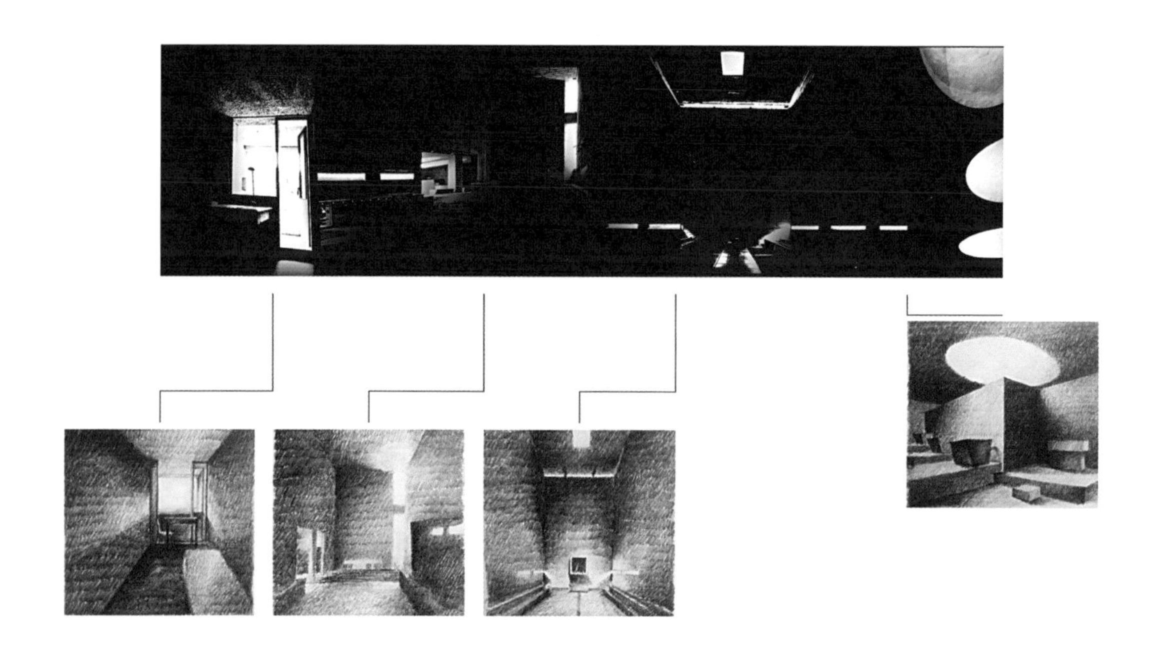

开篇：主要方案图。**本页上图：**拉图雷特修道院分析。**本页下图：**草图。**对页：**初期研究模型分析。

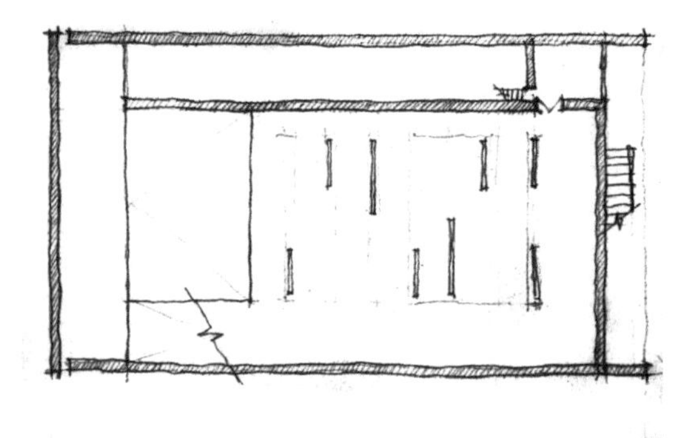

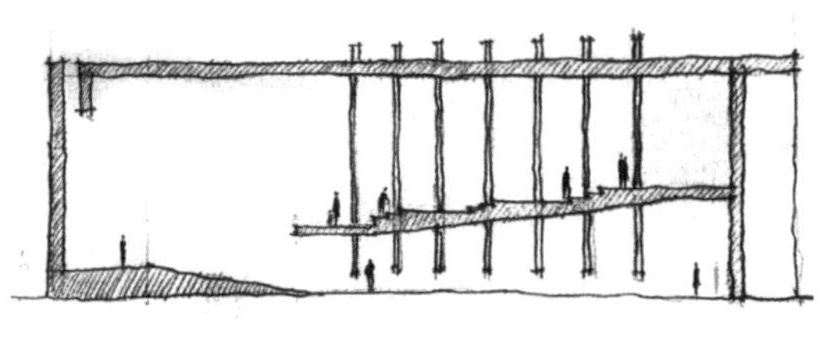

美术馆位于清华学堂北侧院内，体量上以简明的方形与西侧同方部对应，呼应了场地中北侧水利馆的轴线，同时插入原本封闭的清华学堂内院以激发活力。在内部，建筑沿流线营造了由下至上，由自由到理性的空间渐变，从而建构起朦胧到清晰的认知体验。使用者由一层进入展览空间，摸索路径出“洞”，在尽端的“池”停歇转折，而后拾阶上“山”，最后离开展览空间。八片可以移动的展墙贯穿上下，将光从顶至底导引，并作为线索串联起整个空间。

案例分析及概念生成 李天颖 张昊天 2015.03.16

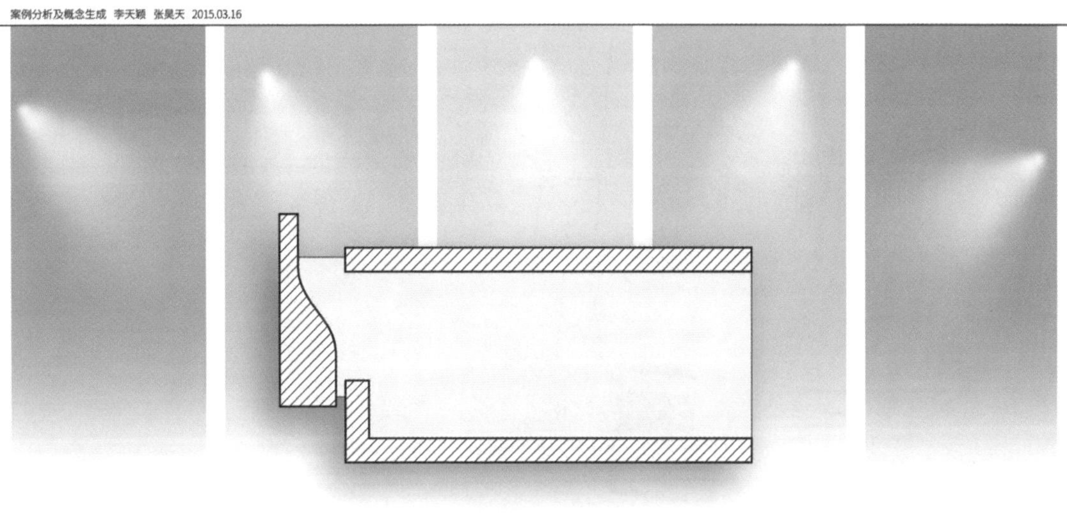

不同角度光线的光照效果

本页上图（顺时针）：总平面图。一层平面图。二层平面图。**本页下图（从上至下）**：剖面图。立面图。**本页上图**：东北角透视。**对页上图**：轴测图。**对页上图**：布展场景。

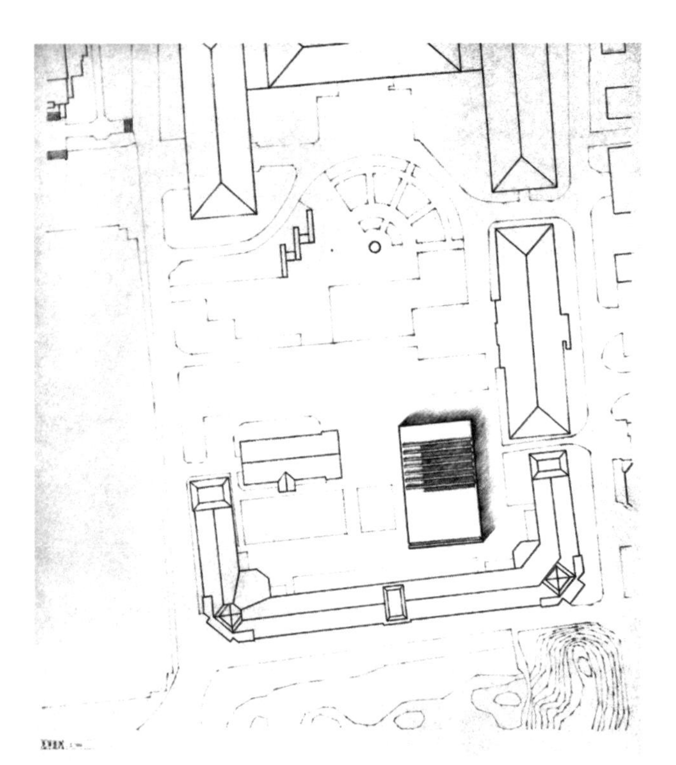

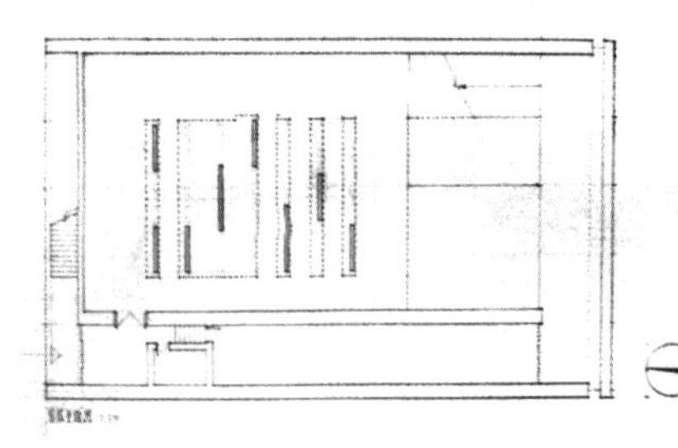

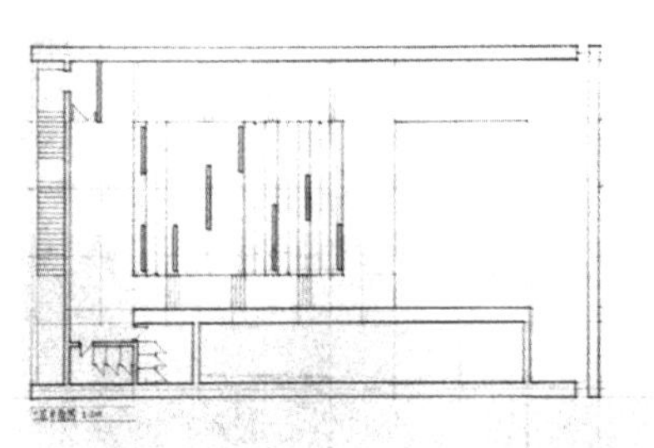

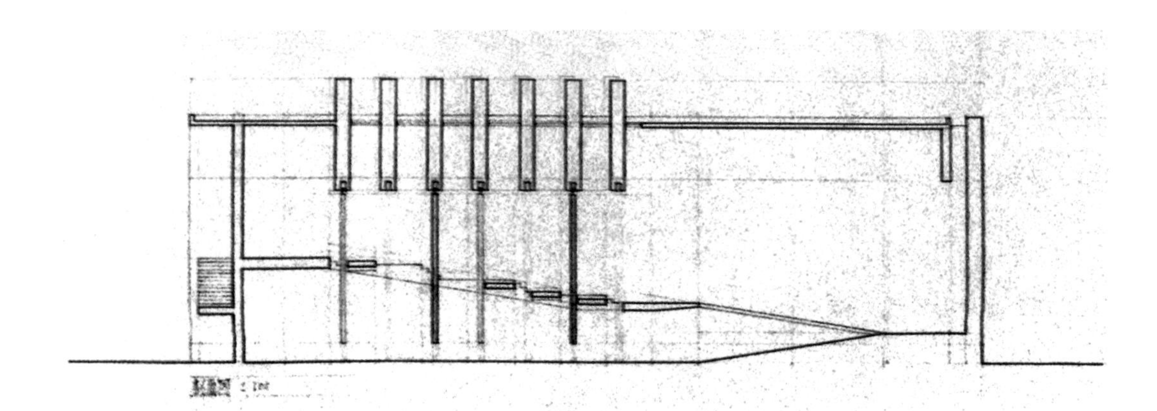

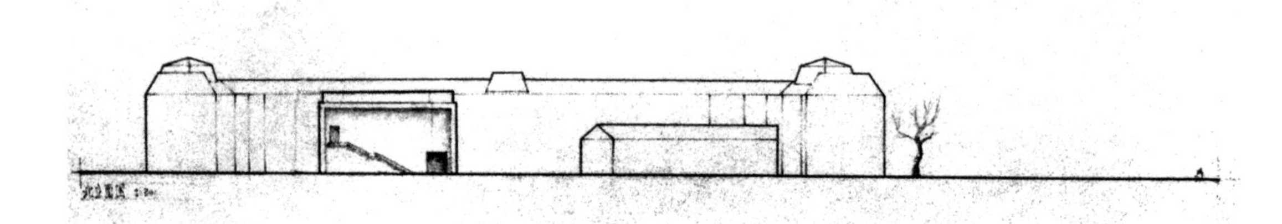

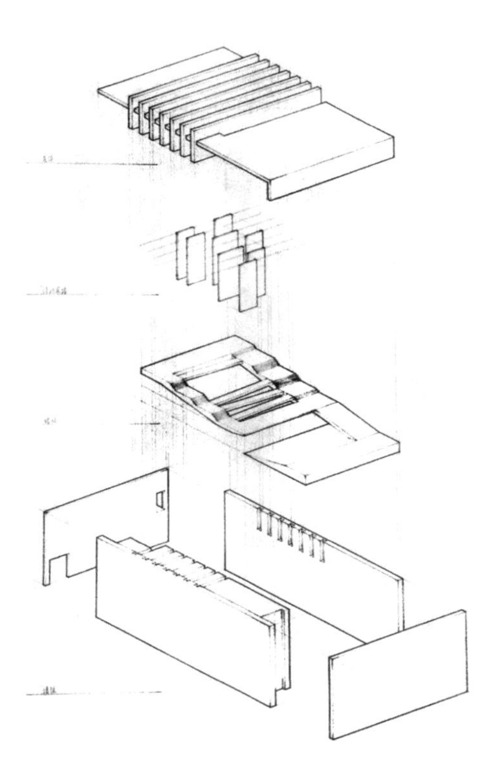

Art Gallery is located in the North yard of Tsinghua School, which corresponds to the same location of west side in volume in a concise manner, it guards the Water Conservancy Museum that serves as the axis in the north side, while penetrating into the inside yard of closed Tsinghua School to stimulate the vitality. In the interior, the buildings create the spatial gradation from down to up and from freedom to rationality along the streamline, thus constructing the cognitive experience from obscure to clear. The users enter into the exhibition space from the first layer, explore the path to go out from the "hole", have a turn in the "pond" rest area in the near end, then pick up the stairs to the mountain, and finally leave the exhibition space. The eight pieces of movable exhibition wall pass through the upper and lower, guiding the light from the top to the bottom, and connect the entire space in series as a clue.

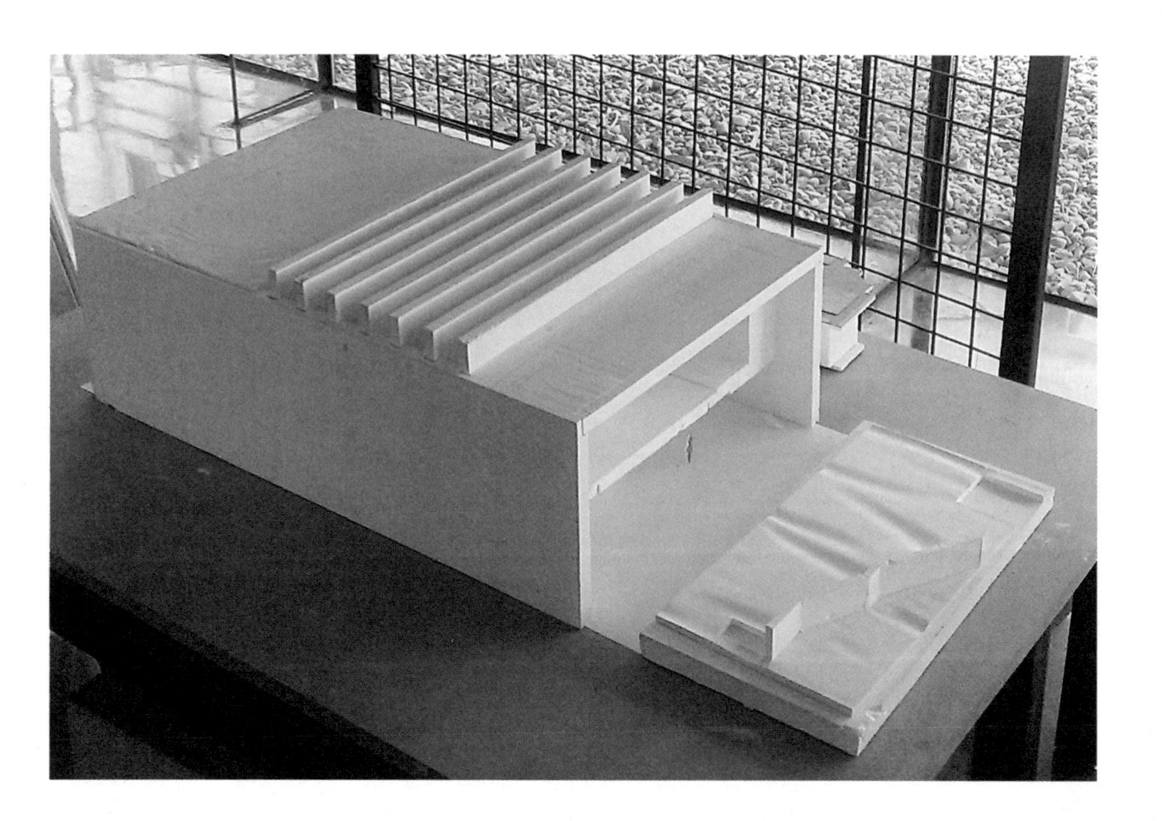

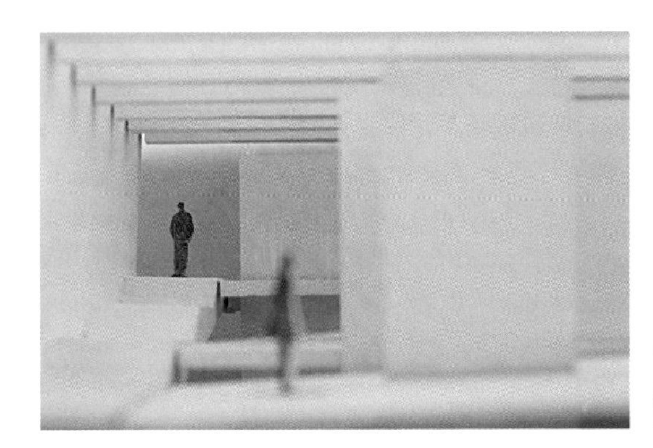

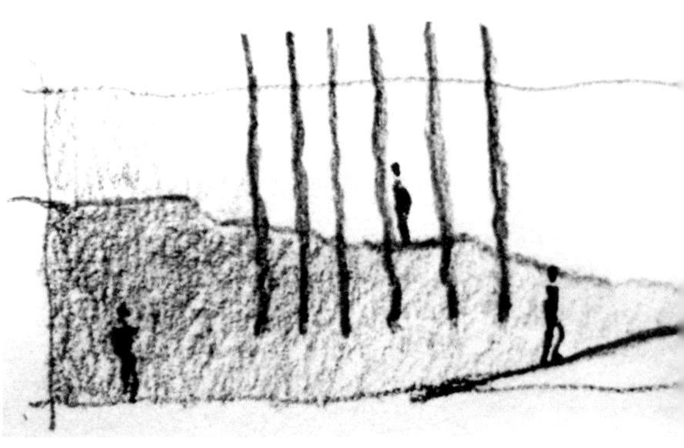

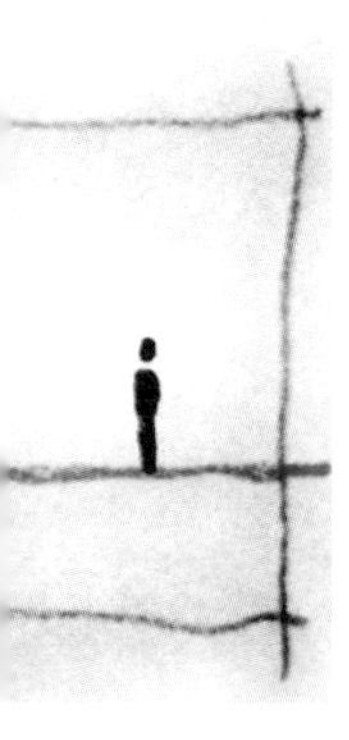

对页上图：局部透视。**跨页上图**：空间结构。**跨页下图**：一层透视。**本页下图**：穿行方式。

教师点评

学生由对拉图维特修道院的研究开始，受到建筑中光线塑造精神空间的启发。学生用大尺度的石膏模型研究了自然光线对空间神秘性和精神性的塑造。在美术馆的设计中，设计结合流线，将不同的光体验组合成为亮度由弱渐强的序列，用光的明暗和韵律营造了建筑的叙事性，使参观者完成了"寻光"和升华的过程。贯通两层的可移动悬挂展板又为艺术品提供了多变灵活的展出方式。

Teacher's comments

The students started from the study of the monastery of Latour Witt, and were inspired by the shape of spiritual space through the light in the building. The students studied the shaping of natural light on the mystery and the spirit of space with large-dimension plaster models. In the design of the museum of fine arts, the design combining with streamline combines the different light experiences into a sequence of light from strong to weak in brightness, the light tone and rhythm created a narrative of the building, which enables visitors to complete the process of "finding the light" and sublimation. The removable hanging panels trough two layers provide a flexible exhibition way for artworks.

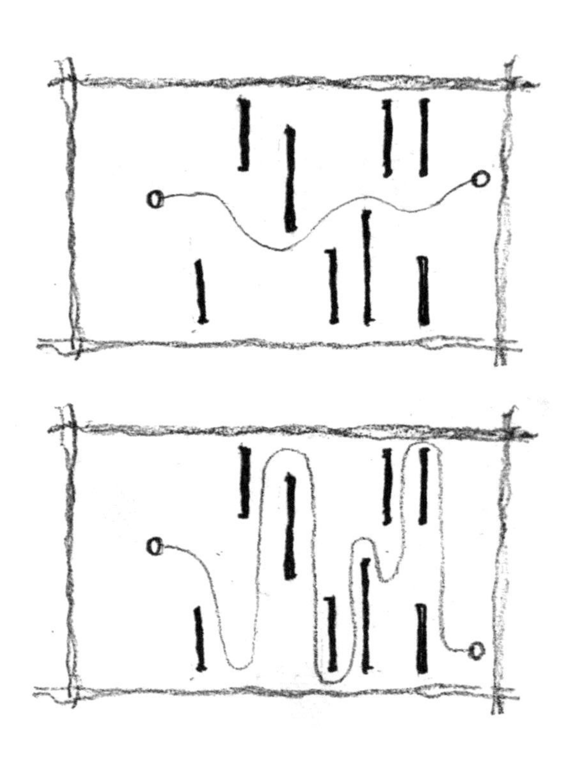

创作展陈空间/ CREATION EXHIBITION SPACE

项目选址：清华学堂以北、同方部以东，包括现在的动振楼和这三栋楼所围成的内部庭院

项目类型：创作、展示、交流的空间

建筑面积：800m² **建筑面积：**1200m²

方案设计：刘宇涵

指导教师：朱锫

完成时间：2015

方案设计：王玉颖

指导教师：朱锫

完成时间：2015

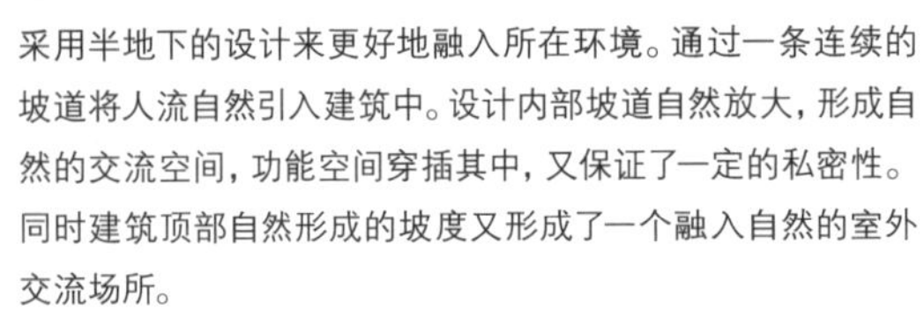

采用半地下的设计来更好地融入所在环境。通过一条连续的坡道将人流自然引入建筑中。设计内部坡道自然放大，形成自然的交流空间，功能空间穿插其中，又保证了一定的私密性。同时建筑顶部自然形成的坡度又形成了一个融入自然的室外交流场所。

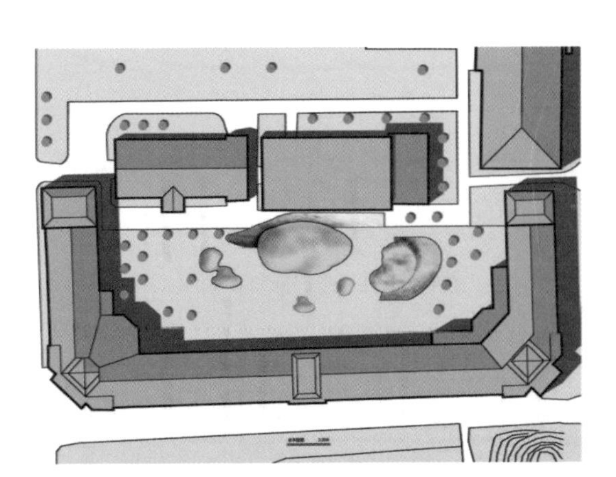

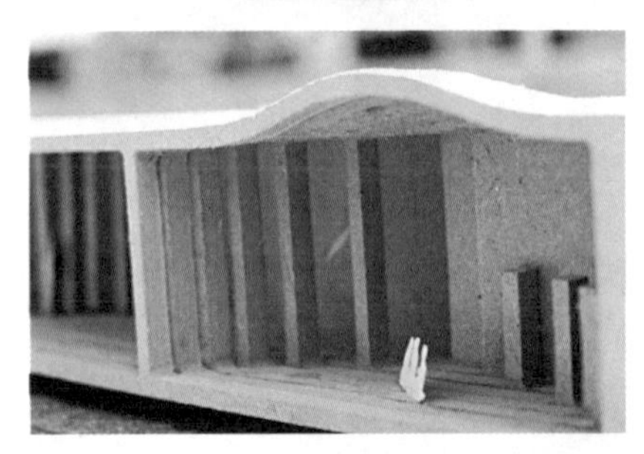

藏一美术馆/
TIBETAN ART MU-
SEUM
项目选址：清华大学清华学堂内院
项目类型：光影美术馆、展览大学生艺术作品
建筑面积：613m²
占地面积：2400m²

方案设计：唐波晗
指导教师：朱锫
完成时间：2015

方案设计：祁佳
指导教师：朱锫
完成时间：2015

设计中我们主要进行了历史地段的应对与光影空间设计。在内院中做了联系清华学堂和北面大广场的下沉庭院。置入五个分散的盒子，通过缝隙进行组织，形成自然与建筑之间不停转换。

我们采取了一个半透弧面墙作为光的载体。针对不同的展厅设置了不同的光环境。如绘画展厅，采取较微弱的采光效果，弧面较小。光还随时间发生变化，在弧面会产生律动的光影效果。

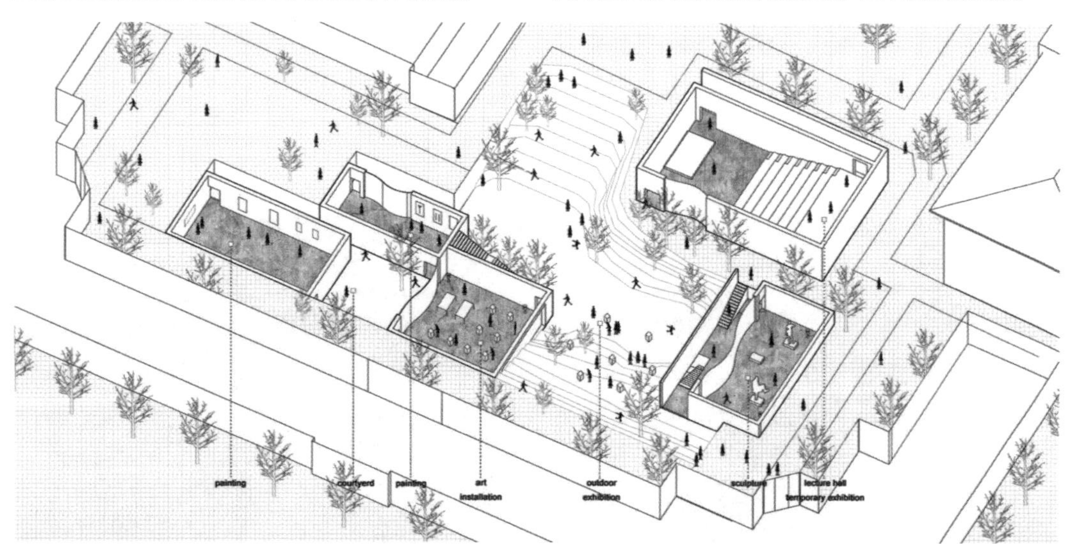

消隐·无重力/ BLANKING AND NO GRAVITY

项目选址：清华学堂以北、同方院以东

项目类型：学生活动中心

建筑面积：1700m²

用地面积：2300m²

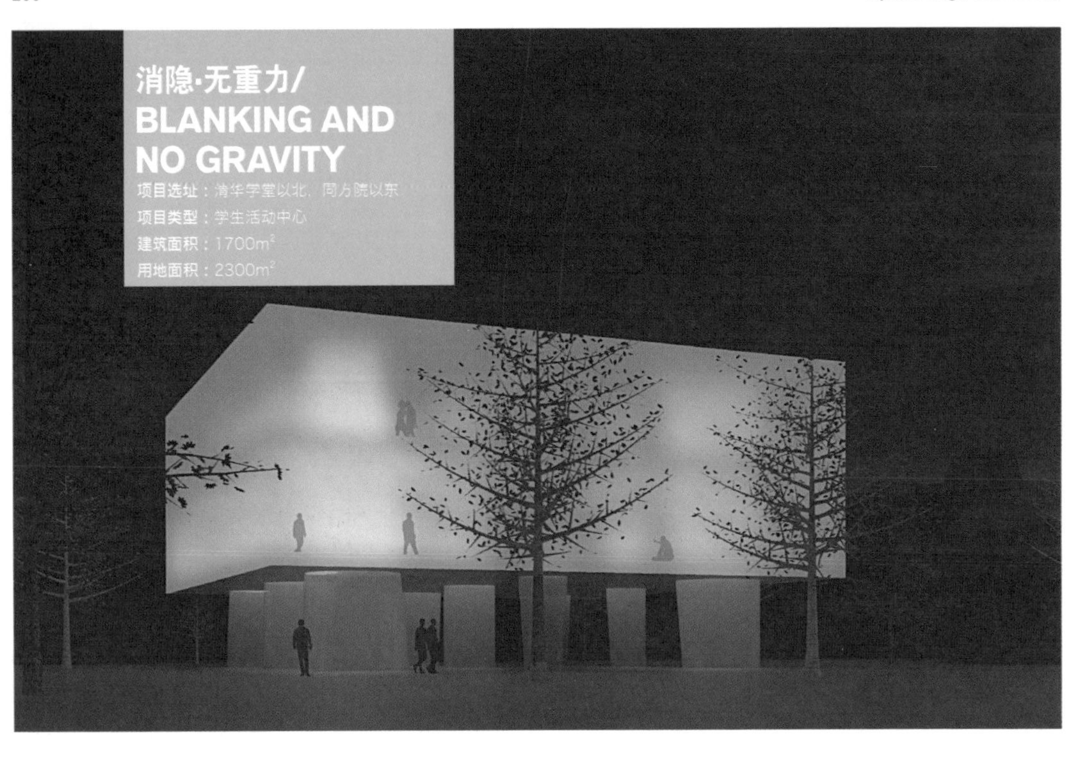

方案设计：唐博

指导教师：朱锫

完成时间：2015

方案设计：宋雨

指导教师：朱锫

完成时间：2015

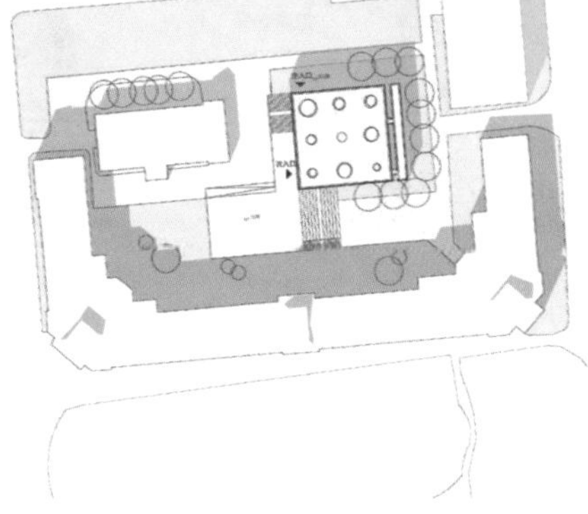

总平面图 1：500

该学生活动中心以“消隐·无重力”为概念。建筑外部采用玻璃幕墙和反光膜，白天反射周边、融于环境，夜晚则如灯笼般点亮地段。

建筑内部利用半透明采光筒，使承重结构模糊；光筒也在各层营造不同光照氛围，呼应不同功能。

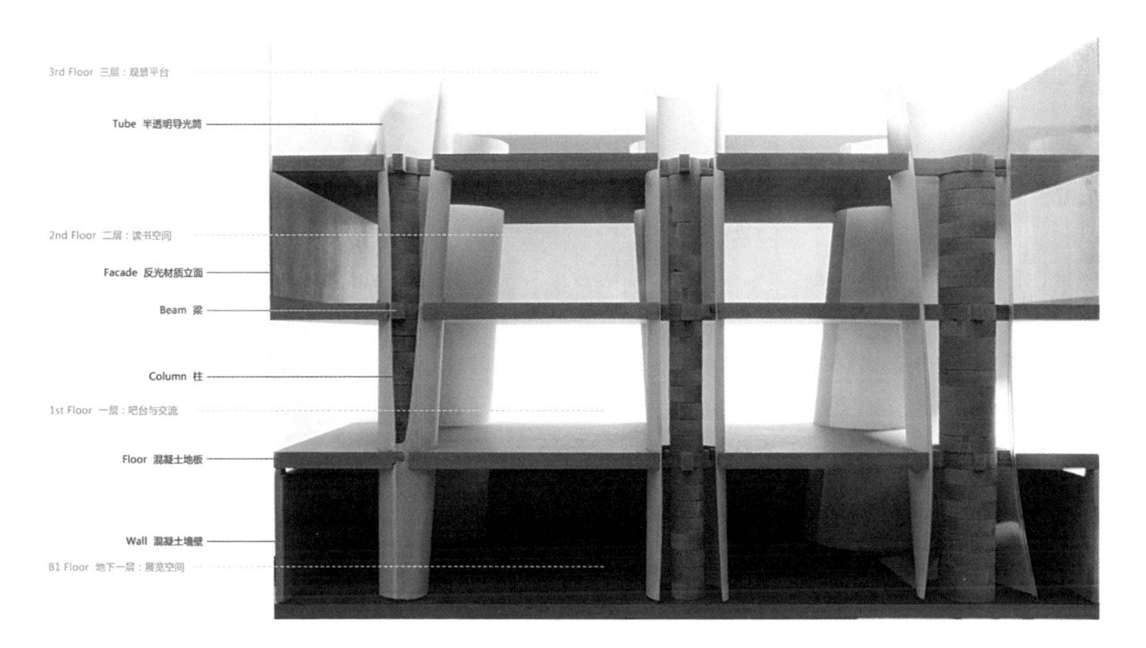
3rd Floor 三层：观景平台
Tube 半透明导光筒
2nd Floor 二层：读书空间
Facade 反光材质立面
Beam 梁
Column 柱
1st Floor 一层：吧台与交流
Floor 混凝土地板
Wall 混凝土墙壁
B1 Floor 地下一层：展览空间

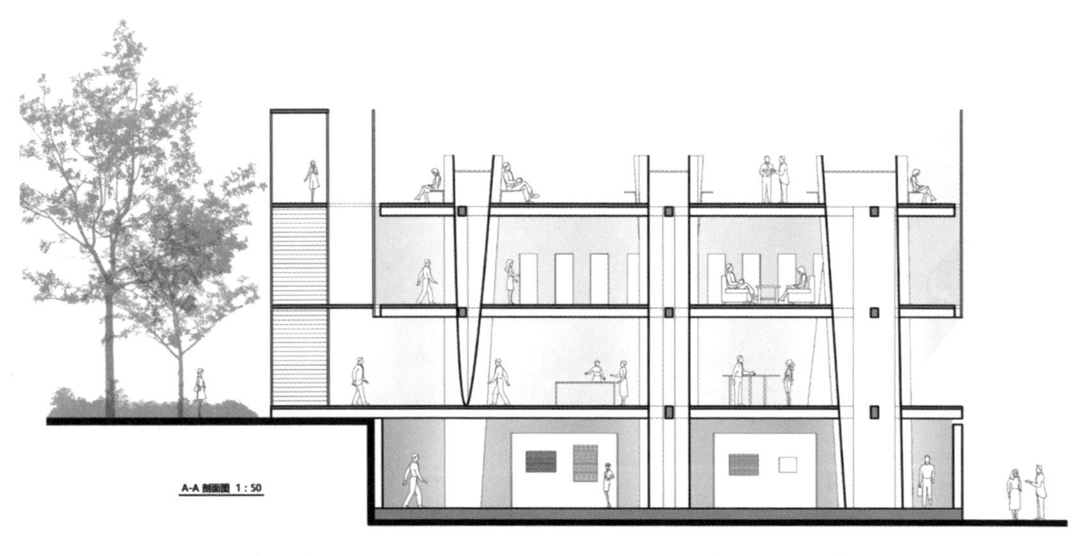
A-A 剖面图 1 : 50

后泡沫城市

李虎

OPEN 建筑事务所创始合伙人

李虎
OPEN 建筑事务所
创始合伙人、主持建筑师

教育背景
1991年 – 1996年
清华大学建筑学院建筑学 学士
1996年 – 1998年
美国莱斯大学建筑学院建筑学 硕士

工作经历
1998年 – 1999年
MICHAEL GRAVES 建筑事务所 建筑师
1999年 – 2000年
PKSB 建筑事务所 建筑师
2000年 – 2010年
Steven Holl 建筑事务所 建筑师、合伙人
2009年 – 2015年
Studio-X美国哥伦比亚大学北京建筑中心 主任
2003年至今
OPEN 建筑事务所 创始合伙人、主持建筑师

主要论著
李虎; 黄文菁.《应力》[M], 中国建筑工业出版社, 2015
李虎. 开放·随想[J].城市空间设计·新观察, 2013(6)
李虎. 开放·建造[J]. 建筑技艺, 2013(2)
李虎; 黄文菁. 成长的礼物——秦皇岛·歌华营地体验中心设计[J]. 建筑学报, 2012(12)

设计获奖
2012 – WA中国建筑奖 优胜奖
2012 – 中国建筑传媒奖最佳建筑奖
2013 – 亚太区室内设计大奖公共空间类金奖
2014 – GQ年度建筑师
2014 – RTFA反思未来奖
2014 – WA中国建筑奖 居住贡献优胜奖
2015 – 伦敦设计博物馆年度设计提名
2015 – 美国建筑师协会纽约分会年度优秀设计奖
2015 – 意大利A'设计大奖赛铜奖

代表作品
歌华营地体验中心（图1）、北京四中房山校区（图2）、HEX-SYS/六边体系（图3）、退台方院（图4）、清华大学海洋中心（图5）、当代MOMA（图6）、万科中心（图7）

后泡沫城市的再设计
——社区中心的生成与植入

三年级建筑设计（6）设计任务书

指导教师：李虎
助理教师：罗韧

未来的建筑师们，当你们毕业的时候，开始你们的建筑实践生涯，你们所面临的大多数任务可能已经不再是宏大规模的设计与建造。这些工作已经在过去的十几二十年里被你们上一两代的建筑师们在无比的匆忙中透支掉了。在高速度的设计与建设中，难以置信的巨大数量的房屋遍地开花，城市面貌被迅速改写，旧城被拆改，新城不断涌现。在物质表面变化的同时，一个城市的精神层面在同步发生裂变。这些瞬间出现的、新的城市空间与结构形态，在潜移默化地、深刻地影响着每一个城市人的生活，包括他们的交往方式，他们的社会关系。这些在短视主义与急功近利驱动下所造就的城市，将是你们不得不去面对，去发现，去改造，去再创造的新的战场，需要你们运用智慧与策略去寻找变革的可能。
这个课程设计就是基于此目标的一次共同探索：去发现城市再设计的可能性；去尝试一种新的团队合作；去学习一种研究和观察城市的方法；去实验一种新的空间设计方法，以及设计表现的可能性。我们在这场针对未来的模拟演习中共同寻找和体验建筑创作的乐趣。

——李虎

场地

泡沫城市速生时期，由于诸多原因而导致了城市土地使用效率低，遗留了大量空闲土地，是城市重生/再建设的基地，东直门香河园，东城与朝阳两区交界处的三角形空地，是一个典型的场地。由于位于两区交界，外加此区域居住形态/人口构成/近期历史变迁也都很复杂/丰富。
基地面积：约3500m^2
建筑面积：2500~3000m^2

建筑功能

社区中心/邻里中心/社区服务中心，是未来城市重生/社区营造的一个催化剂。

课程结构

1. 研究：研究的方法+理论研究和阅读（横向及纵向历史，相关理论）
所有学生会分成三个小组，各组拥有各自针对性的研究内容。两周的研究结束后，所有的研究内容需汇总并编撰成为全组成员共享的一本研究文献。具体分组方法如下：
1.1：基地研究：该小组主要以课程设计的指定基地为研究对象，对基地的自然和社会信息进行调研和整理
基地现状：基地尺度研究及对比；基地的竖向及垂直空间分析；基地及周边交通分析；基地及周边设施分析基地所属社区调研与分析：东直门香河园及左家庄一带社区历史沿革（至少应包括行政区划变迁，路网

变迁，城市功能及空间变迁，建筑群落变迁等）；该社区现阶段人口构成及特征基地周边社区居民对公共服务设施诉求调研（调查问卷方式）。

1.2：中国社区及社区中心历史/现状研究：研究内容：该小组主要以“社区”及“社区中心”这两个概念在中国的历史发展为对象展开调研，并剖析中国社区中心运营的现状。

1.3：世界社区及社区中心历史/现状研究：在两周的研究结束后，全组会根据研究成果、在导师指导下、共同讨论制订出建筑设计的任务书；每个学生在接下来的时间里会依据该任务书进行建筑设计。

2. 设计：任务书的创造；空间影响社会行为的潜力；发展设计能力

在两周的研究结束后，全组会根据研究成果、在导师指导下、共同讨论制订出建筑设计的任务书；每个学生在接下来的时间里会依据该任务书进行建筑设计。

在前两周城市研究的基础上，发现各自建筑设计的入手点，发展清晰的设计概念，探究"空间"与"行为"的相互关系，并进行“从内到外的”建筑设计。与此同时，学生还应运用不一而足的设计方法及手段（包括实体模型、手绘、电影、电脑等）来推动设计的进行。

3. 表达：概念/空间/材料的表达方法

设计表达也是这次设计课的训练重点。将强调对建筑概念的清晰表达，并期待每位同学在规定成果之外、运用对各自概念适合的创造性手段去充分表现其设计的深度与细节（如大比尺模型，动画，影像或手绘），以及对设计生成过程的记录。

4. POST-STUDIO：研究及设计成果的整理

在studio结束之后的暑期里，我们希望有志愿者参与，把全组的研究及设计成果整理出来，编纂成书，作为建筑师对社区中心与城市现有问题的回应、及对两者之未来的想象。

成果要求

最终设计应着力于满足以下5点特征：

（1）动人的 spectacular

（2）人本的 humanistic

（3）在地的 site-specific

（4）有趣的 playful

（5）概念清晰的 conceptual-clarity

艺术过滤器/ ART FILTER

项目选址：机场高架下三角地块

项目类型：城市艺术中心（社区中心）

建筑面积：2400m²

用地面积：3000m²

总 装 招

方案设计：苏天宇
指导教师：李虎
完成时间：2015

开篇：主要方案图。**跨页图：**地段分析。**对页上图：**尺度分析。**对页下图**地段场景照片。

中国的高速城市化为世界瞩目，而这其中暴露出的城市问题同样由于城市化进程的压缩而格外凸显。北京，作为中国大都市的代表，同时集中了高速发展带来的红利和弊端。城市中艺术空间的发展速度远远赶不上城市的蔓延，造成城市艺术性匮乏。这种匮乏包含两个层面：实体上的，以及精神上的——如果艺术空间是高高在上而非触手可及，人们也很难从中找到自己的位置和归属感。此设计中，我通过喇叭状的空间原型，尝试使不同类型的艺术空间重新与人发生联系。在高架桥下，我布置了“触、演、读、食”四个平民化的艺术空间，使之服务社区。在布置充分的公共空间同时，我也将强有力的体量冲破高架桥限制，发表着昂扬的艺术宣言。

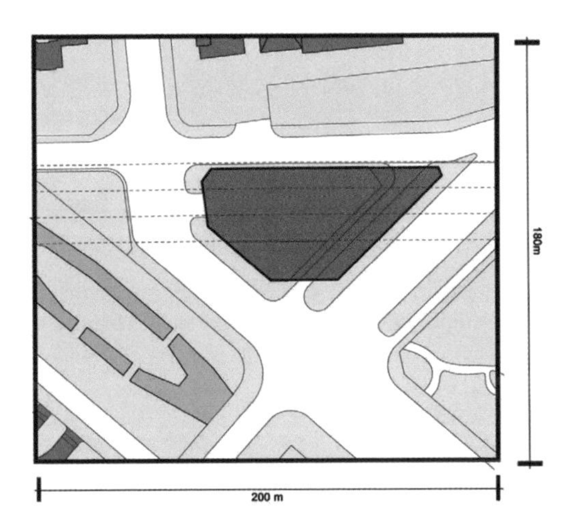
180m
200 m

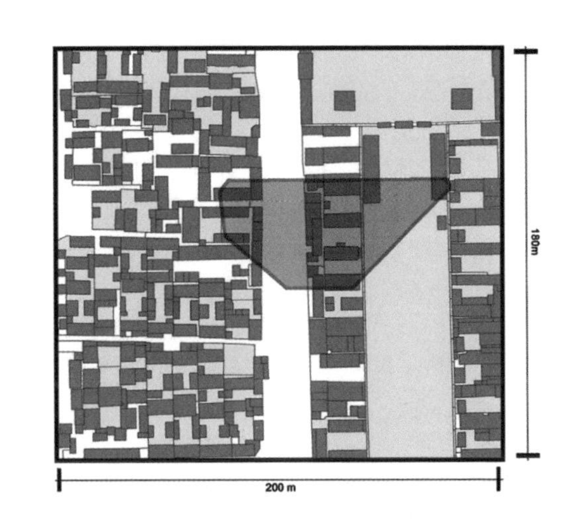
180m
200 m

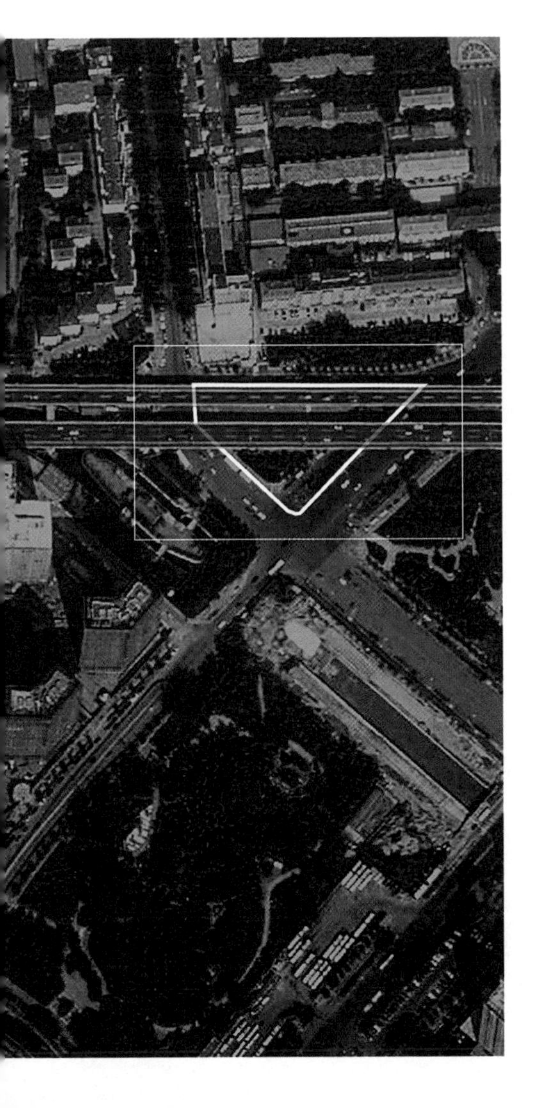

China's high-speed urbanization is the world's attention, and the urban problems exposed are especially prominent due to the compression of the process of urbanization. Beijing, a representative of the Chinese metropolis, collects the bonus and drawbacks as the result of rapid development. The development speed of art space in the city is far behind the spread of the city, causing the lack of urban art. This shortage includes two aspects: physical and spiritual — if art space is superior and is not within touch, people are difficult to find their location and ownership. In this design, I try to make different types of art space to reconnect with people through the trumpet shaped space prototype. Under the viaduct, I arrange the "touch, play, read and eat," the four civilian art spaces to serve the community. In arranging the full public spaces, I will make the powerful volume break through the viaduct constraints, publishing a spirited art declaration.

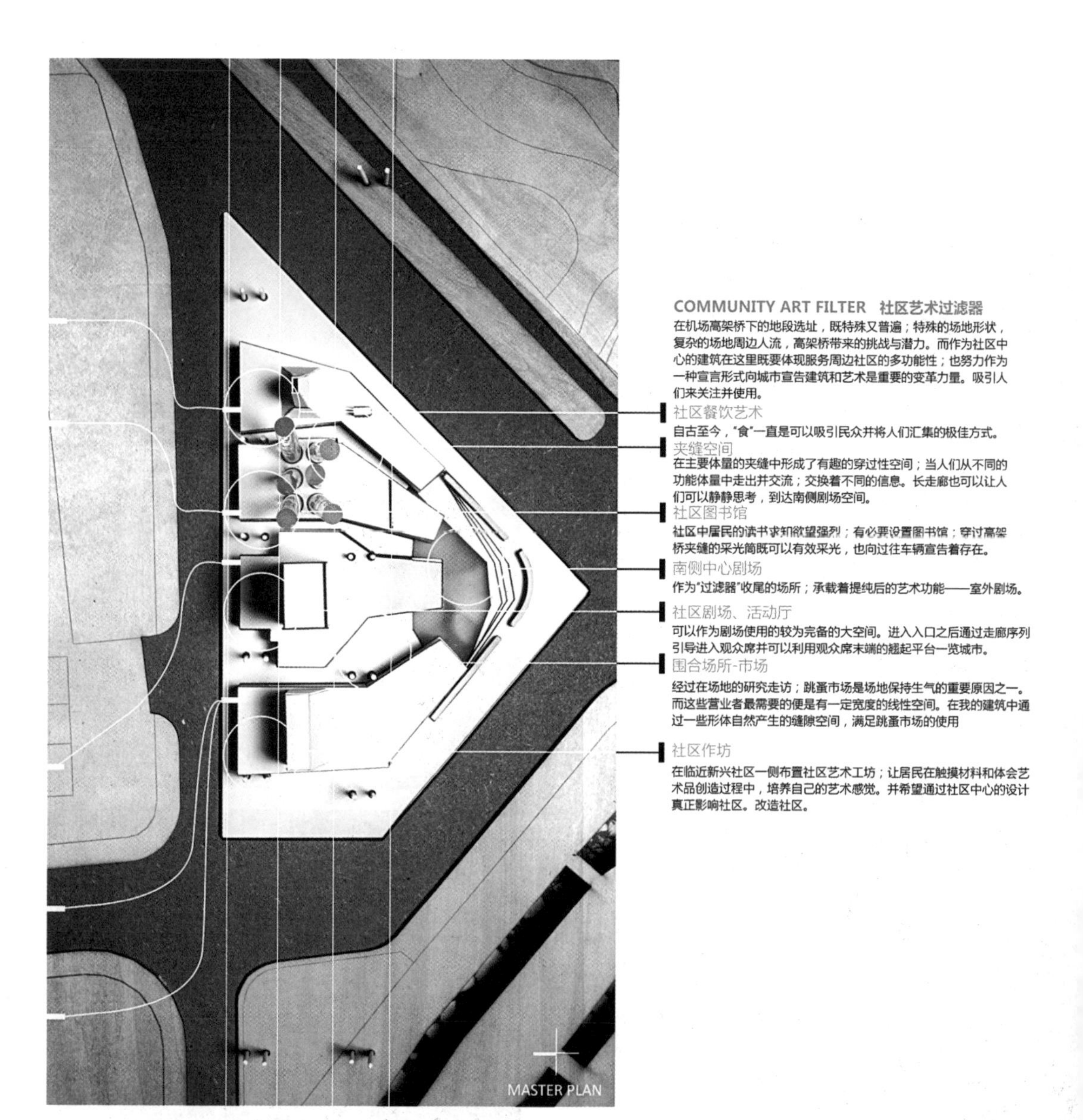

COMMUNITY ART FILTER 社区艺术过滤器
在机场高架桥下的地段选址，既特殊又普遍；特殊的场地形状，复杂的场地周边人流，高架桥带来的挑战与潜力。而作为社区中心的建筑在这里既要体现服务周边社区的多功能性；也努力作为一种宣言形式向城市宣告建筑和艺术是重要的变革力量。吸引人们来关注并使用。
社区餐饮艺术
自古至今，"食"一直是可以吸引民众并将人们汇集的极佳方式。
夹缝空间
在主要体量的夹缝中形成了有趣的穿过性空间；当人们从不同的功能体量中走出并交流；交换着不同的信息。长走廊也可以让人们可以静静思考，到达南侧剧场空间。
社区图书馆
社区中居民的读书求知欲望强烈；有必要设置图书馆；穿过高架桥夹缝的采光筒既可以有效采光，也向过往车辆宣告着存在。
南侧中心剧场
作为"过滤器"收尾的场所；承载着提纯后的艺术功能——室外剧场。
社区剧场、活动厅
可以作为剧场使用的较为完备的大空间。进入入口之后通过走廊序列引导进入观众席并可以利用观众席末端的翘起平台一览城市。
围合场所-市场
经过在场地的研究走访；跳蚤市场是场地保持生气的重要原因之一。而这些营业者最需要的便是有一定宽度的线性空间。在我的建筑中通过一些形体自然产生的缝隙空间，满足跳蚤市场的使用
社区作坊
在临近新兴社区一侧布置社区艺术工坊；让居民在触摸材料和体会艺术品创造过程中，培养自己的艺术感觉。并希望通过社区中心的设计真正影响社区。改造社区。
MASTER PLAN

对页：总平面图。**本页上图**：剖面空间草图。**本页下图**：高架桥人视图。

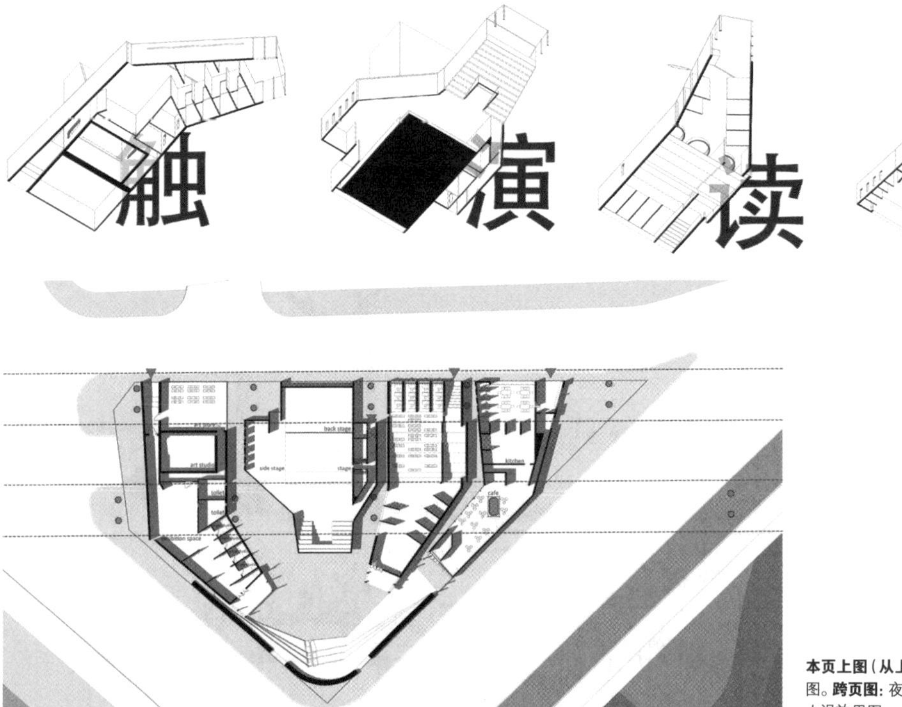

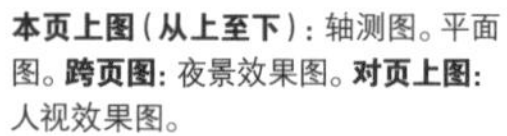

本页上图（从上至下）：轴测图。平面图。**跨页图：**夜景效果图。**对页上图：**人视效果图。

教师点评

苏天宇很早就定下了“社区艺术工坊”这样的想法，希望能够鼓励社区居民参与到创造性的艺术行为中来凝聚社区。出色的动手能力帮助他以实体模型的方式不断地深化方案的推进。着眼于艺术创作行为本身的多样性和丰富性，并通过对建筑中人的行为进行组织和拼贴，他成功地使“艺术改变社区”这一概念落地，并以建筑空间将这种组织表达出来。通过将自然光以不同的形式引入室内空间，方案提供给不同艺术活动以相适应的光环境。而这些形式各异的光装置，也成为建筑在视觉上的宣言。

Teacher's comments

Su Tianyu determined the idea of "community art studio" in an early time, and hoped to encourage community residents to participate in the creative art activities to enhance community cohesion. The excellent hands-on ability helped him to continue to deepen the development of the scheme in the form of solid model. Focusing on the diversity and richness of artistic creation itself, the organization and collage were carried out through the human behavior in the building, so that he successfully implemented the concept of "art changes a community", and expressed this organization through architectural space. By introducing natural light into the indoor space in different forms, the scheme can be provided to different artistic activities in order to adapt to the light environment. And these various forms of light devices also become a visual declaration of the building.

“一时兴起，脏手偶得”/ UNKNOWN PLEASURE

项目选址：东直门二环立交桥下三角地

项目类型：社区中心

建筑面积：4000m²

用地面积：3000m²

方案设计：项轲超
指导教师：李虎
完成时间：2015

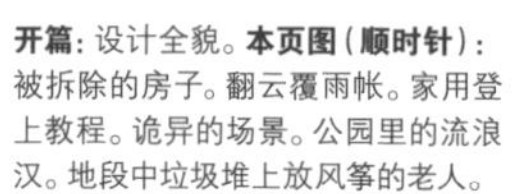

开篇：设计全貌。**本页图（顺时针）：**被拆除的房子。翻云覆雨帐。家用登上教程。诡异的场景。公园里的流浪汉。地段中垃圾堆上放风筝的老人。

“一时兴起，脏手偶得”假想了这样一个场景，由于地段中原城中村的拆迁疏忽，几户人家被遗忘在地段中无人问津，起初为了夺得城管部门的注意，同时改善生活条件，地段中居民从周边地区收集垃圾来进行自宅的加建，但是并没有人管。到了后来，随着垃圾建筑物的增加，来自各方的人群聚集到了地段中来，本来用于“报复”社会的私搭乱建现在变成了能够为社区提供各方面服务的另类的社区中心，随着时间的推移，地段中的垃圾建筑不停地被拆除，回收，而后建起新垃圾建筑，生生不息地发展下去不断为城市和社区倾注活力。

“去找最能打动自己的的东西。”——李虎老师

一个处于高架桥下噪音不断被人遗忘的寂寞地段，睡在公园的流浪者的帐篷，站在堆着建筑废料的垃圾堆上放风筝的老人，拾荒者利用捡来的材料装饰建造起来的房子，随处可见的废弃三轮车，自行车……

古谷实的漫画中无依无靠在城市中流浪的年轻人，用垃圾做艺术的家伙们：Joseph Cornell，劳申伯格，赤瀬川原平……在树上盖茶室，在屋顶种韭菜的藤森照信，William Health Robinson的漫画中使用"高科技"炮制的各种生活装置，使用奇怪的材料和方式盖房子的Smiljan Radic，《庇护所》和《没有建筑师的建筑》中的大自然造就的房屋，《东京制造》中描述的“寡廉鲜耻”的滥建筑……

个人的爱好和在地段中的观察研究不断积累，最后筑成了垃圾社区这样一个表现形式。

这个明显脱离常规的设计从最开始并不是叙事性的，当然也不是关于社会贫困或者人文主义，只是想用充满趣味的方式给自己苦大仇深让人情欲尽失的传统建筑设计学习状态画上句号。

问答：（1）设计中遇到的困难和转折点

8周的设计进程中第4周和第6周的时候我的设计都发生了翻天覆地的变化。

第4周面临着严峻的问题，真的垃圾建造是什么样的过程呢？这些过程又应该如何表现出来呢？用手绘的方式总是欠缺说服力，这时恰好遇到了因社团活动收集到了20辆废旧自行车的殷婷云同学，我们连夜在全建馆收集各种板材废料并将他们和二十辆废旧自行车在第二天要进行中期评图的场地上搭建成了一个自行车的"宝座"，在搭自行车的时候试验各种自行车之间的交接方式，无论多么微薄，这样的建造经验是任何图纸软件模拟不了的。

第六周时苦于找不到组织垃圾城市的方法，想象不出垃圾城市的样子，李虎老师告诉我一个在民旺胡同附近用垃圾建造起来的房子。拎着四瓶燕京啤酒的我兴奋地敲开了用铁钉固定在木头梁架上的塑料板门，与热情的房子主人交谈使我受到了极大的启发，这座房子直接摧毁了我的建筑观，房子原来可以是这样的东西！随后又走访了附近的垃圾场，废品收购站，最后我终于知道什么东西能够让自己感到满意和兴奋了。

问答：（2）对于建筑的认识的转变

认识到了学生的建筑设计必将演变成一场与自身的严肃问答，愧对自己的部分和自己的成长都会在设计中闪现。

对于建筑恒定的结构和明确的意义必将会失效，在考虑事物之间联系的同时也决不能忽略事物的本质。

问答：（3）通过与建筑师的接触得到的知识

建筑师必将是高度严谨和自律的同时又时刻充满激情的。刚刚通宵交过图的李老师和罗老师出现在课堂上的时候，仿佛看到了建筑师身上沉重而孤独的责任。

问答：（4）吐槽/点赞

李虎老师和罗韧老师是两位非常酷的老师和建筑师，我时常幻想有一天成为他们的样子。

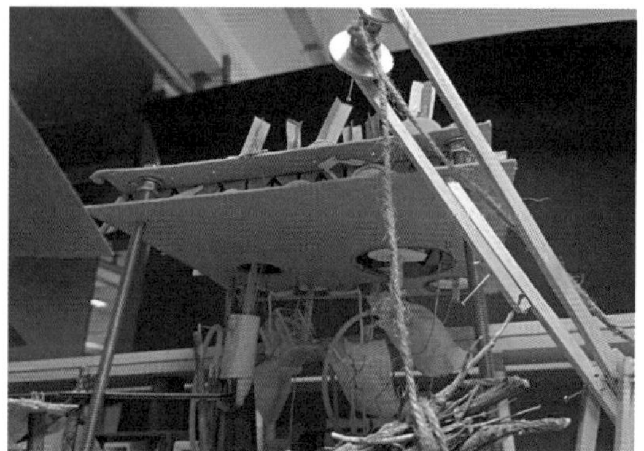

本页上图（从左至右）：风力发电搅拌豆浆糖炒栗子旋转木马机。街边场景。**本页中图（从左至右）**：垃圾房子。满地垃圾。**本页下图（顺时针）**：模型总图。木塔。桥上风景。棚子之下的旋转木马。

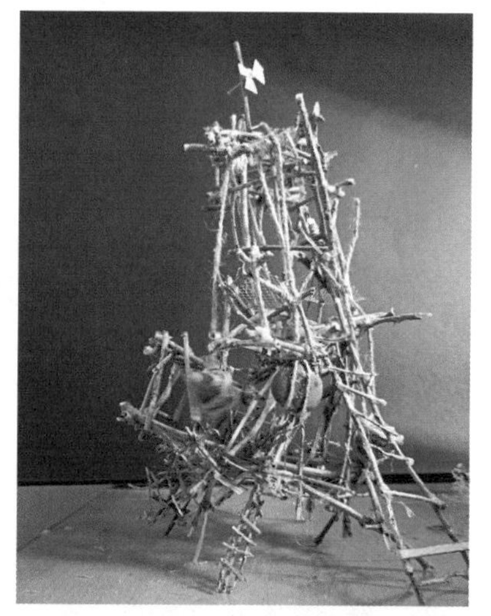

"One does something out of interest and gets something with dirty hands". He imagines such a scene: Because of the negligence of demolishing the original urban village in this area, several households are left here and nobody cares for them. At first, in order to attract the attention of urban management department and improve living conditions, the residents here collected rubbish from surrounding areas to build their own houses, but nobody stopped them. Later, along with the increase of rubbish buildings, the people from all directions gathered here. The originally private and disordered buildings for "revenging" the society become a weird community center for providing the community with all-round services. As time goes on, the rubbish buildings here are demolished and recovered continuously and the new rubbish buildings are built successively for injecting the city and community with vitality.

"Find things which can move you most" - Teacher Li Hu

The lonely and noisy area under the viaduct being forgotten by people, the tents of vagrants sleeping in the park, the flying-kites old people standing on the garbage heap of building wastes, the houses built by junkmen with discarded decoration materials, and the abandon tricycles and bikes everywhere...

The helplessly young people wandering in the city in cartoons of Minoru Furuya and the guys making art works by using rubbish: Joseph Cornel l, Robert Rauschenberg, and Genpei Agasegawa..., Terunobu Fujimori building the teahouse on the tree and planting fragrant-flowered garlics on the roof, William Health Robinson making various devices in daily life by using "high technologies", and Smiljan Radic building the house with strange materials and methods... The houses built by the nature in Sheltered and Architecture without Architects, and the "shameless" architecture described in Made in Tokyo...

The personal hobby and the continuous accumulation of observational study on areas form such a rubbish community. The design deviating from conventions was not narrative at the beginning and not about the social poverty or humanism. He only wanted to draw a full stop in an interesting way for the traditional architectural design learning which made him suffer bitterly and have no passions.

教师点评

项柯超的设计恐怕是全组中一个最有争议的一个。他是一个对社会问题有着强烈看法、对弱势群体有着强烈同情心的同学，这个社区中心的题目对他是一个独特的机会。在前半程中，他一直在摸索如何切入，直至他开始观察到场地周边的一些实际非常规的民间建造而受到了启发。当我看到他开始动手尝试建造的时候，开始鼓励他完全通过手工建造的方法来设计，如同制作一场戏剧的舞台道具。他在最后一周内爆发出了惊人的热情和能量，通过他充满细节和故事性的模型表达了出来。

Teacher's comments

Xiang Kechao's design is probably the most controversial design among the designs of the whole group. He is a student who has a strong view on social problems and a strong sympathy for the vulnerable groups. The topic of community center is a unique opportunity to him. In the first half, he did not know how to start the design until he was inspired by the actual non-conventional civil construction around the site after observation. I encouraged him to do it by the hand-made method, just as he was making stage properties for a drama when I saw him starting to construct. He showed tremendous enthusiasm and energy through his model full of details and stories in the last week.

对页图：桥下小剧场。**本页图：**自行车宝座。

后泡沫时代城市再设计——深夜食堂/MIDNIGHT DINER

项目选址：北京市北二环机场高架下

项目类型（功能）：社区中心

建筑面积：2300m²

用地面积：约 3500m²

方案设计：徐菊杰
指导教师：李虎
完成时间：2015

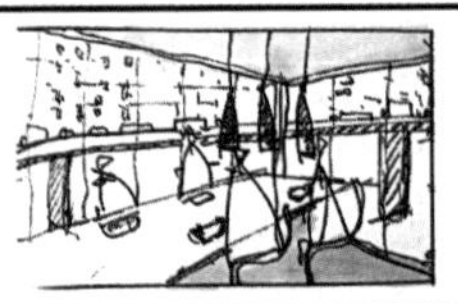

又是堵车的工作日早晨，高架夹在两侧高楼中，每辆车夹在每辆车中间

坐在出租车中百无聊赖的年轻白领姑娘，漫不经意的抬头向外看去。

“为什么不走下高架试试看？”从两桥之间建筑的井口中隐约透出的光轻轻的邀请着她，她发觉这个想法难以抗拒。

她付过车费，拉开车门，在众多从车内发出的惊异目光中跃下高架，敏捷的翻过护栏，

打开门，却是各色各样的人，热热闹闹挤在一起，聊天、吃饭。他们看见她，露出了熟悉的微笑。

Narrator 1: 高架来人

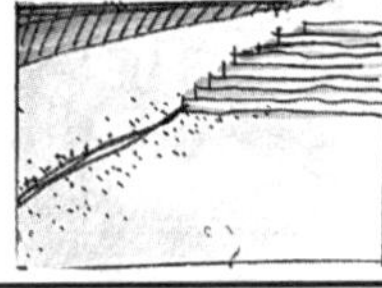

她走在之间，高架桥之间，天和地之间，欣然的熟悉和未知的紧张之间。

下午两点，昏昏欲睡的蝉鸣，桥下分明的阴影。

孩子的视角，高架桥落在白墙上的影子。

孩子蹲下来，把自己的影子也投在白墙上，模仿着各种小动物的样子。

影子随着太阳在墙上缓缓变换位置。

Narrator 2: 孩子

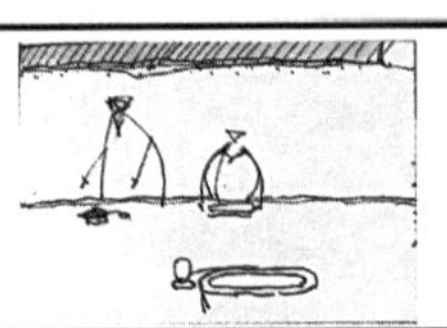
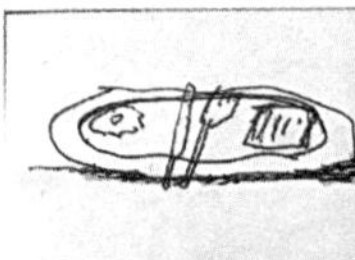

或者在桥下，无忧无虑的荡起秋千。

好奇的窥探着大人的世界。

餐食是城市浪漫车轮，台阶层层，越过斑驳的阳光，青年急匆匆推门而入，为了上班前的早餐。

餐桌旁，几个相熟的社区居民安静的吃着饭。

安静的早餐。

Narrator 3: 社区青年

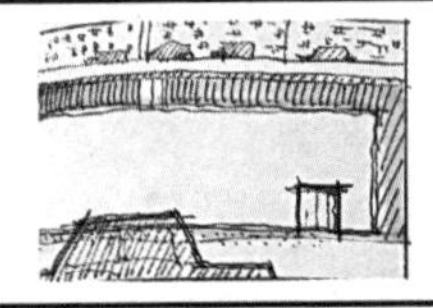

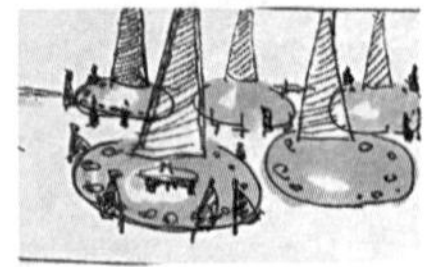

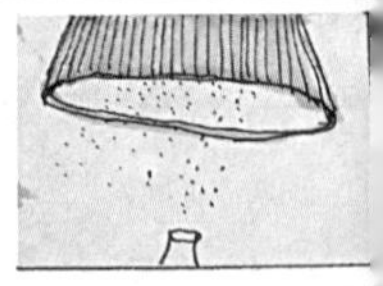

夜深了，两侧高楼逐渐拉上了帷幕，但高架上依旧是喧闹的灯带，桥下有一个小小的入口，固执的点着一盏灯。

与沉默的黑墙形成对比的，是温暖的灯光和进出的居民。

围炉而坐，巴山夜话。

热气腾腾的火锅。

光暗明灭，在灯光筒中嬉戏、回响。

Narrator 4: 深夜食堂

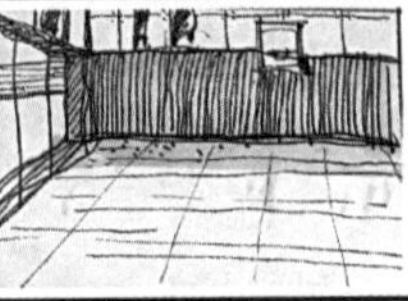

下沉的广场聚集着跳蚤市场，人群熙熙攘攘。

与之一门之隔的是半地下活动场地，走廊让路过的居民好奇的注视着场馆内的活动。

坐在活动场地顶上，看着远处的高架和底下的跳蚤市场。

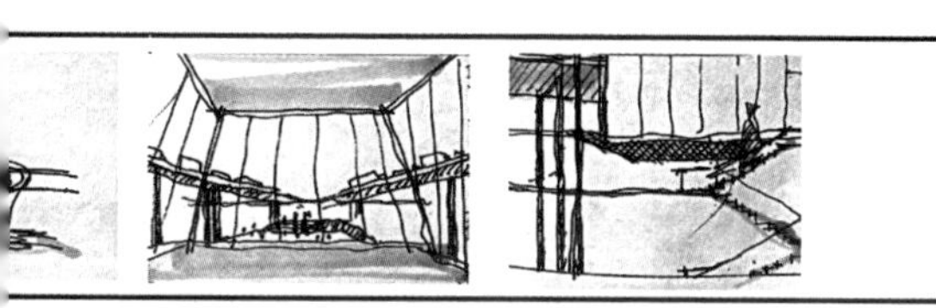

后泡沫时代城市再设计，关注于城市问题，落脚在人的生活。怎样激发废弃场地的潜能？社区中心的精神何在？人的生活状态又是怎样？

日剧《深夜食堂》中，一方小桌承载了一组人物的悲欢离合。社区中心的核心应是人和人相遇、发生故事的场所。但这样温情的联系却也随着城市的高速发展日益稀薄。吃则不分文化不问阶层，可以作为联结的纽带。以社区食堂为主，辅以其他公共空间的建筑功能由此而生。而对空间使用场景的想象，以文字、影像的方式对生活的关注，亦带来设计方法上的再思考。在老师的鼓励下，我尝试用电影的方式创作一组戏剧化的建筑空间，希望能提高人与人、心和心之间的相遇浓度，重构温情纽带。在设计过程中，1:50的手工模型用以推敲空间，分镜草图帮助探索人的活动，文字则编织可能性，进而形成故事再作用于空间。村上春树的《1Q84》中，青豆意外从高架走下，整个故事由此展开。设计中借鉴了这个开头，考虑到地段位于两条并列高架之下，如果能使建筑体量从中挤出，堵车的人与社区居民相遇，也该是有趣的故事吧？除了异质人群的相遇，社区中的不同与相似的人呢？他们又能发生怎样的故事？

空间、故事两向合拢，场地与活动相互激荡。整个建筑就在空间、时间、人的活动的三种维度上反推而成。最后呈现出一个内部空间丰富、外部立面神秘的黑盒子。设计成果表达时也借助了电影叙事手法，利用一系列黑白透视讲述了一组在建筑中人和人相遇的故事。最初从对城市问题的好奇和对人的一点热忱开始的设计，也最终终于此。

开篇：主要方案图。**跨页图**：设计过程中的草图—中期评图。

The city redesign in the post bubble era focuses on the urban problems, which are settled in the lives of the people. How to stimulate the potential of abandoned sites? What is the spirit of community center? What about people's living conditions? In the Japanese drama "Midnight Canteen", a square table carries the vicissitudes of life of a group of characters. The core of the community center should be a place where people meet and have stories. But the warm link becomes increasingly thin with the rapid development of the city. Eating, regardless of culture and class, can be used as a connection of link. The building functions, take the community canteen as the center, supplemented by other public space produced from this. The imagination of space usage scenarios, with the attention to the life with text and image, also brings the thinking of the design method. Encouraged by the teacher, I try to use the film to create a set of dramatic architectural space, hoping to improve the concentration of people and people, heart and heart, and reconstruct the warmth bond. In the design process, 1:50 manual model is used to hammer out space, the storyboard sketches help explore the human activities, the text weaves the possibility, then forms the story and then acts on the space. In 1Q84 of Haruki Murakami, Qingdou accidently walked down from the overhead, the whole story unfolded. The design borrows this beginning, taking into account that the location is under two parallel overheads. If the construction volume can be extruded, it will be an interesting story when the people in traffic jam encounter with community residents, isn' t it? In addition to the meeting of heterogeneous population, how different and similar people do in the community? What kind of stories can they have? The space and stories gather from two directions, the site and interaction stimulate each other. The whole building is derived from the three dimensions of space, time and human activities. At last, a black box with rich inner space and mysterious outer façade is presented. The expression of the design results also draws support from narrative techniques of film, and a series of black and white perspective explain a group of stories of the encounter among people in the building. The initially design starting from the curiosity of the urban problems and the enthusiasm to people is derived from this finally.

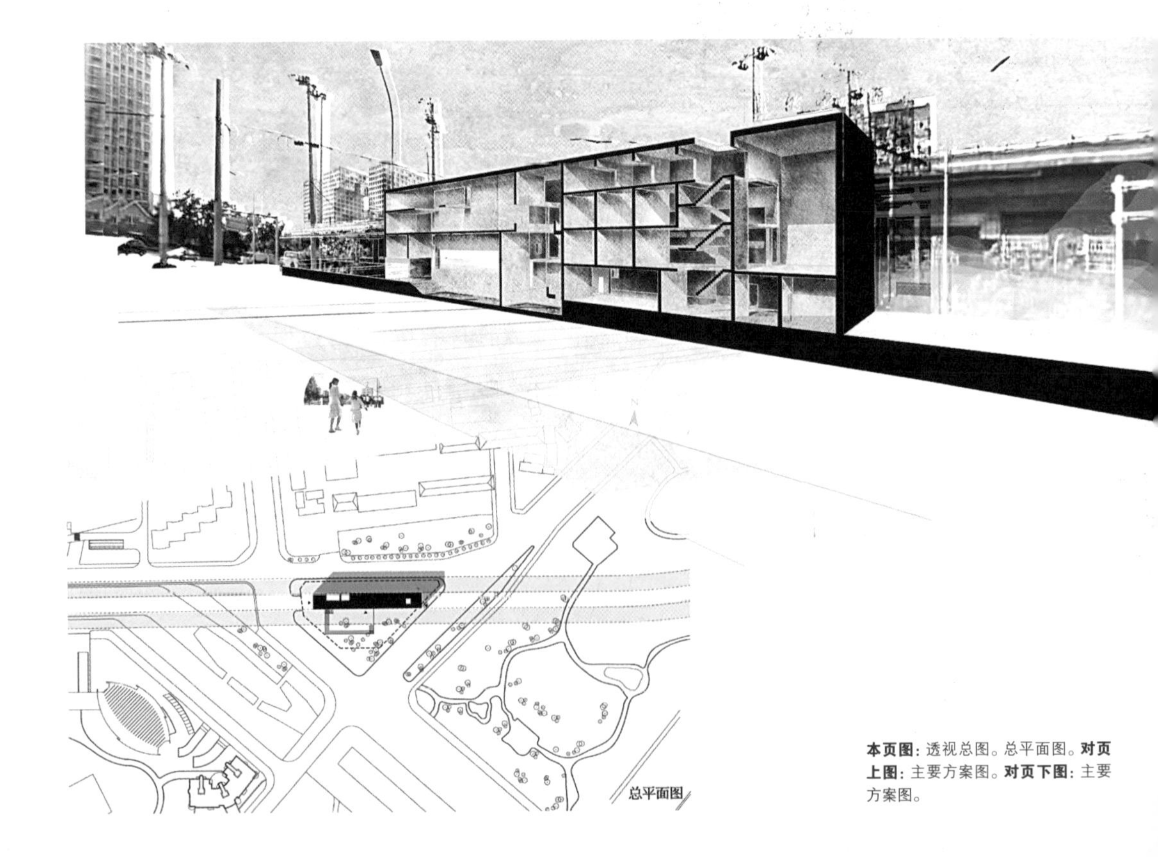

本页图：透视总图。总平面图。**对页上图：**主要方案图。**对页下图：**主要方案图。

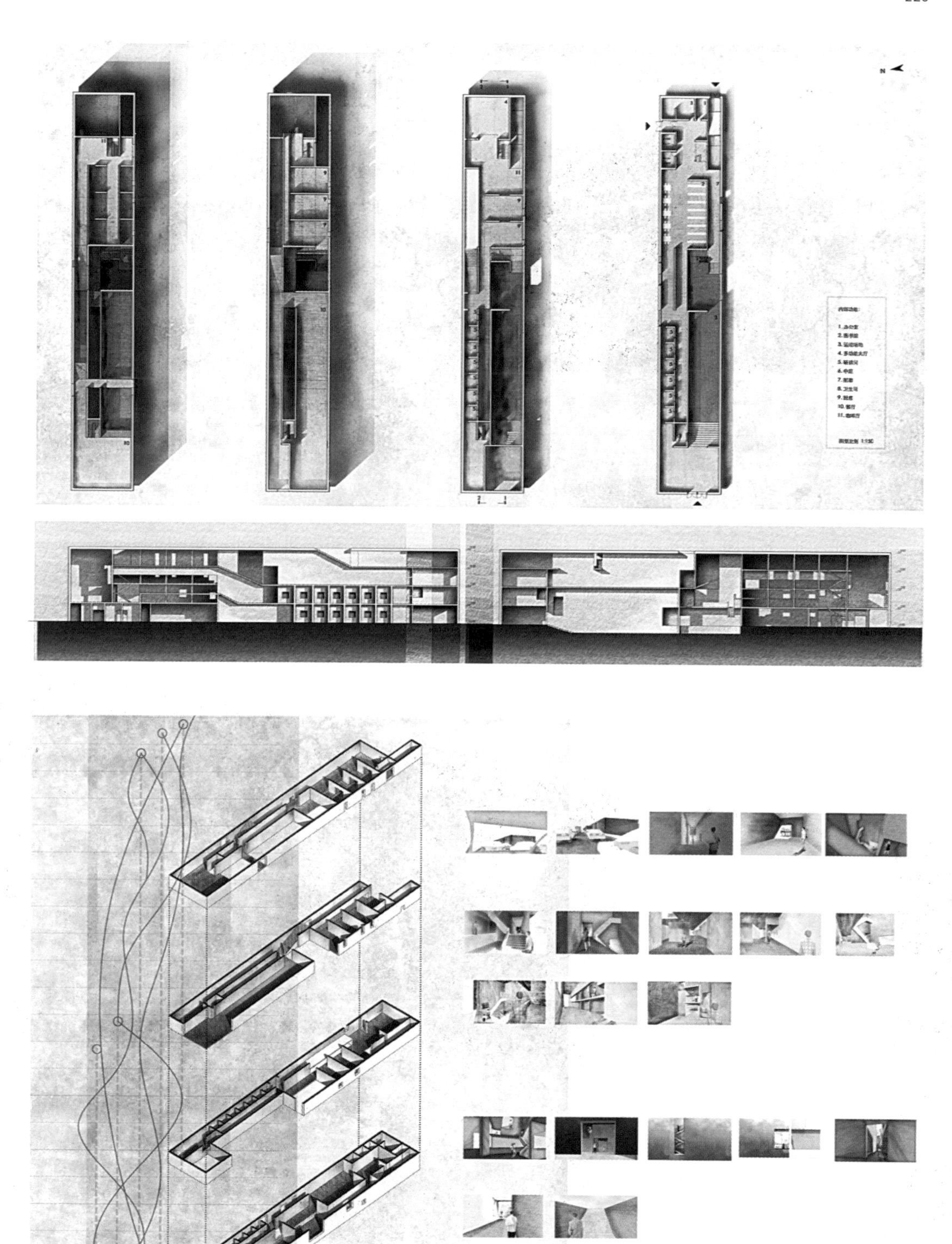
N
内部功能:
1.办公室
2.图书馆
3.运动场地
4.多功能大厅
5.研读间
6.中庭
7.展廊
8.卫生间
10.餐厅
11.咖啡厅
图纸比例 1:150

对页上图：最终成果模型。**跨页图**：最终成果模型。**本页上图**：最终成果模型。

教师点评

徐菊杰是一个很有想象力的同学，同时手绘能力很强，她从设计的开始即在想象一些空间使用的场景。这让我联想起电影与建筑的关系，以及通过电影的方式来创作一组建筑空间的可能性，并且开始鼓励她大胆地去实验一下这样的设计方法。她在设计过程里，通过编写故事，通过设想空间的片段，从空间和时间以及人在空间的活动中反推出一个建筑。最后呈现出的内部空间丰富的拆解黑盒子模型以及特殊的黑白透视系列，表现力很强烈。是一个“电影”设计方法的很好尝试。

Teacher's comments

Xu Jujie is a very imaginative student, with very strong freehand sketch ability. She started to imagine some scenes of space uses at the beginning of her design. This made me associate the relationship between films and architecture, as well as the possibility of creating a set of architectural space in the film mode, thus I encouraged her to boldly test such a design method. During the design, she derived reversely a building from the preparation of the story, from the fragments of imagined space, from space and time, and form the activities of people in space. At last, the decomposed black box model rich in interior space and the special black and white perspective series presented finally were powerful in expression, which is a good attempt to a design method of a "film".

际遇/ CROSSROADS ENCOUNTER

项目选址：北京当代 moma 东南角

项目类型（功能）：社区中心

建筑面积：$3000m^2$

用地面积：$1500m^2$

方案设计：白若琦

指导教师：李虎

完成时间：2015

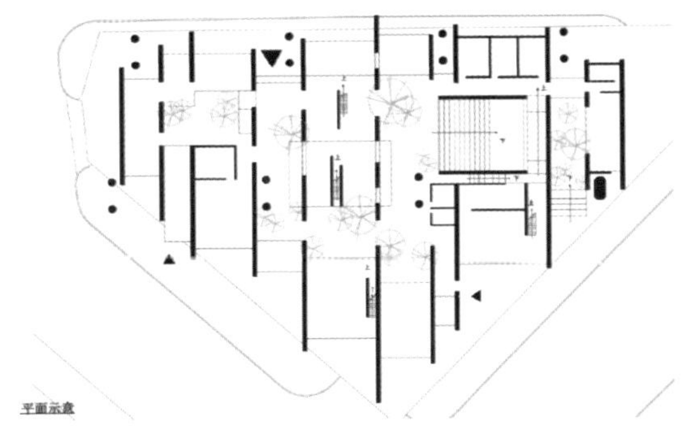

平面示意

这个设计从一开始到一年后的今天，我一直在思考：在这样的一个特殊地段，做一个设计中心，真的合适吗？这个问题其实没有答案。所以设计中我试图专注于空间，而不是城市。

场地本身有两条高架穿过，对下方的空间性质作出了很强的限定。我选择的回应方式是用垂直的片墙在下方穿过，营造一个三维正交体系。片墙和片墙之间，高价与下方的空间，在几个方向是独立又联合——如果场地本身在城市中不具备吸引力，就用单纯的空间趣味来吸引人群。

社区集市/ BAZAAR COMMUNITY

项目选址：当代 MOMA 三角地，北京
项目类型（功能）：社区中心
建筑面积：3200m²
用地面积：2000m²

方案设计：程瑜飞
指导教师：李虎
完成时间：2015

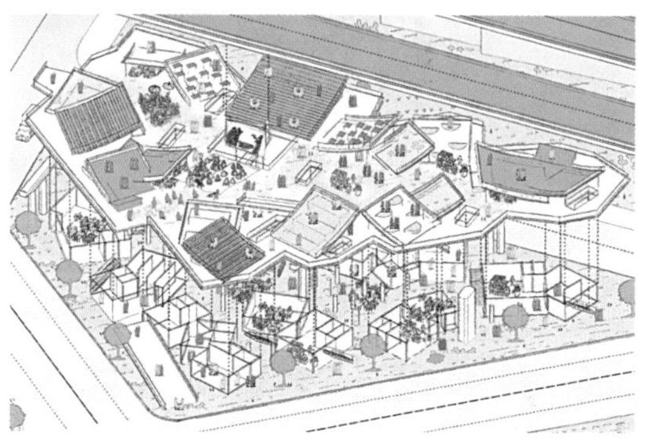

从场地本身存在的跳蚤市场现象出发，一方面希望为跳蚤市场本身提供一处可以聚集、交流的场所，另一方面营造一个如同跳蚤市场一样自由、活跃的社区中心。同时，二者之间又可以相互交流，以期形成一个聚落与集市共存、自主性与包容性共存的社区中心。由跳蚤市场体现出的自发组织的特性出发，社区中心应当成为一个自由、自发的居民行为的集合地。目的是让居民在社区中心里所进行的行为能像跳蚤市场自由碎片组织的摊位一样分布在场地里，居民根据自己的行为选择活动进行的功能盒子，同时盒子之间又是通过自由组织的街道过渡空间联系起来。由此，社区中心成为一个自由组织、自发性与包容性共存的居民交流场所，它对于居民是自由、无缝的场所。

吊装市集/ LIFTING THE MARKET

项目选址：北京香河园桥下三角地空间

项目类型（功能）：社区中心

建筑面积：3100m^2

用地面积：4200m^2

方案设计：钟程

指导教师：李虎

完成时间：2015

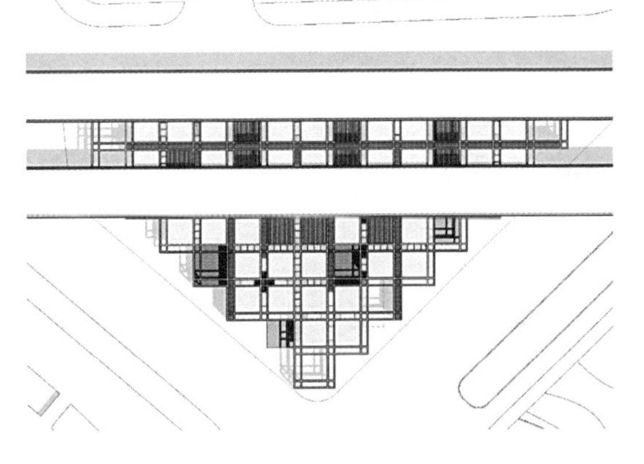

方案保留场地原有的跳蚤市场，将社区中心实体部分悬挂起来，把底层近人尺度的空间留给市集。在社区中心中，小单元提供活动空间，单元与单元之间以“交换”作为主题和联系纽带，建筑作为背景和画布被人们改写功能，形成场所特性。

城市天空/ SKY IN THE CITY

项目选址：东直门香河园，东城与朝阳两区交界处的三角形空地
项目类型：社区中心
建筑面积：3300m²
用地面积：3500m²

方案设计：黄也桐
指导教师：李虎
完成时间：2015

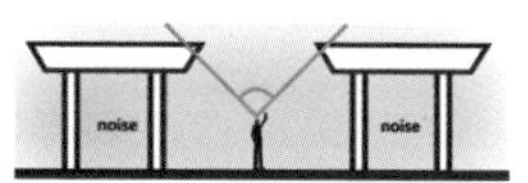

noise noise WALLS FOR QUIETNESS

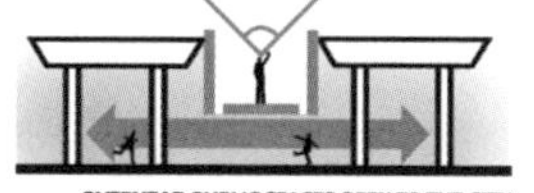

方案从场地特有的高架桥元素出发，将后泡沫城市中高架桥下的失落空间转化为狭缝看天的精神空间，强化场所印记。建筑底层架空流通，向城市开放，南北两侧置入丰富的艺术类功能空间以丰富居民的文化生活，同时为内部狭缝空间隔绝喧闹的城市环境。

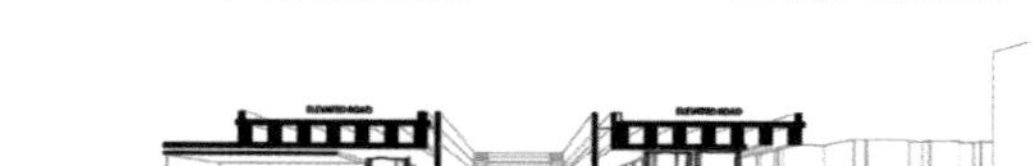

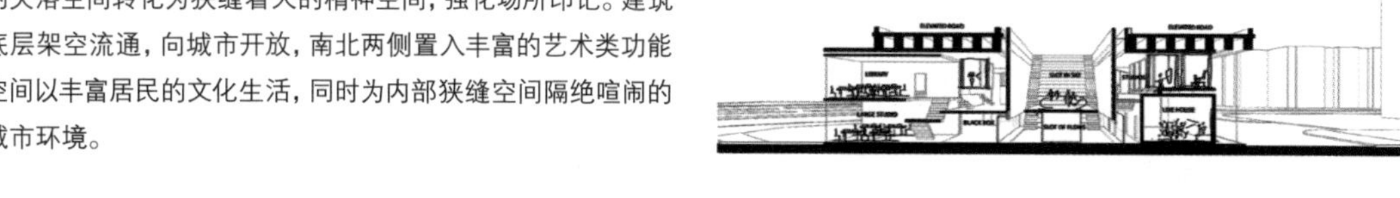

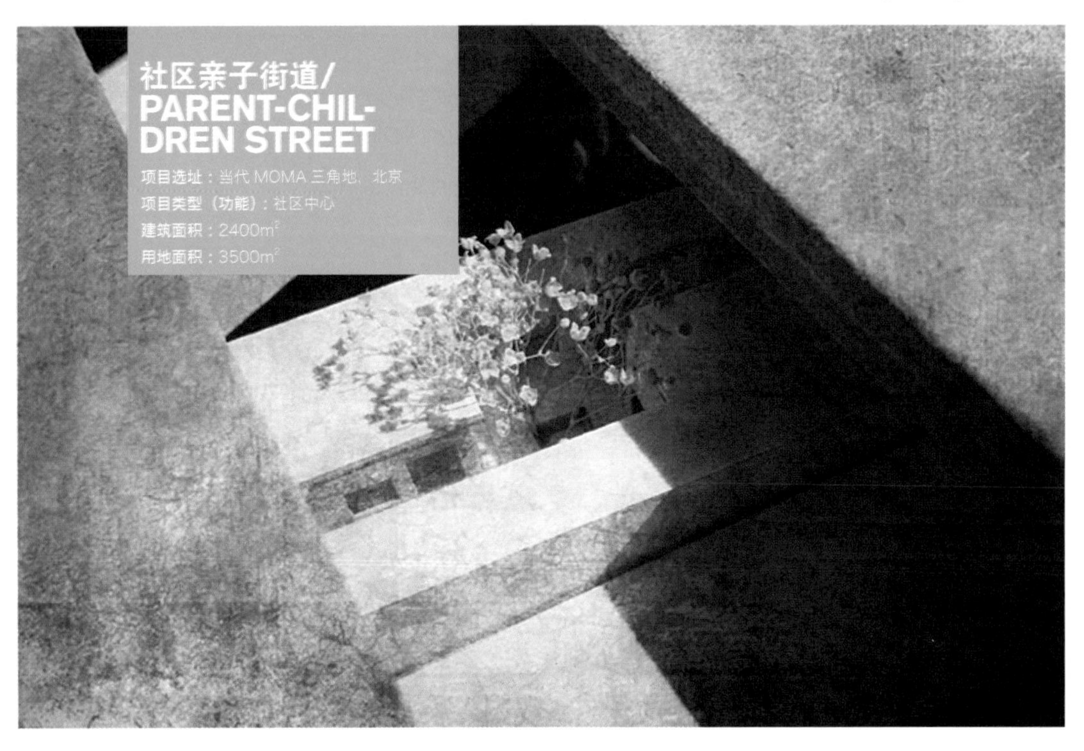

社区亲子街道/ PARENT-CHILDREN STREET

项目选址：当代 MOMA 三角地，北京
项目类型（功能）：社区中心
建筑面积：2400m^2
用地面积：3500m^2

方案设计：李秀政
指导教师：李虎
完成时间：2015

社区中心应当提供联系各个人群行为的空间。而我希望选取亲子生活成为联系这些生活模式的桥梁。街道样的空间模式将使得固定的活动与儿童自由移动的模式线性共存而又相互交织。白色的建筑体量又将如同雕塑一般改变整个场所而形成不同的空间影响社区生活。

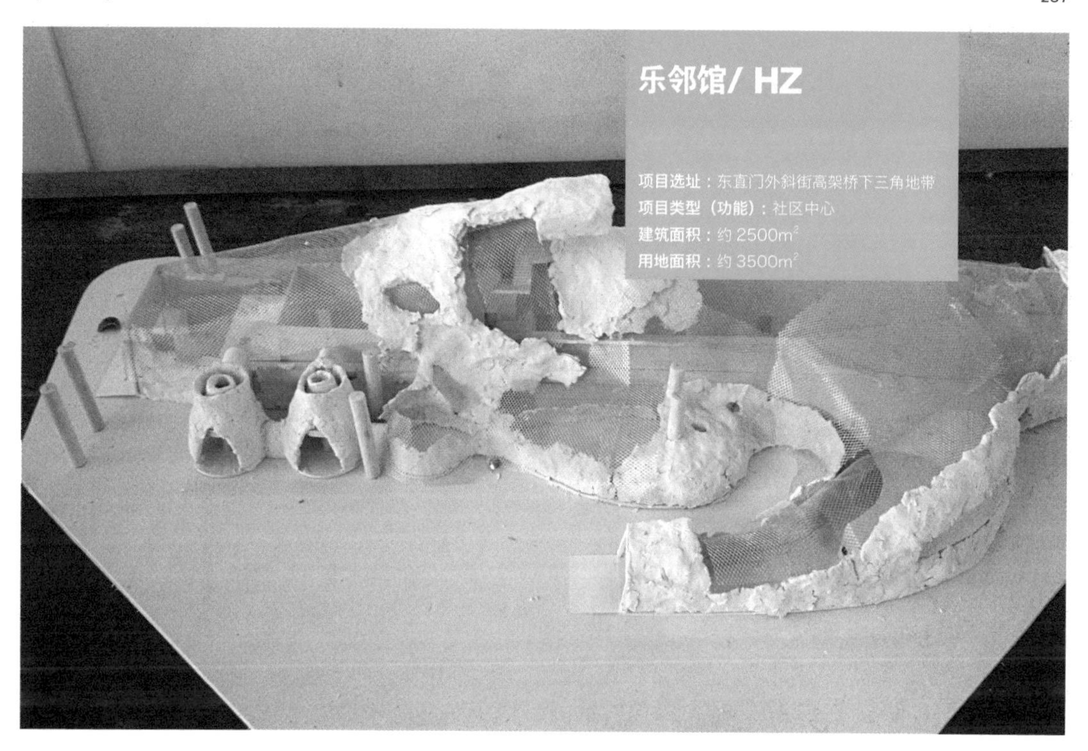

乐邻馆/ HZ

项目选址：东直门外斜街高架桥下三角地带
项目类型（功能）：社区中心
建筑面积：约 2500m²
用地面积：约 3500m²

方案设计：林浓华
指导教师：李虎
完成时间：2015

社区中心是一个关于“关系”的建筑，而“关系”存在于交流中：人与地。探讨城市"声命"的我，想象着这样的一个社区中心，人们未必相见、谈话，却在某种层面上相遇、交流，共同倾听着、创造着城市的声音。

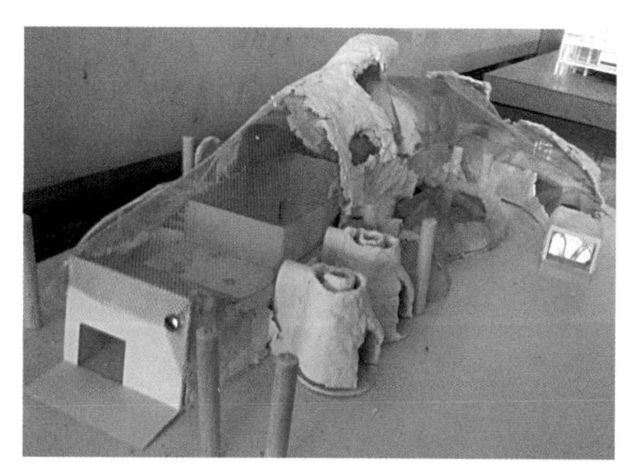

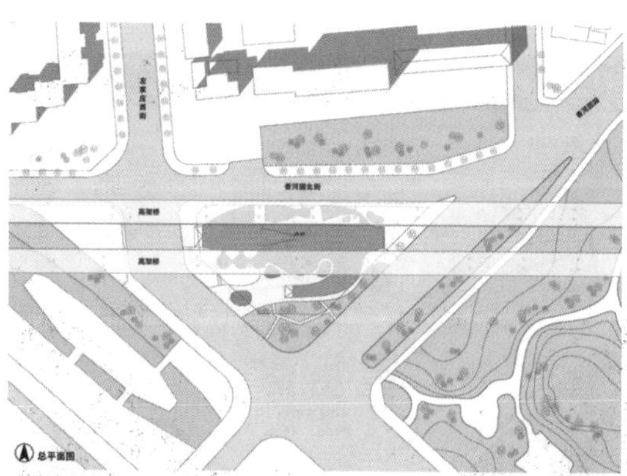

超市作为社区中心/ SUPERMARKET AS COMMUNITY CENTER

项目选址：北京东直门机场高速高架桥下三角地
项目类型：社区活动中心
建筑面积：2700m²
用地面积：2200m²

方案设计：马志桐
指导教师：李虎
完成时间：2015

居民对于社区中心的需求是不确定的，且他们的需求会因时而变。故本方案提供了一个大型的“开放可变空间”：通过模拟超市空间中后勤+可变空间的模式，设计了一个提供多种空间选择的大型开放室内空间，其空间模块亦可通过直接置换后勤空间的方式整体改建，最终使得社区中心如同超市一般灵活，满足不同人群的不同需求。

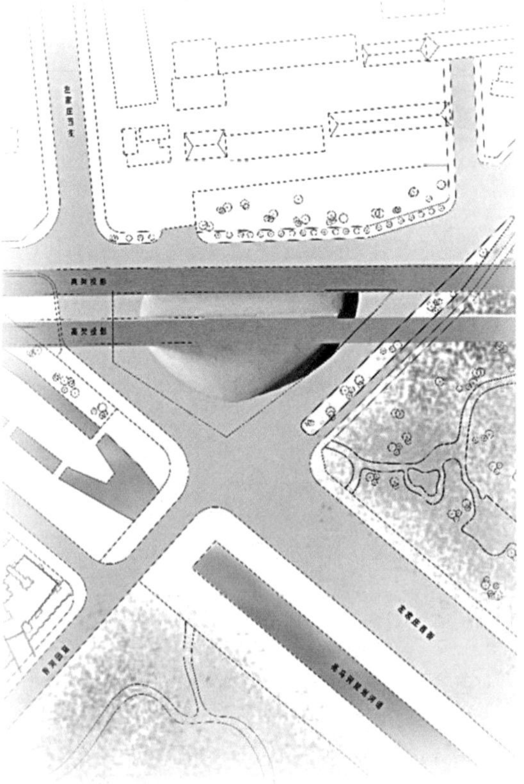

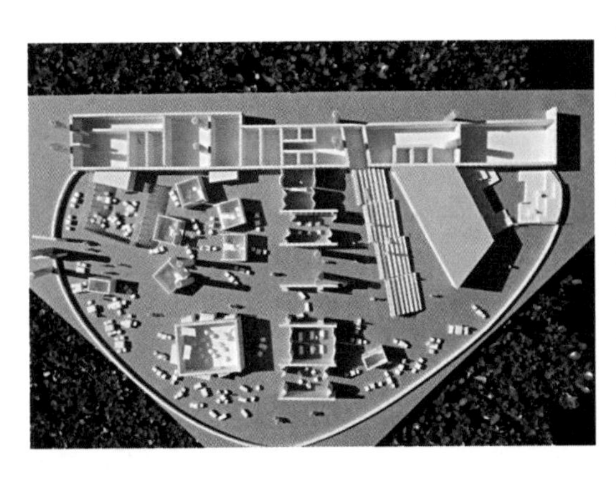

社区集装箱工坊/ COMMUNITY CONTAINER WORKSHOP

项目选址：北东直门首都机场高速高架桥下
项目类型：社区活动中心
建筑面积：固定部分：$1424m^2$
用地面积：$3760m^2$

方案设计：王澜钦
指导教师：李虎
完成时间：2015

本设计试图探讨一种可随着时间和个人需求的变化而有机生长的福利性社区中心模式。设计充分挖掘场地的“廉价”属性，营造一种开放式的无身份标签的垂直立体场地，并通过可随时插入和搬出的集装箱改造工坊。

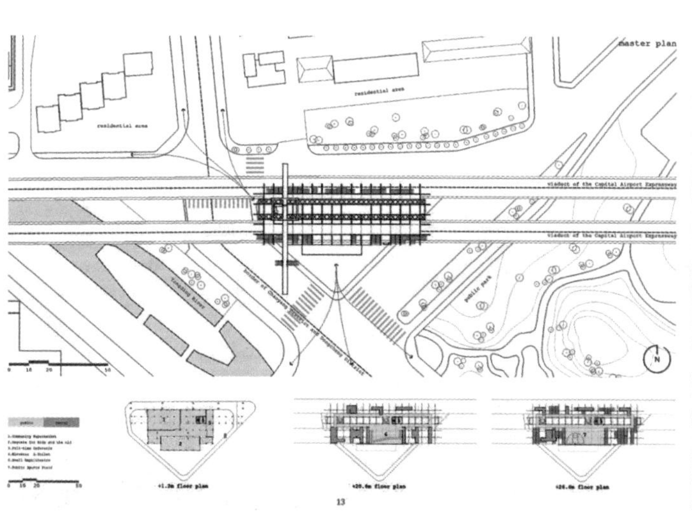

连续·剧/ SERIESCENE

项目选址：当代 MOMA 南侧高架桥下三角地
项目类型（功能）：社区中心
建筑面积：3500m²

方案设计：许达
指导教师：李虎
完成时间：2015

现行社区中心通常使用情况是让同龄人聚集，非同龄人分开。但是在一个循环社区中，建筑师通过连续不间断的流动空间创造一个能够消解年龄差异，匀质并且无限循环的社区空间。充分发挥地段所处位置的“岛屿”性质，漫步其中的过程，可以和来自各个方向社区的居民进行交流和互动，克服高架桥下空间的负面效应。

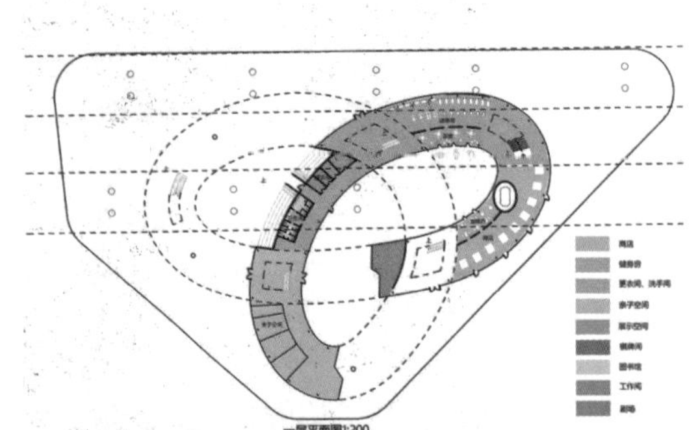

一层平面图1:200

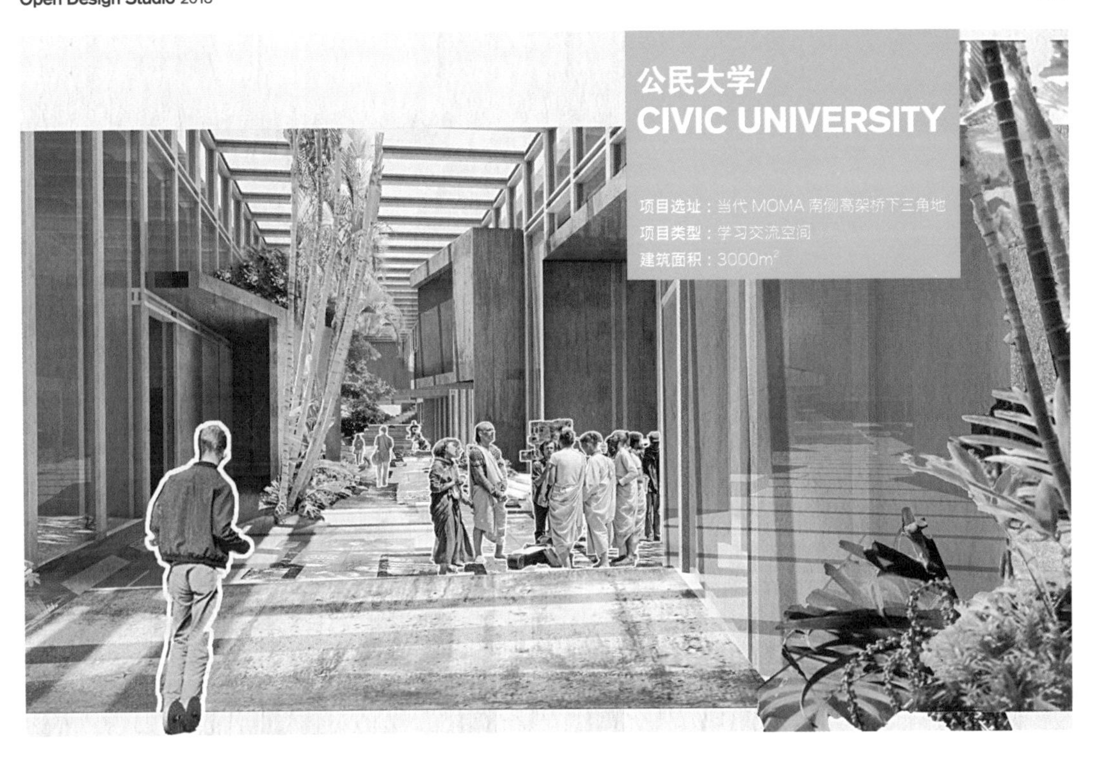

公民大学/ CIVIC UNIVERSITY

项目选址：当代 MOMA 南侧高架桥下三角地
项目类型：学习交流空间
建筑面积：3000m²

方案设计：杨明炎
指导教师：李虎
完成时间：2015

这个设计的理念“公民大学”（people's university）来自1918年Henry Jackson的 A Community Center，书中提到：人们对“进步”与“民主”的追寻源自“人内在对于知识的渴望”，而社区中心应当是一所“公民大学”。

另一方面，设计前期对日本与中国台湾的研究中，我发现其公民馆在鼓励居民终身学习的同时，也在公民社会的转型中散发能量。这种"解放思想，改善社会"的社区精神值得大陆学习。

当我追问“公民大学”的场所氛围时，我看到“大学”从古代城市学院发展聚合为当代完整的校园过程中，“分享”是重要的精神连结。大学之所以不是学院的累积，是因为各学院在一个时空中分享场所、知识。“分享”如同“树下的智识交流”，恰恰是迈向公民社会的必要觉知。

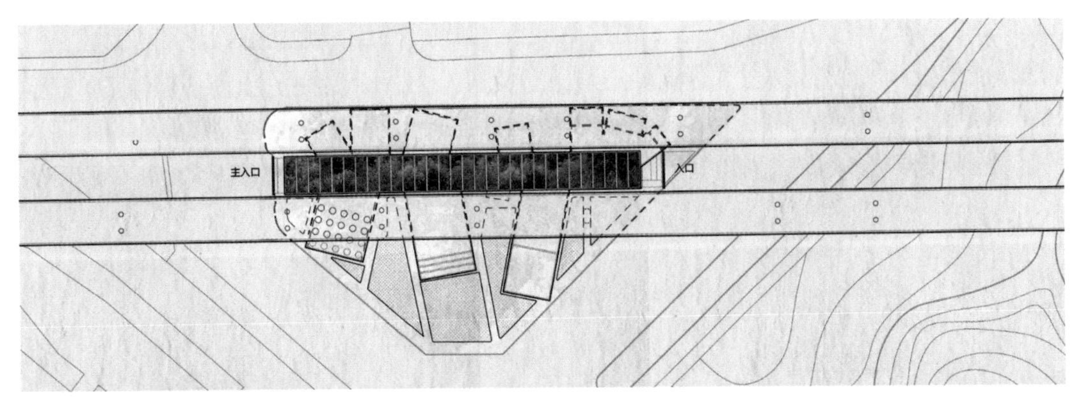

注重形式的

徐全胜

北京市建筑设计研究院有限公司
总经理

徐全胜
北京市建筑设计研究院有限公司
总经理

教育背景

1987年 – 1992年
清华大学建筑学院建筑学　学士
1996年 – 1999年
清华大学建筑学院建筑学　硕士
2001年
意大利格里高蒂事务所学习
2010年
美国芝加哥大学学习公共管理

工作经历

1992年 – 2012年
北京市建筑设计研究院建筑师、副所长、所长、副总建筑师、副院长
2012年至今
北京市建筑设计研究院有限公司党委副书记、董事、总经理、副总建筑师

主要论著

徐全胜.《设计随笔-关于北京恒基中心》,《建筑学报》, 1998
徐全胜.《建筑创作是一个解题的过程—北京地震局海湾科技减灾中心创作体会》,《建筑创作》, 2002

设计获奖

1994 – 第四届中国青年建筑师杯优秀方案奖–深圳特区报社方案
2000 – 九十年代北京十大建筑奖–北京恒基中心
2001 – 首都规划设计展专家组优秀方案奖群众组十佳方案奖–北京市高级人民法院审判业务用房
2002 – 首都规划设计展专家组优秀方案奖群众组十佳方案奖–北京电视中心
2007 – 北京市优秀工程一等奖–北京市高级人民法院审判业务用房
2007 – 国家优质工程奖银奖–北京市高级人民法院审判业务用房
2008 – 第五届中国建筑学会建筑创作优秀奖–北京市高级人民法院审判业务用房
2003 – 中国青年建筑师奖设计竞赛佳作奖
2003 – 北京市优秀工程三等奖–中国航天大厦
2004 – 中国建筑学会第四届青年建筑师奖
2012 – 北京市优秀工程一等奖–全国人大机关办公楼
2012 – 绿色建筑创新单项奖–全国人大机关办公楼
2012 – 北京市优秀工程一等奖–福建公安指挥情报中心
2012 – 北京市优秀工程三等奖–北京百子湾1号小区体育休闲配套设施
2012 – 北京市优秀工程三等奖–呼和浩特大唐喜来登大酒店

代表作品

北京恒基中心(图1)、中华女子学院(图2)、北京市高级人民法院审判业务用房(图3)、北京电视中心(图4)、全国人大机关办公楼(图5)

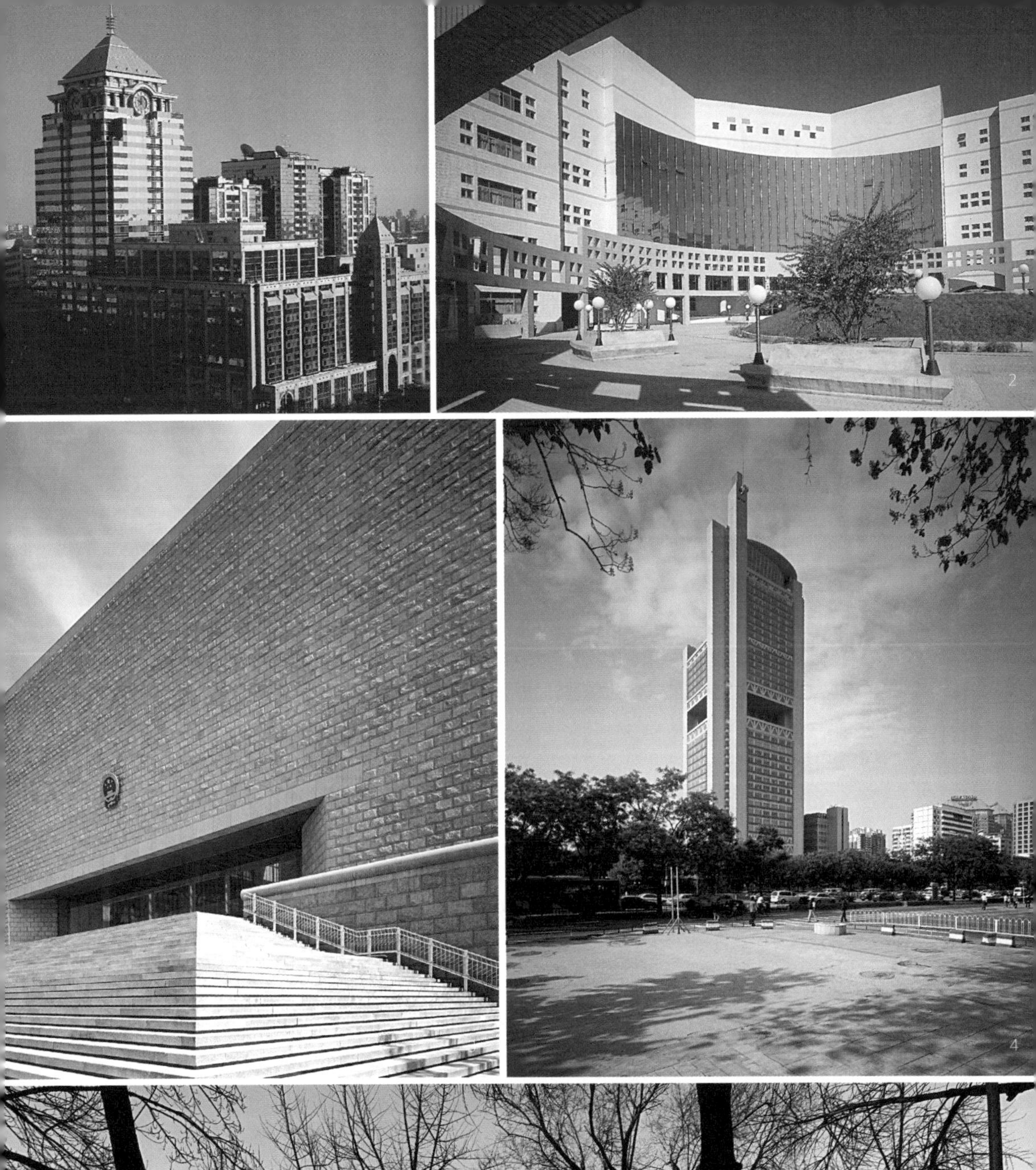

注重形式的建筑设计

三年级建筑设计（6）设计任务书

指导教师：徐全胜
助理教师：段华楠　廖思宇　刘伯宇

建筑的形式—FORM

“形式”是建筑的基本元素，建筑以“形式”存在，是建筑的表象，是城市“千城一面”或者“千变万化”的基本组成单元。
建筑设计是一个解题的过程，“形式”与功能的关系，是建筑设计的经典问题，“形式”还与技术、经济、文化等等有关，形成的答案以“形式”表现出来。
建筑“形式”是结果，但有其形成的逻辑和过程。
目前国内对建筑形式有越来越重视的趋势，最近很流行的一些关于建筑的讨论，基本也是在建筑"形式"的层面。
我们这里探讨的建筑“形式”包括：建筑的室外空间、建筑的室内空间、建筑的体形、建筑的内外立面等。

课程描述

先从案例分析学习开始，研究知名建筑师经典的建筑设计作品，分析环境、地段、功能等对建筑的客观要求以及其理论、个人特点对设计完成作品的影响，体会完成作品体现出的解题的过程、答案，建筑"形式"的形成过程和代表的意义。
课程的选题是设计B公司园区内的一幢办公楼A，园区由9栋现状建筑组成。设计要梳理园区的中心空间，并进行“注重形式”的城市设计和单体建筑设计。

课程目标

1. 案例分析：教师会以贝聿铭为例，做一个30分钟有ppt的演讲，分析作者个人特点以及他的建筑设计理论，并以其作品—北京中银大厦为例分析“贝氏建筑”形式的特点。要求同学每3~4人为一组，学习一个知名建筑师和他（她）设计的一座建筑。通过了解这个建筑师的生平，阅读理论书籍，详细分析学习一个建筑设计作品。建议：柯布西耶、密斯凡德罗、路易斯康、伊东丰雄、隈研吾、UNSTDIO、OMA等。目的是体验、了解最终建筑“形式”形成的过程和逻辑，列出建筑“形式”所代表的定义、围度和内涵。
2. 建筑设计：项目用地在北京，通过现场踏勘，了解地段的情况，熟悉设计任务书，设计要从城市设计入手，分析梳理整个园区的空间、环境，研究园区使用者对园区整体的使用要求和行为模式，统一设计办公楼A南北广场空间和首层空间的形式。然后按照设计任务书提出的功能和面积要求，满足适用的规范、标准等的限制条件，进行建筑单体—办公楼A的建筑设计，重点探讨建筑的形式。

建筑设计要求

1. 建筑选址：办公楼A位于北京市西城区南礼士路62号，B公司园区内。

2. 规划设计条件：建筑限高40m，并满足对北侧3#住宅楼的日照间距要求，板楼1：1.7，塔楼1：1.1。地上建筑面积5000m^2，地下2层，建筑面积1000m^2。地上二层和地下二层与园区其他建筑地上二层和下二层建筑联通。

3. 功能要求：1~2层设门厅，咖啡厅300m^2，B公司博物馆500m^2。其他层设置集团会议室500m^2，满足150人的开放式办公室（15m^{2}建筑面积/人），其中含15间玻璃隔断的单间办公室，15m^2/间，2个建筑名人设计工作室300~500m^2/个。每层设带独立盥洗区的公共卫生间，男厕设2个小便厕位、2个座便厕位；女厕设3个座便厕位。设1350kg电梯2部。

4. 造价要求：建筑层高不限，单方造价不限，以适宜为准。

5. 设计要求：对B公司园区进行城市设计，设计梳理办公楼A的南北广场，对贴临的办公楼C提出建筑“形式”的设计意向。设计办公楼A，注重建筑的“形式”，可以体型、空间、立面之一作为设计、创作的重点，达到一定的设计完成度和表现的深度。

6. 创意要求：有所不同-Something special。

设计成果表达

1. 案例分析阶段：小组工作，10页电脑演示文件，1人代表小组做10分钟演讲。集体完成关于建筑“形式”的定义、围度和内涵描述。

2. 设计过程：课题统一制作1/500的B公司办公园区体块模型。每位同学制作1/500以建筑用地为底盘轮廓的单体体块模型，用作过程推敲讨论方案之用。

3. 中期评图：概念设计深度。1/500总平面图，1/200首、二层平面图，标准层平面图，剖面图，三维电脑概念模型，1/500以建筑用地为底盘轮廓的实体单体体块模型。

4. 终期评图：20页电脑演示文件。1/500总平面图，1/200各层平、立、剖面图。Sketchup深度的电脑渲染图2张，其中人视一张、鸟瞰图一张

参考书目

《建筑：形式空间和秩序》程大锦著，天津大学出版社

《像建筑师那样思考》豪鲍克斯著，姜卫平、唐伟译，山东画报出版社

《总体设计》凯文林奇、加里海克著，黄富厢、朱琪、吴小亚译，中国建筑工业出版社

转折/TRANSITION

项目选址： B 公司园区内的办公楼 C 座，位于北京市西城区南礼士路 62 号

项目类型： 办公建筑

建筑面积： 6700m^2

用地面积： 1663m^2

方案设计：杨子瑄
指导教师：徐全胜
完成时间：2015

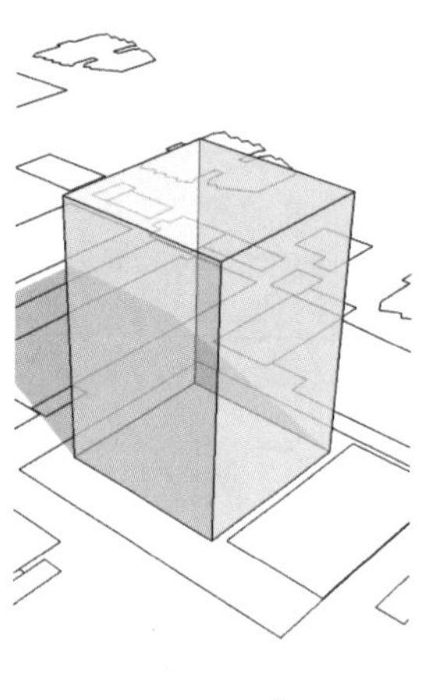

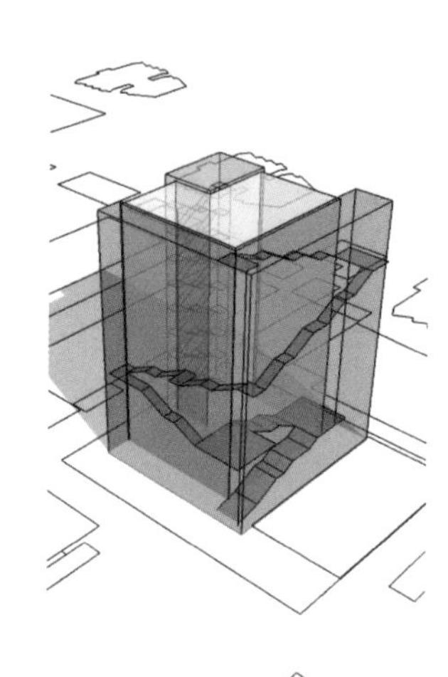

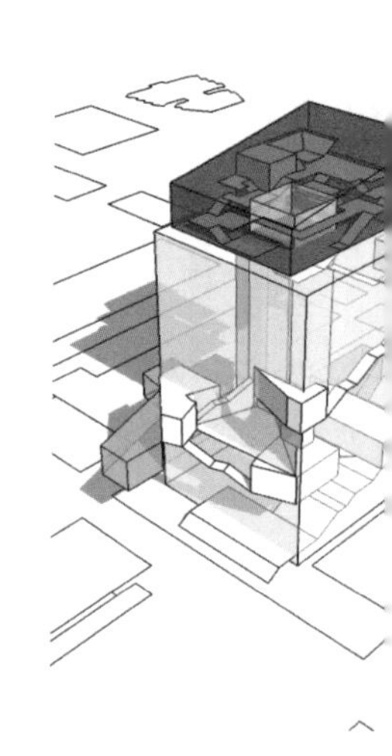

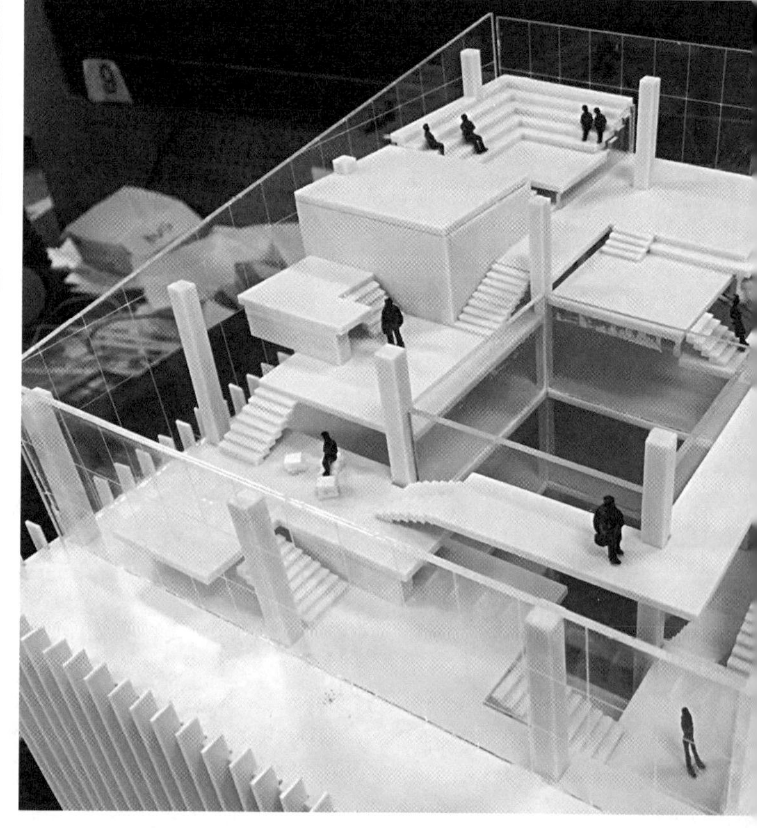

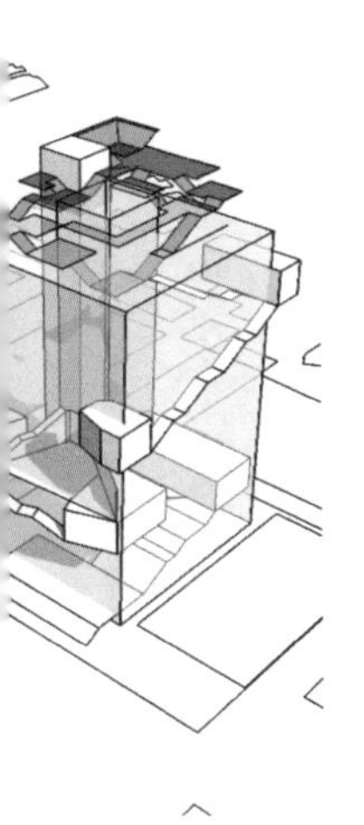

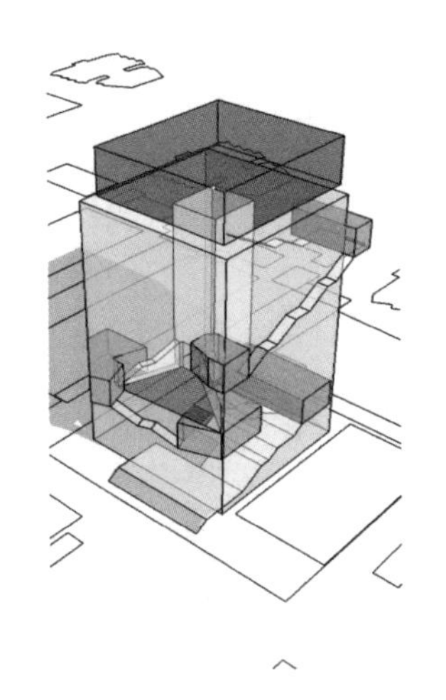

这次设计的主题是“注重形式的建筑设计”。经过解读，我认为“形式”分为“形”和“式”。“形”是一个可见的结果，想要得到好的结果，就需要找对问题，于是我们要思考为了什么设计，出发点是什么，要用什么“形”来回答初始的问题。而“式”表达了一种创造结果的过程或机制，本身是名词，但表达出动态的发展过程。

设计任务是北京院院区内的一栋办公楼重建，即为建筑师们设计办公建筑，于是我尝试将自己代入使用者，从调研开始，想象自己在其中工作的场景，从中寻找切入点。第一次去北京院调研地段的时候听老师介绍得知，北京院的组织架构分为院、所、室，各自独立运行，也许共用同一栋楼，甚至同一层，但是民间交流很少，仿佛在一个无形的盒子里。然而，随着BIM的发展，也许以后这种人员固定的组织方式不是最有效的，而是不同专业的人在不同项目中任意搭配、合作，人员组合更多样，建筑师之间、以及与其他设计师之间的接触交流更多。因此我希望在方案中营造出宜人的办公空间和丰富的交往空间。

开篇：主透视。**对页上跨图**：生成分析图。**对页下跨图**：最终成果模型。**本页下图**：总平面图。

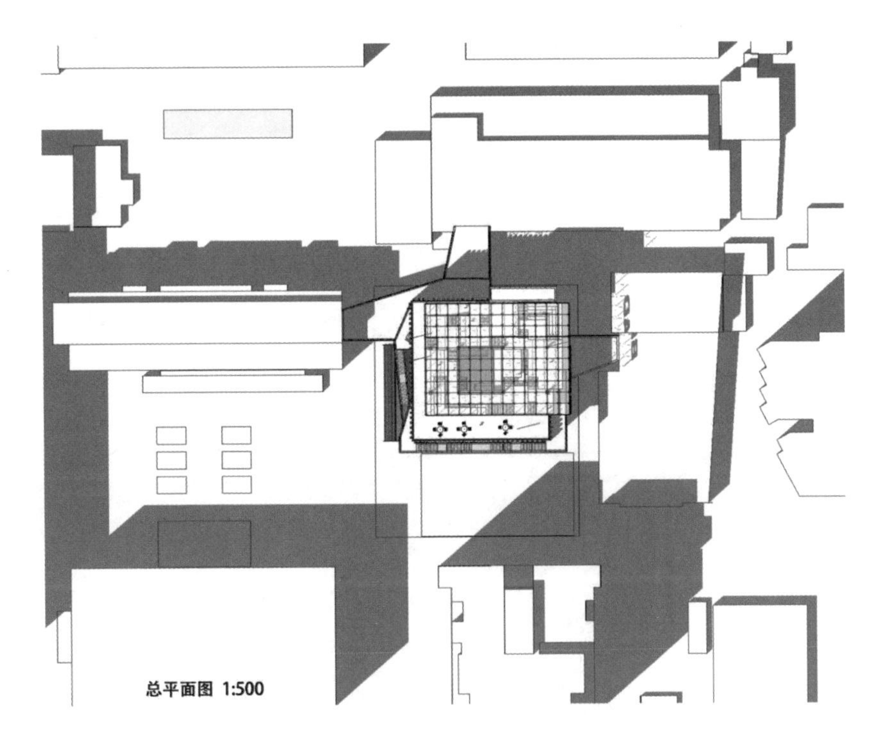

The theme of this design is "the architectural design paying attention to the form". After interpretation, I think the "form" is divided into "shape" and "style". The "shape" is a visible result, if you want to get a good result, you need to find the problem, so we have to think about what the design is for, what is the start point, what kind of "shape" should be used to answer the initial question. The expression of "style" is a process or mechanism of a creation result, it is a noun, but presents a dynamic process of development.

The design task is the building reconstruction in the yard of Beijing Institute, namely the design of office building for architects, so I try to put myself in the user, from the beginning of the research, I imagine myself in the working scene to look for the breakthrough point. I learned from the introduction of the teacher in the first time when I went to investigate the section in the Beijing Institute that the organizational structure of Beijing Institute is divided into the department, division and section, which operates independently, maybe they share the same building, or even the same layer, but people exchange few, as if in an invisible box. However, with the development of BIM, perhaps the organization mode of fixed staff of is not the most effective, but people from different professions may match randomly and cooperate in different projects, the combination of personnel will be more diverse, the contact and communication among architects and other designers will be more frequent. Therefore, I hope to create both pleasant office space and rich communication space in the scheme.

对页图：最终成果模型整体。**本页图：**最终模型入口效果。

教师点评

杨子从设计之初，就在探索以一条连续、转折的人行实体动线，联结建筑所有楼层的不同功能和所有相对应的空间，形成并逐渐完善这个设计概念，专注于此，贯彻设计的始终。通过创造性地努力和探索，最终解决了设计问题，圆满完成了课程设计。她最终的设计成果，建筑功能合理，建筑空间丰富灵动，由于这条立体转折的空中步道，使建筑的体型和外立面生动、独特，使常规的写字楼建筑唯美并与众不同。

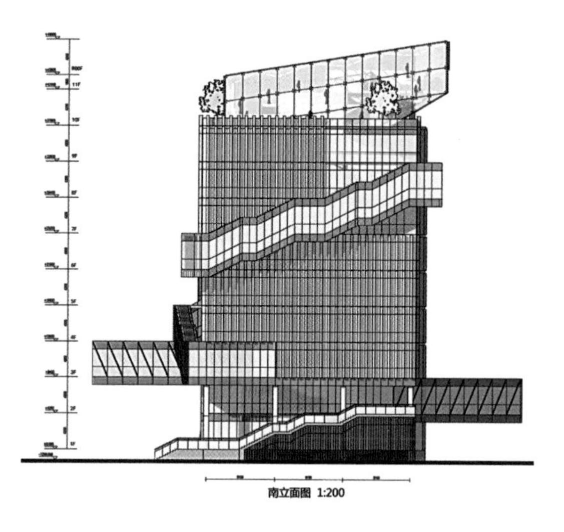

Teacher's comments

Yang Zixuan is exploring a continuous and turning generatrix at the beginning of the design to connect different functions and all corresponding spaces of all floors in the building, forming and gradually improving the design concept. She is focusing on this and implementing it into the design. Through creative efforts and exploration, the design problems are ultimately solved, and the curriculum design is successfully completed. Her final design result is featured by reasonable construction functions and rich architectural space, the three-dimensional turning skywalk makes the building shape and facade vivid and unique, which is different from the aesthetics of conventional office buildings.

本页上图：立面图。**本页下图：**最终成果模型。**对页图：**夜景鸟瞰图。

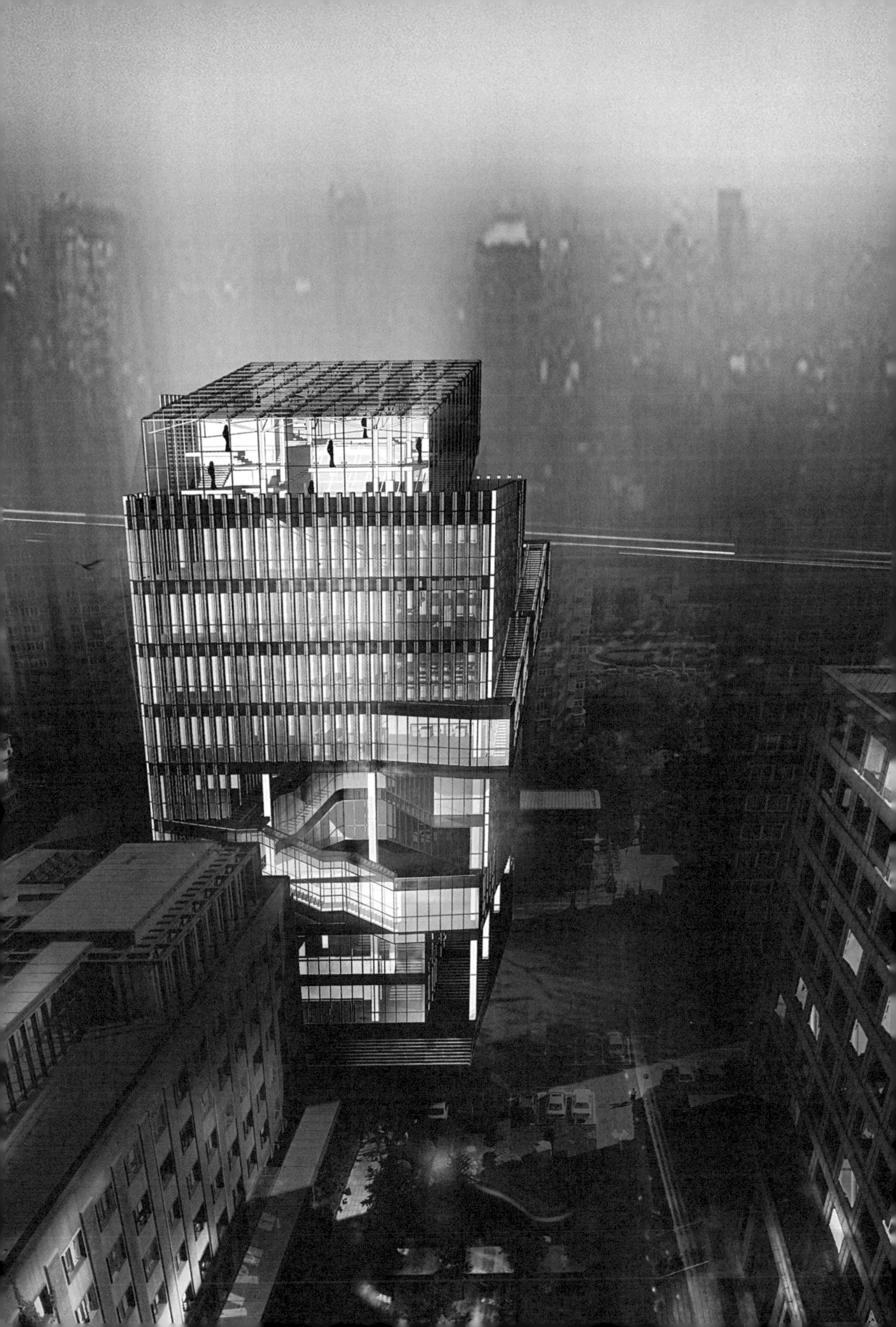

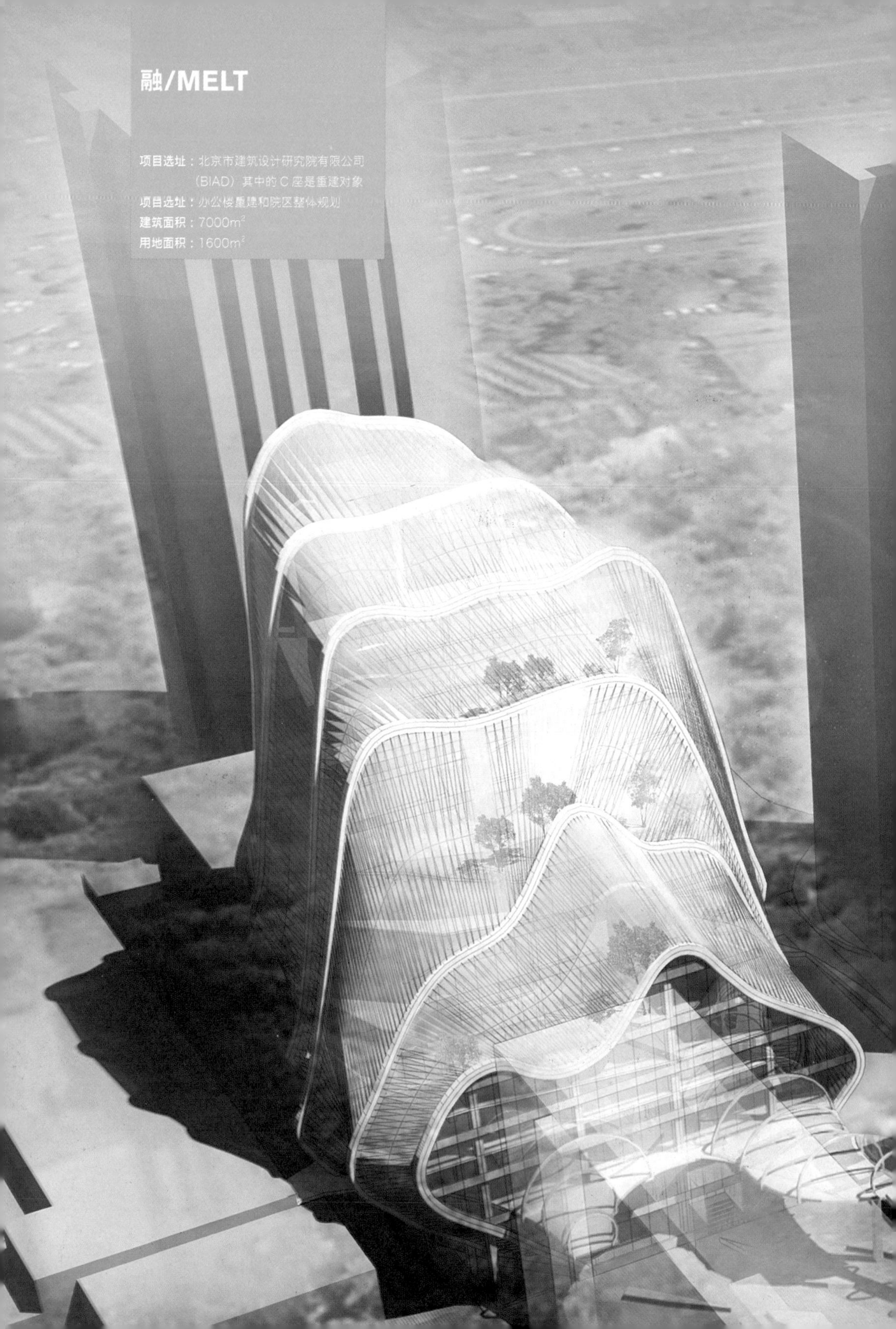

融/MELT

项目选址： 北京市建筑设计研究院有限公司（BIAD）其中的C座是重建对象

项目选址： 办公楼重建和院区整体规划

建筑面积： 7000m^2

用地面积： 1600m^2

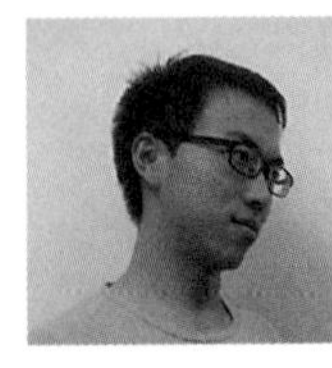

方案设计：周皓
指导教师：徐全胜
完成时间：2015

方案设计：孙越
指导教师：徐全胜
完成时间：2015

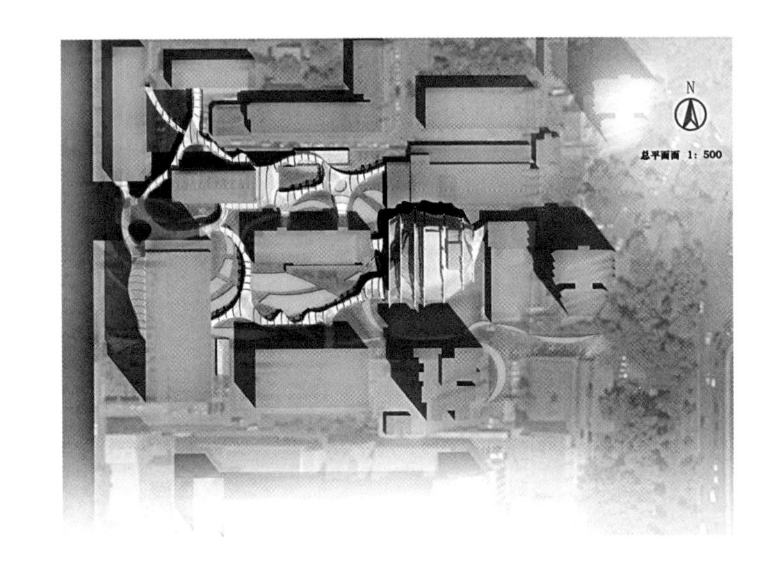

▪周边休闲景观

距院区有一定距离

南礼士路公园

街角休闲场所

▪大院边界

围和程度高，对社会开放度低

大院的边界具有较高程度的围和，内部形成自己的圈子，却难为外界熟知，对社会开放度低。

▪院区内交通与停车

人车混行、车辆占道停放

院区内人车混行，空间紧张，车辆除停放在地下停车场中很多都随意停放在道路两侧

▪建筑师工作特点

团队的沟通交流结合独立的设计

建筑师每天到达上班地点要先与其他各工种沟通交流，解决问题，然后安排个人工作，开展独立的设计工作。

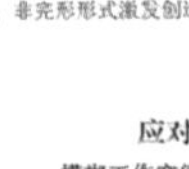

▪创意

应对创意类工作

新形式要具有差异化，激发创造活力

建筑设计行业需要建筑师时常涌现创意和想法，与老建筑差异化的曲线非完形形式激发创造活力。

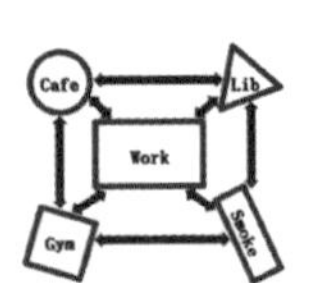

▪未来

应对建筑师工作特点

模糊工作空间与其他空间的界限

利用云技术，实现随时随地工作，模糊工作空间与其他空间，包括交通空间的界限。

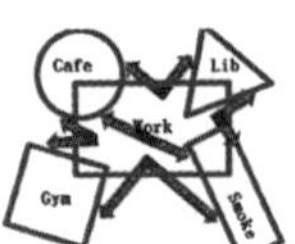

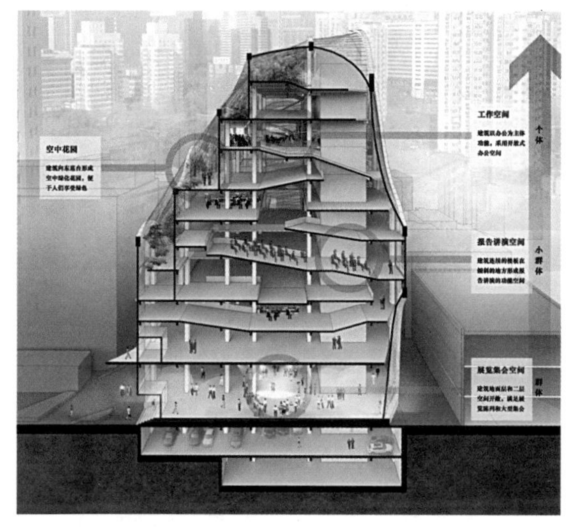

北京市建筑设计研究院有限公司（BIAD）作为北京众多“大院”之一，具有悠久的历史底蕴和深刻的文化内涵，但大院模式隔绝的边界带来的封闭性十分突出。在北京院内，整个公司被划分为众多规模不同的设计院、所、室和其他功能空间，它们彼此独立、高效地运转，但功能分离的模式却不利于建筑师们交流沟通。

我们的理念是通过架空地面形成城市公园消解北京院与城市之间的边界，通过创造连续的多功能空间消解院、所、室和其他功能空间的边界，将它们融合起来。

最终我们提出了“融”的院区和建筑概念，并将这种全新的、融合的“功能体”以一种"自然形"的建筑造型加以实现，彰显新态度，为院区注入新活力。

开篇：院区整体主要透视。**对页上图**：总平面图。**跨页图**：分析图总汇。**本页上图**：建筑剖透图。

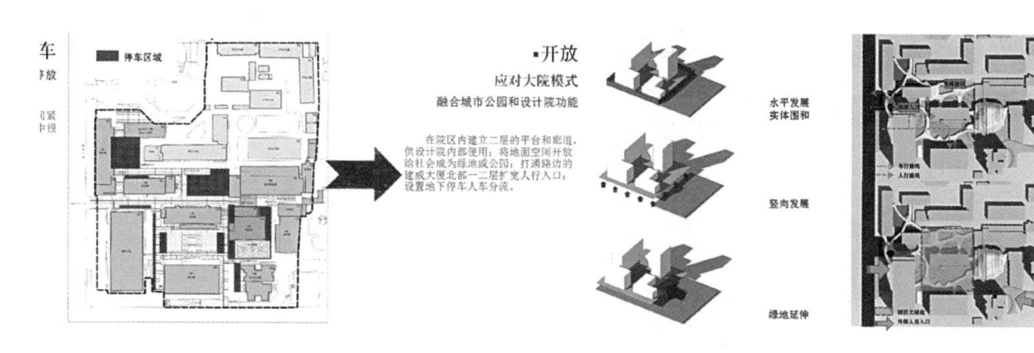

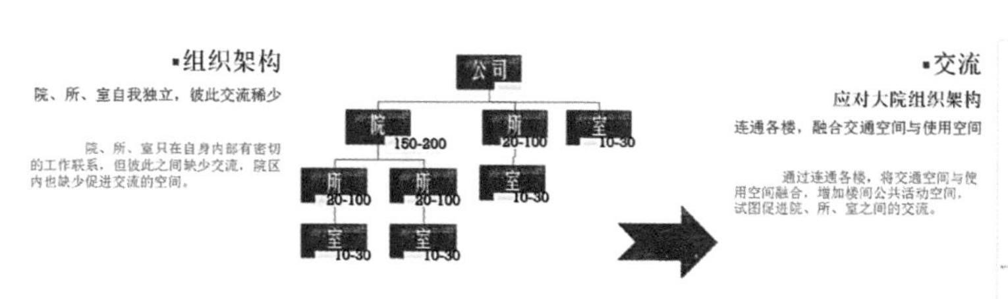

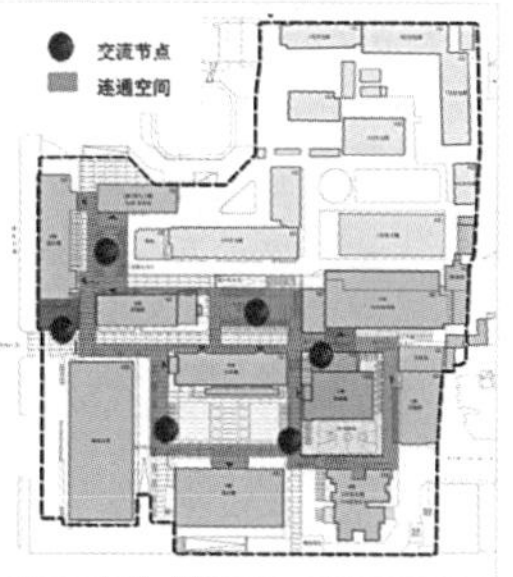

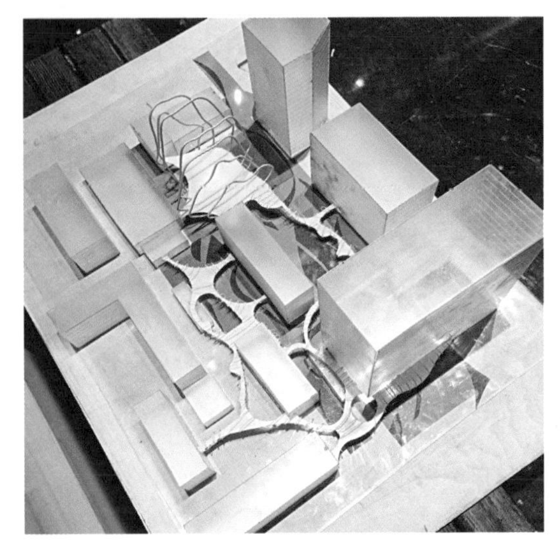

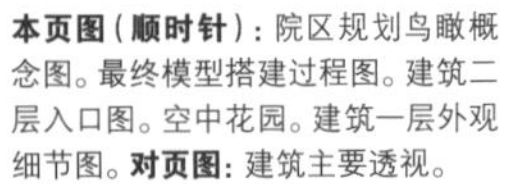

本页图（顺时针）：院区规划鸟瞰概念图。最终模型搭建过程图。建筑二层入口图。空中花园。建筑一层外观细节图。**对页图：**建筑主要透视。

Beijing Institute of Architectural Design Co., Ltd. (BIAD), one of numerous "institutes" in Beijing, has long history and profounds cultural connotation, but the closed mode brought by the isolated boundary of the institute is very prominent. In Beijing Institute, the company is divided into design institutes, divisions and other functional spaces of many different scales, they are independent from each other and operate efficiently, but the separated function model is not conducive to the communication among architects.

Our idea is to create city gardens on overhead ground to decompose the boundary between Beijing institute and the city, decompose the boundaries between the institute, division, section and other function spaces through creating continuous multi-function space and then integrating them. Finally, we propose the "integrated" institute yard and architectural concepts and implement the new integrated "function body" with a "natural" architectural modeling, highlighting new attitude and injecting new vitality into the institute.

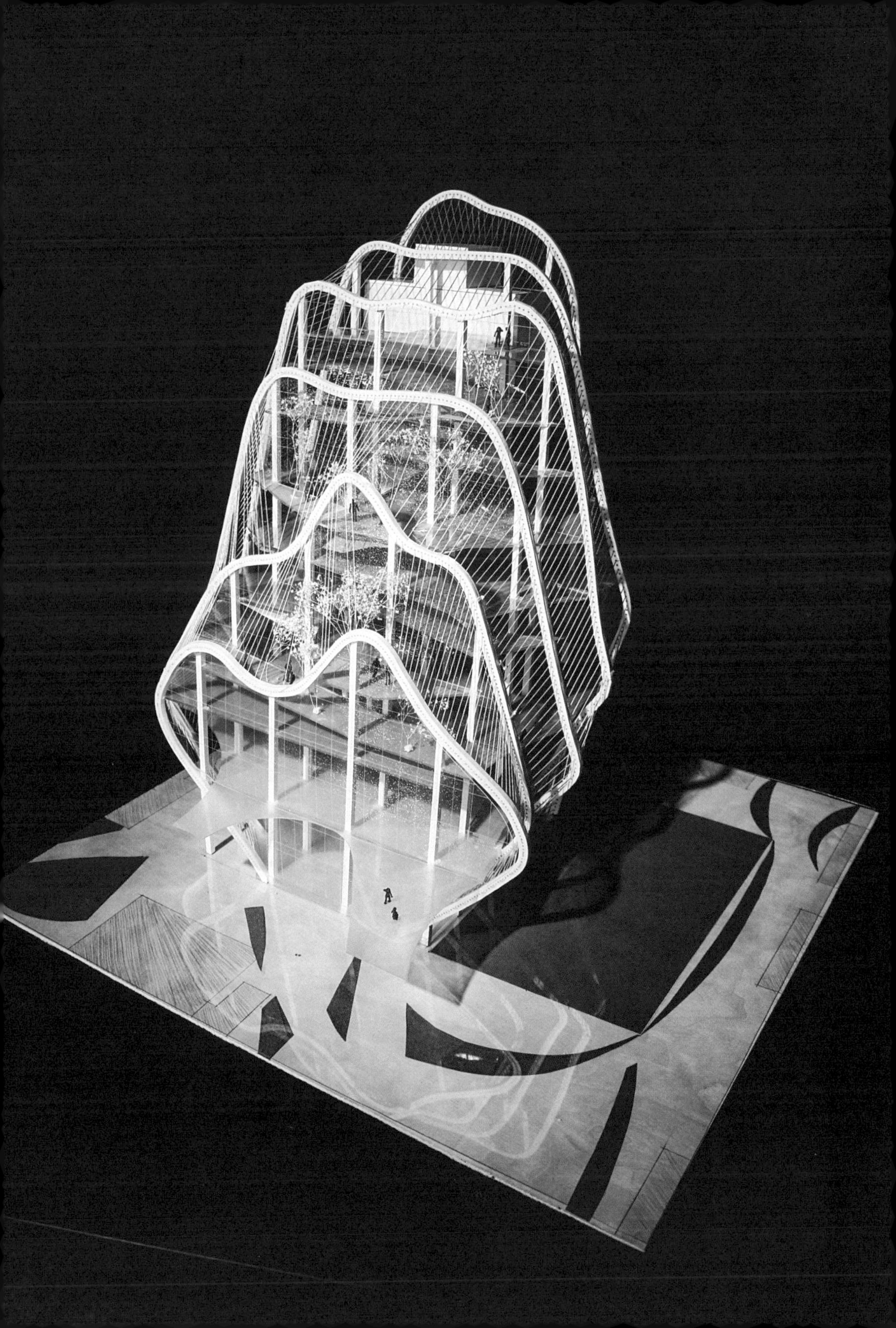

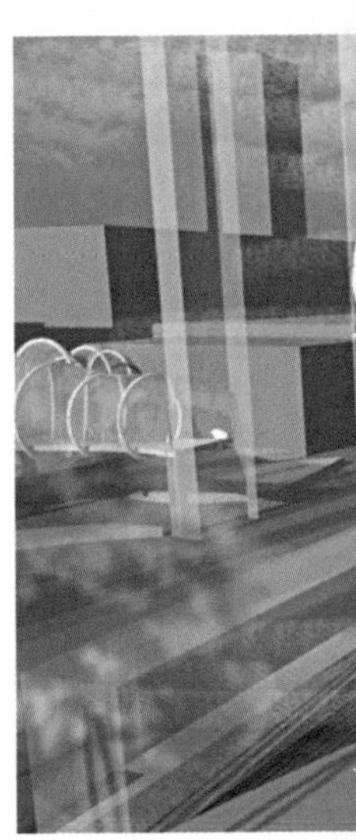

BIAD

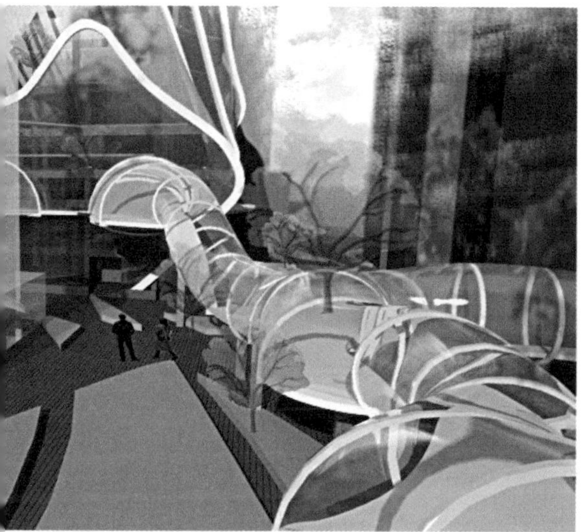

对页上图：院区廊道透视图。**跨页上图：**建筑入口图。**跨页下图：**院区入口图。**本页上图：**建筑室内图。

教师点评

两位同学从设计伊始，就有强烈、鲜明、独特的从"建筑形式"出发的设计理念，试图以他们理想中的"形式"，来解决B园区所面临的城市设计和建筑设计问题，实现他们创意中的B园区最终的意向和氛围。他们认真进行现场调研、分析，梳理园区的功能、交通组织和员工的行为模式，探讨新办公楼的使用与建造，研究形式的意义、实现的手段等等问题。最终，正如我们所见，他们的设计，满足了课程设计任务书的要求，并通过他们不懈的坚持，最终圆满实现了他们原创的、自我的追求形式的梦想，成果精彩并与众不同。

Teacher's comments

The two students had strong, vivid and unique design concept starting from "architectural style" at the beginning of the design. They tried to solve the urban design and architectural design issues of Park in their ideal "form" and achieved their final intention and the climate diagram of Park B in their creation. They carefully conducted field research and analysis, arranged the functions, traffic organization and employee behavior patterns of the park, explored the use and construction of new office buildings, and studied the significance of form, the means of implementation and so on. Eventually, as we have seen, their design meets the requirements of course design task book, and they finally accomplish their original and self-pursuing dream through their tireless persistence. The result is wonderful and unusual.

连·钻/ LIAN DIAMOND

项目选址：北京市建筑设计院大院（北京市西城区南礼士路62号）

功能定位：办公楼设计与园区规划

建筑面积：5284m²

用地面积：1575m²

方案设计：商宇航
指导教师：徐全胜
完成时间：2015

方案设计：李乐
指导教师：徐全胜
完成时间：2015

开篇：主要方案图。**本页左图：**模型。**本页右图：**总平面图。**对页图：**北京院办公楼设计图纸。

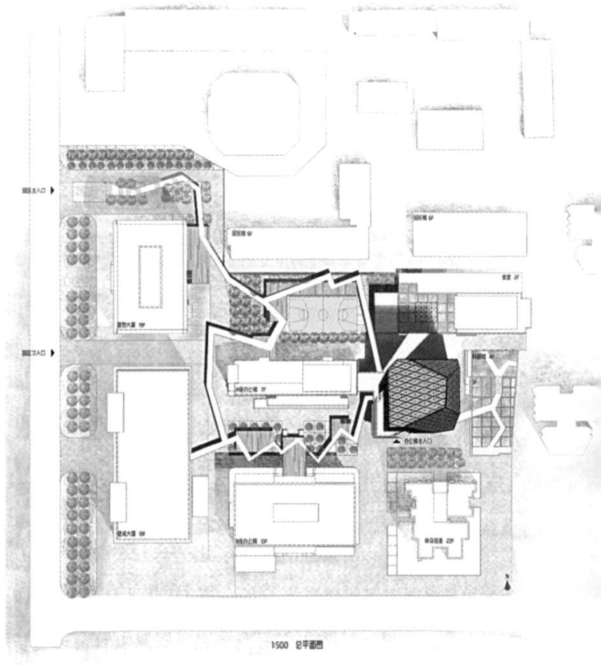

设办公楼单体以钻石为概念，充分考虑日照、景观、交通等环境要素，通过斜线切削，形成晶莹剔透的有冲击力的形体，与园区原有平直形象对比，富有标志性。园区规划以连通为概念，以办公楼为原点，与周围楼连接，依据交通流线，公共空间系统分析，向外放射出二层室外漫步道加休息平台系统，如同钻石的光芒闪耀，将园区各楼串连成一个整体，营造出优美精致的园区公共休闲空间。

楼的表皮通过三角形幕墙强调钻石的晶莹感，以8.1m柱网与4m层高之间的关系作为基本模数形成幕墙龙骨，再进一步细分，部分做为开启扇，挑台上设置部分三角形绿化，栏板角度倾斜与楼体平行。办公楼楼顶设计9m通高大空间，360度观景，可以作为交图后庆祝，聚会的场所。

连·钻

北京建筑设计院园区
规划及办公楼设计

学生姓名：建21 商宇航 2012010030 建22 李乐 2012010047
指导教师：徐全胜 林卫 段华楠 屠思宇 交图日期：2015.4.25

体形虚实与周围环境的关系

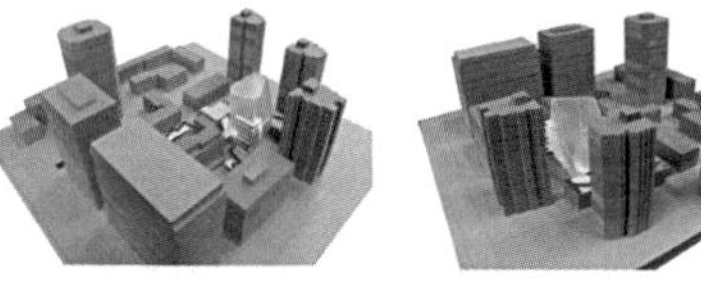

钻石体形生成过程

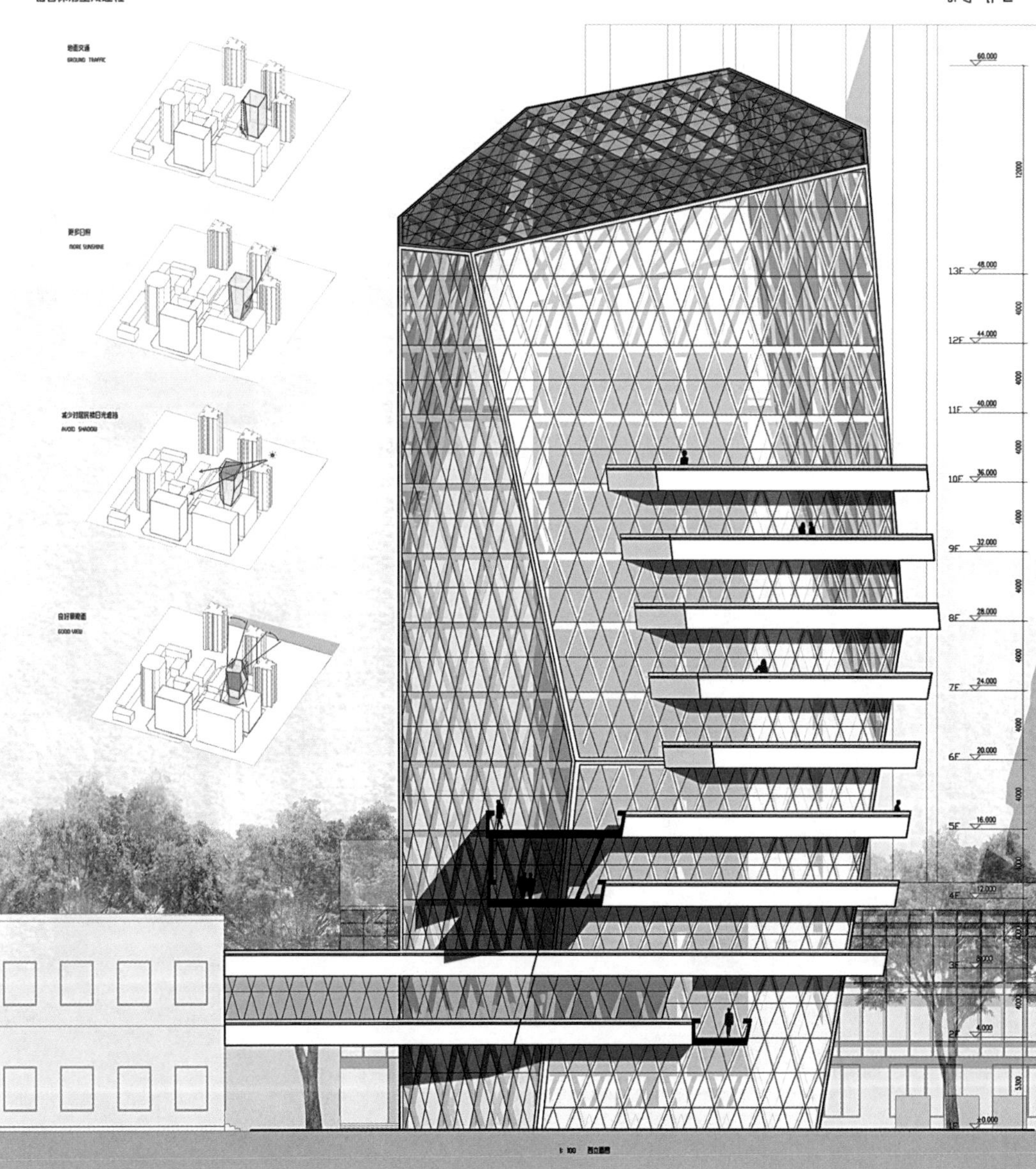

1:100 西立面图

The office building monomer, taking the diamonds as concept, gives full consideration to the sunshine, landscape, traffic and other environmental factors, through the diagonal line cutting, it forms the crystal clear impact form, compared with the original flat image of the park, it's symbolic. Park planning takes the connection as a concept and the office building as the origin, with the surrounding buildings connected. According to the traffic flow and analysis of public space system, the outdoor walking channel and rest platform system are radiated outwards from the second floor, which are like the shining diamonds, connecting the buildings in the park serially into a whole, and thus creating exquisite public leisure space in the park.

The skin of the building highlights the crystal feeling of the diamond through triangular curtain wall. The relationship between 8.1 column net and 4m floor height serves as the basic module to form curtain wall keels. By further subdivision, part of them can be opened. Triangular greening is set on the part of the cantilever platform, and the angle tilting of breast board is parallel to the building. The 9m large space is designed at the top of the building, which can realize 360-degree viewing, and can be used as a place for celebration after the delivery and a place for gathering.

本页上图：模型顶层效果。本页下图：模型入口效果。**对页图：**最终模型全貌。

本页图：模型半鸟瞰效果。**对页图：**
模型入口局部效果。

教师点评

商宇航和李乐的设计，从城市设计入手，分析园区的空间特征对新建单体建筑的限定，生成了建筑“钻石”的体量，在精心设计建筑单体和园区的前提下，对建筑立面进行重点设计，建筑表现精彩，提交了精致完美的成果。

Teacher's comments

The design of Shang Yuhang and Li Le started from the urban design to analyze the park spatial characteristics' limits on newly constructed individual buildings and generate the massing of the building "diamond". On the premise of elaborate design of single buildings and parks, they focused on the design of the building facades. Their building design performance is wonderful, presenting exquisite and perfect results.

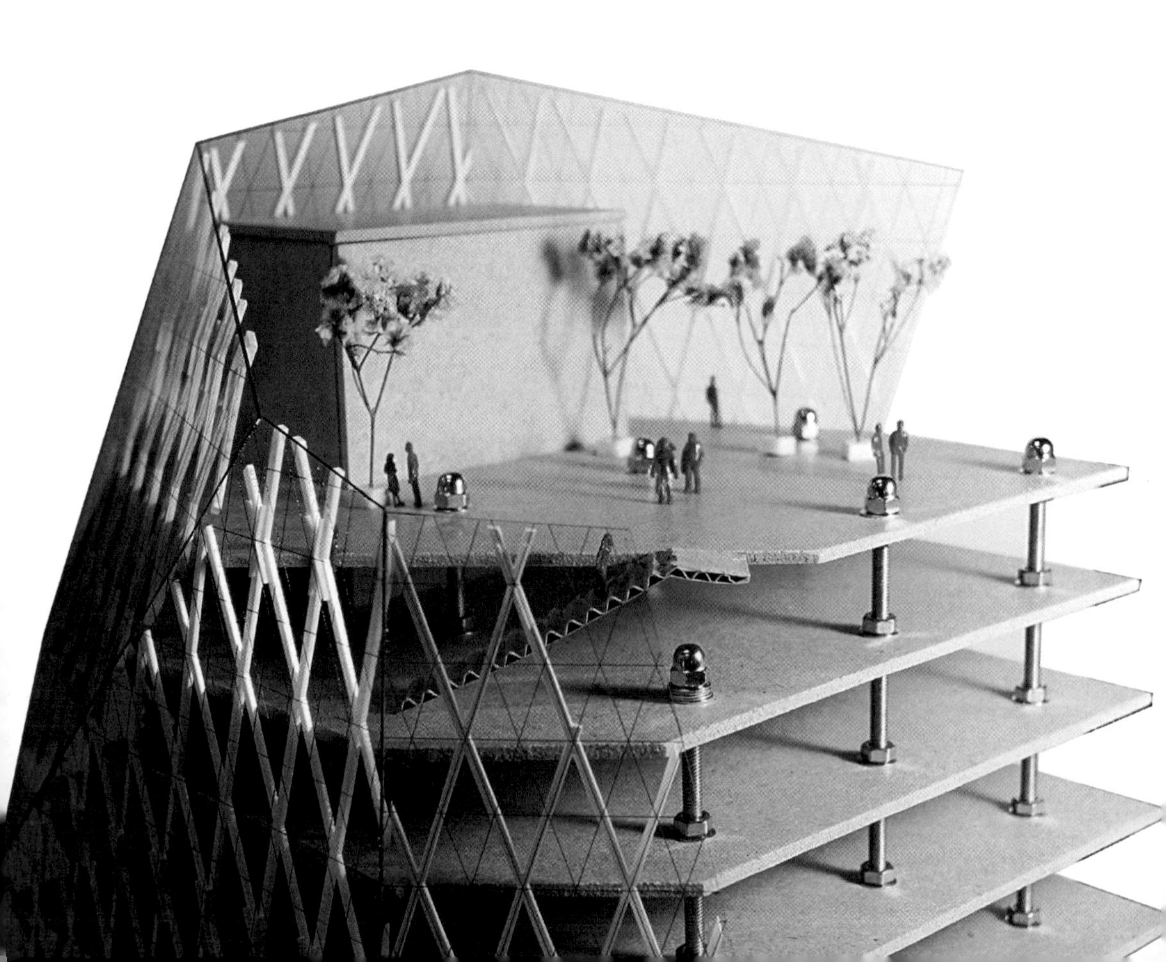

BIAD办公楼/BIAD OFFICE BUILDING

项目选址：北京市建筑设计研究院内
功能定位：办公楼
建筑面积：7068m² 基地面积：1694m²
占地面积：874.4m² 绿化率：52.1%

方案设计：高浩歌
指导教师：徐全胜
完成时间：2015

本次设计从办公楼将多个个人与工作单元集中在一起的状态中提取出概念“集成”，以正立方体作为单元，结合对邻近建筑及环境的分析，对小单元的形态及排列做出变化，营造出各单元彼此独立，同时共享公共空间的办公环境。

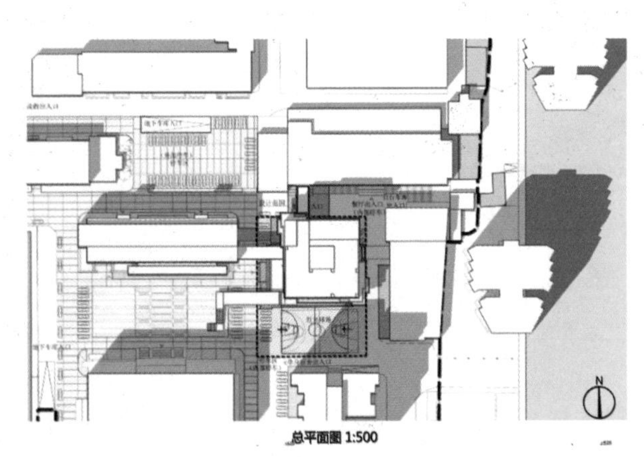

总平面图 1:500

BIAD空中乐园/ BIAD AIR PARK

项目选址：北京市建筑设计研究院
项目类型：办公楼
建筑面积：5366m² 容积率：3.18
绿化率：35.3% 建筑高度：42m

方案设计： 郭琳
指导教师： 徐全胜
完成时间： 2015

方案设计： 龚怡清
指导教师： 徐全胜
完成时间： 2015

针对BIAD园区现有情况，通过以下几点切入设计：1. 创造可更换的工作环境，激发灵感。 2. 创造交互性强的灰空间，与室外广场结合，形成园区核心景观。3. 解决拥挤问题。具体设计中引入了大型通高空间和绿化系统，内嵌小型独立单元，并以格栅统一立面效果。

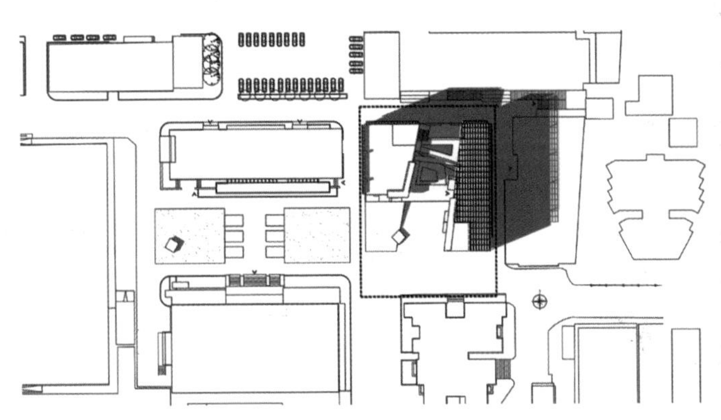

BIAD办公楼/BIAD OFFICE BUILDING

项目选址：北京市建筑设计研究院内

功能定位：办公楼

建筑面积：6000m² 占地面积：900m²

容积率：1.5 绿化率：0.35%

标准层面积：767m²

方案设计：胡毅衡

指导教师：徐全胜

完成时间：2015

这次设计的始发点主要是透过分析建筑设计师的日常工作模式，得出建筑师需要个人的空间之余还需要公共交流空间。因此设计了每两层中间会有一个较大的通高空间，以交流讨论之用，而大空间的两旁就是一些小单元块，为建筑师个人空间。

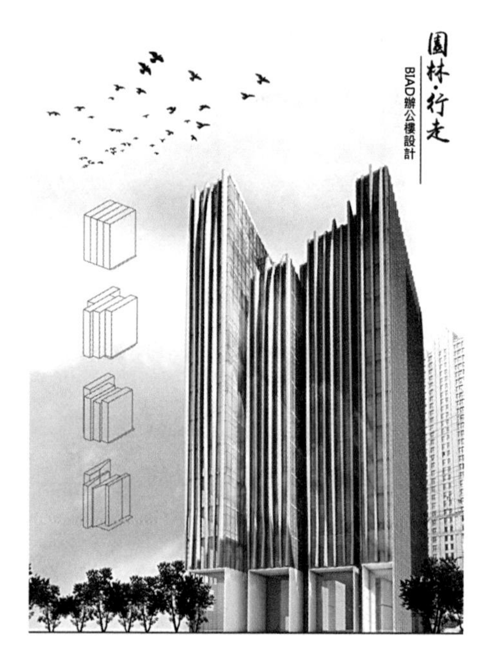

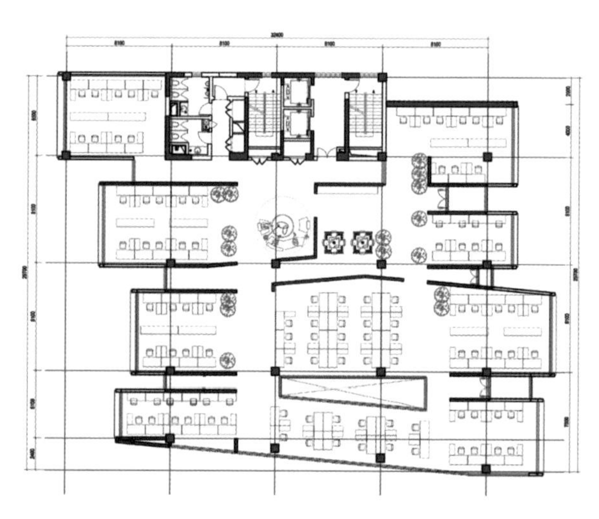

一方/A PARTY

项目选址：北京建树设计研究院
项目类型：办公楼
建筑面积：6500m²
用地面积：1680m²

方案设计：黄孙杨
指导教师：徐全胜
完成时间：2015

设计题目为：注重“形式”的建筑设计。设计的主要关注点是“形式”和“功能”之间的关系，形式相对于功能可以是脱离的，也可以相互关联。设计中通过在建筑中置入新的功能——篮球场，来探索建筑形式的可能性。

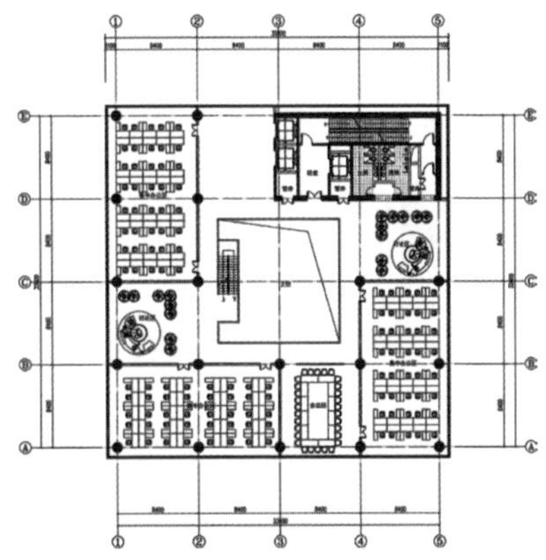

错动的办公室/ WRONG OFFICE

项目选址：北京市西城区南礼士路 66 号
功能定位：办公楼
建筑面积：6554m²
用地面积：755m²

方案设计：金兑镃
指导教师：徐全胜
完成时间：2015

设计的出发点是对普通办公空间的问题意识。大多数办公空间形式单一、缺乏活力。本方案从“错动”的形式出发，利用楼层的横向错动引发纵向的空间属性变化，试图营造办公与休闲混合的活跃的办公室。

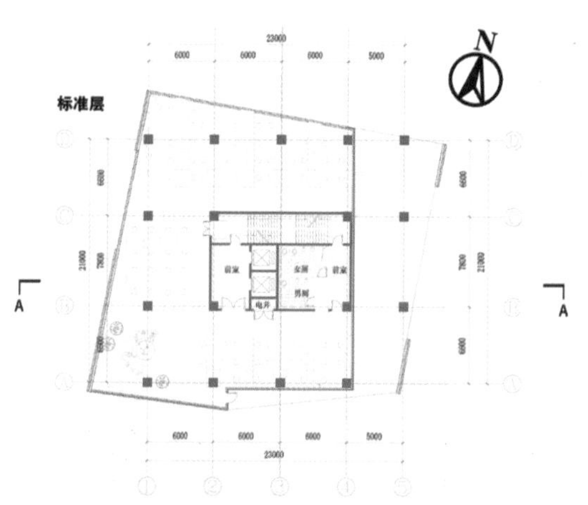

光之/YUAN

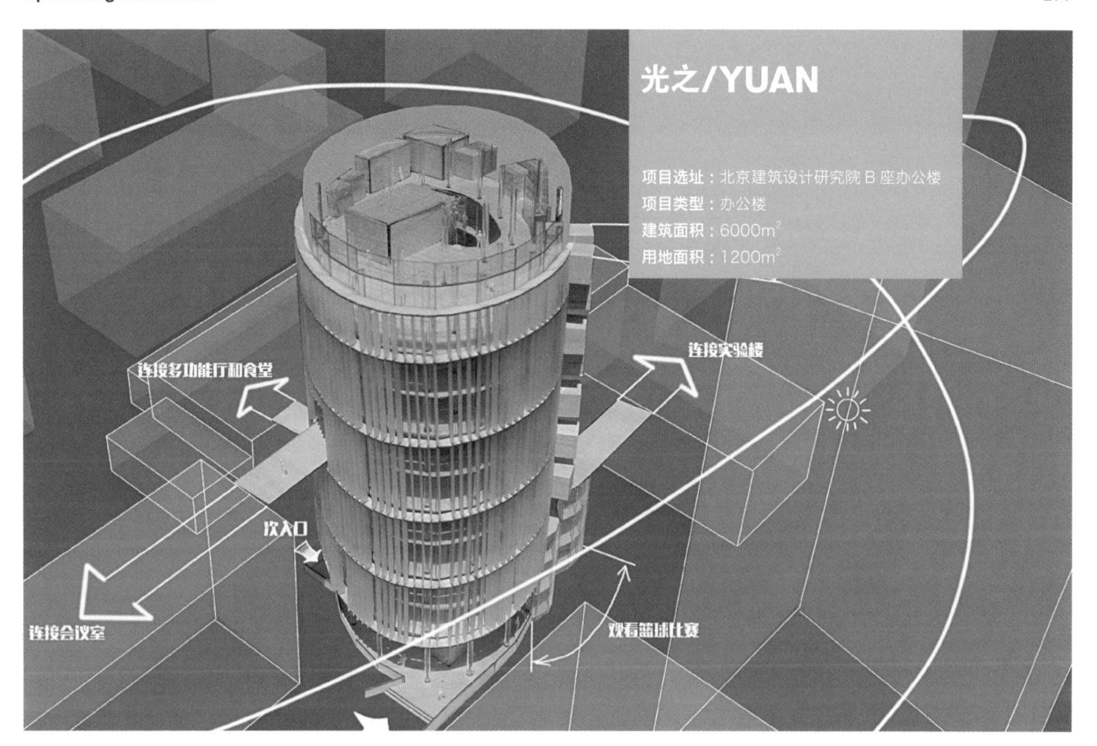

方案设计：李智
指导教师：徐全胜
完成时间：2015

方案设计：李姝琳
指导教师：徐全胜
完成时间：2015

本设计方案立足园区环境，通过椭圆形体量介入拥挤局促的场地。同时为了营造舒适光环境的办公氛围，贯穿吹拔作为采光筒，将开敞办公空间和私密的会议、休闲空间的光环境联系起来，实现"光"之乐园。

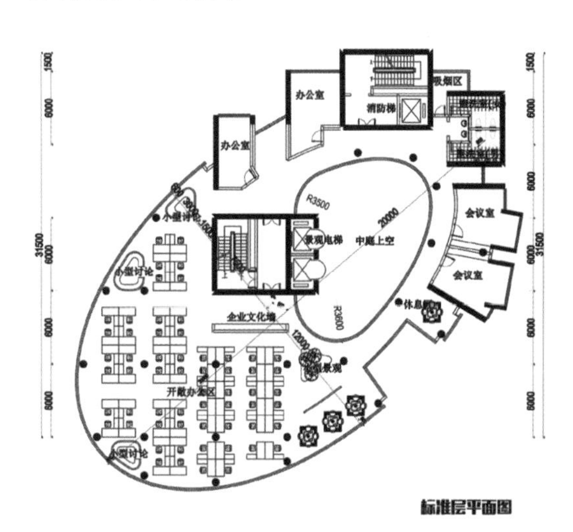

标准层平面图

空间单元

崔彤

全国工程勘察设计大师
中科院建筑设计研究院 副院长 总建筑师
中国科学院大学建筑中心 主任 教授 博导

崔彤

全国工程勘察设计大师

中科院建筑设计研究院 副院长 总建筑师

中国科学院大学建筑中心 主任 教授 博导

教育背景

1981年 – 1985年

内蒙古科技大学　工学学士

1994年 – 1997年

清华大学　建筑学硕士

工作经历

1997年至今

中科院建筑设计研究院　副院长、总建筑师

2012年至今

中国科学院研究生院建筑研究与设计中心

主任、博士生导师

主要论著

《当代建筑师系列 崔彤》[M]中国建筑工业出版社

《生长的秩序—泰国曼谷·中国文化中心设计思考》、《建筑学报》2013（3）：100-105

《源于场所的建构》、《新建筑》2012（6）：16-23

《山水之间：中科院研究生院新校区》、《建筑创作》2010（5）：188-197

《景观生态学原理指导下的校园规划设计》、《建筑学报》2009年（5）：90-93

《新与旧—重构过去中的未来》、《建筑学报》2007（6）：72-75

《重构平衡—外交使馆作为一种建筑类型》、《世界建筑》2006（8）：100-103

设计获奖

2005年 – 中国科学院图书馆获全国优秀工程设计金奖

2005年 – 崔彤获国务院政府特殊津贴

2007年 – 崔彤获第一届全球华人青年建筑师奖

2009年 – 中国科学院图书馆获中国建筑学会建筑创作大奖

2010年 – 国家动物博物馆&中科院动物研究所获全国优秀工程设计银奖

2012年 – 崔彤获当代中国百名建筑师称号

2015年 – 国家开发银行获全国优秀工程勘察设计一等奖

2015年 – 泰国曼谷中国文化中心获全国优秀工程勘察设计一等奖

2016年 – 崔彤获全国工程勘察设计大师称号

代表作品

泰国曼谷中国文化中心（图1、图2）、中国科学院国家科学图书馆（图3）、北京林业大学学研中心（图4）、国家开发银行（图5）、通辽科尔沁文明之光博物馆（图6）、京东商城总部办公楼（图7）

“空间单元”建筑设计

三年级建筑设计（6）设计任务书

指导教师：崔彤
助理教师：兰俊

课程概况

课程题目“设计单元”以新视域、新方法整合传统的设计方法，旨在引发学生对模件单元的重新认知，使学生初步掌握作为建筑师必须拥有的设计方法和建筑观。

在八周的学习和设计实践中，通过历时性的线索追踪，呈现出一系列从亘古聚居单元到当今人居环境背景下的模块单元的多样性和复杂性、直到指向未来的基于人脑与电脑“复合参数”的模件单元的重构；通过共时性的多学科和跨领域的模件单元比较研究，以激发学生的灵感，最终应用"空间单元"完成建筑设计。

关于模件单元

模件体系是一种以标准化零件组装成物品的生产体系。这些标准化的零件可以被大量地预制，并能以不同的组合方式迅速装配在一起，从而以有限形式的构件创造出无限形式的单元。而这些最基本的构件被称为“模件”。

中国古代很早就出现了模件体系，传统模件体系的理念不仅影响了中国的文学、绘画、建筑等领域，也影响了青铜器、雕塑、漆器、瓷器等制造业的发展，自唐以后的大量木构建筑更是让模件体系的优点尽显无遗。

伴随东西文化的交流和西方工业革命的开始，传统模件体系也深刻影响了西方人的生产、制造业的发展。德国历史学家雷德侯（Lothar Ledderose）在其著作《万物》中提到的“欧洲人热切地向中国人学习并采纳了生产的标准化、分工和工厂式的经营管理……”正是传统模件体系被欧洲接纳并追捧。今天重提模件，更多着眼于模件化生产、制造、建造过程背后的一种方法或哲学观，最终转化为一种建筑设计方法或智慧。

其中，传统木构建筑在其演进过程中完善了对模件体系的应用，也发展出一整套繁杂华丽的模件单元，我们甚至也可把中国木构建筑“翻译”为“木构高技建筑”。然而，也正是由于对模件系统的广泛应用导致中国建筑几千年来稳定和持续的发展而缺少创造性。因此，如何批判性的学习和应用模件成为本课程的重点。在模件单元标准化和可装配化的基本要求之外，本次建筑设计还需要关注个性化、丰富性、人情味、自然性和场所感。

设计题目

学生可依据对模件单元的认知选择以下两个地段中的一个作为设计题目开展设计。

地段一：清华大学美院北侧“设计聚落”；

（1）区位：清华大学美术学院北侧与博物馆南侧空地（现为停车场）。

（2）基本技术经济指标要求：用地面积约10000m^2，容积率1.5，建筑高度控制在15m之内，绿化率>30%，用地边界可依据设计需要自行确定；

（3）功能要求：

a.“设计聚落”，为建筑学院、美术学院、新闻传播学院、汽车系、机械系等院系相关专业师生提供“设计研发”“工坊制作”“产品销售”等一系列的科教融合，学、研、产一体化的“模件单元”，并通过单元的自我循环和新陈代谢构成一个联系校园又向社会开放的平衡体系；

b. 公共空间，既是联系和整合“设计聚落”模件单元的空间和功能纽带，又是承载大型展示、跨院系交流等的外向性空间系统；

c. 注意地下空间的开发。

（4）其他要点：地段南北的博物馆和美术学院均为清华大学校内体量较大的建筑，设计时需考虑新旧建筑的协调；地段东侧即为城市道路，需考虑与城市开放空间的相互影响；“设计聚落”模件单元应依据实际功能选取不同尺度，可采用垂直组合的空间形态等。

地段二：北京王府井大街“商务会所单元”；

（1）区位：地段位于王府井大街北段，东临王府井大街，西侧为四合院历史建筑群；北侧为金茂万丽酒店、南侧为世纪大厦；东南侧为首都剧场和商务印书馆。

（2）基本技术经济指标要求：用地面积约8900m^2，容积率_1.8，建筑高度_30m，绿化率>30%；

（3）功能要求：

a. 商务会所单元，建筑的主体部分为相对独立性和私密性的会所单元集合，这些会所单元在空间和功能上均满足模件单元的特征；

b. 公共交流空间，是联系和整合模件单元的开放性空间体系；

c. 对外商业，是会所功能的补充，也是对王府井大街地段特征的回应；

d. 考虑充分发掘地下空间的可能性。

（4）其他要点：地段内部西南角有保留古树一棵；地段西侧的四合院建筑群与地段东侧王府井大街沿线的建筑在尺度上差别较大，设计时应考虑新建建筑尺度与它们的关系；故宫位于四合院建筑群西侧，距离地段少于300m，应作为设计时考虑要素。

课程结构

第1~2周：每名学生需要根据教师讲述、集体讨论和课下思考，完善对模件概念的认知，归纳出几种模件单元组合的几种方法，并用通过某种媒介（文字、草图、模型、影像或上述方式组合）进行描述。

第3~6周：在对每种模件单元的特性分析了解的基础上，结合下述设计题目给出的两个设计地段选择自己喜欢并认为适合发展的模件单元深入研究，并将模件单元与地段特征融合，形成初步方案。

第7~8周：方案完善和细部设计。在设计的全过程强调模件单元的基本理念，并将所选择模件单元特征贯彻到主要建筑细部。

重生:光与模组/ REBIRTH:LIGHT & MODULE

项目选址：清华大学艺术博物馆南侧

功能定位：设计聚落

建筑面积：8000 m^2

用地面积：10800m^2

方案设计：王昭雨
指导教师：崔彤
完成时间：2015

开篇：主透视。**对页图**：模型。**本页上图**：单元剖面。**本页左下图**：总平面图。**本页右上图**：一层平面图，**本页右下图**：地下一层平面图。

“模件”准确来说不仅仅是一种设计手法，在这之前，它是一种制造的手法，雷德侯的《万物》将“模件化”和“规模化生产”并列，展现模件最神奇的一面：相似甚至相同诞生万物。这同样是课程最吸引我的一点，同时“模件”的训练将考虑从生产到组装这系列常常被自己忽略的问题，也令我分外期待。

设计的起点是对于自然与人工模件的研究，我选择的研究对象一个是蜂巢，对于这个自然模件，我更多的关注的是空间的构成逻辑、这种组织逻辑对于飞行的适应以及适应人的需求的改造可能性；另一个是可口可乐瓶“Coca-Cola 2nd Lives”（通过置换瓶盖使得废弃可乐瓶有着多重用途）这样的人工模件，通过“可变”与“不可变”带来丰富的适应性，是我的设计适应多重功能的重要思路。

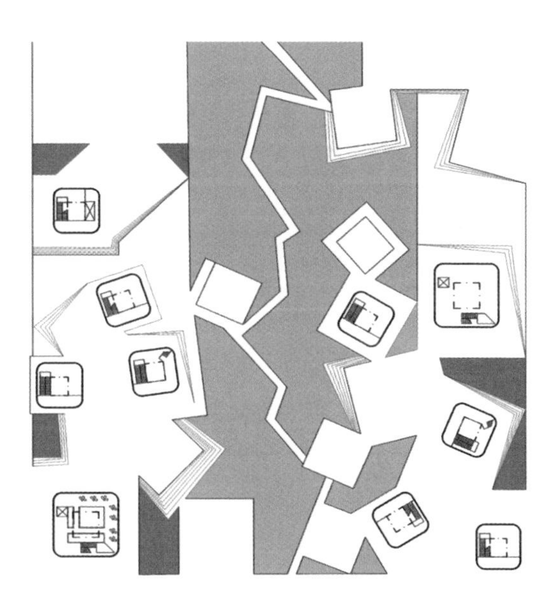

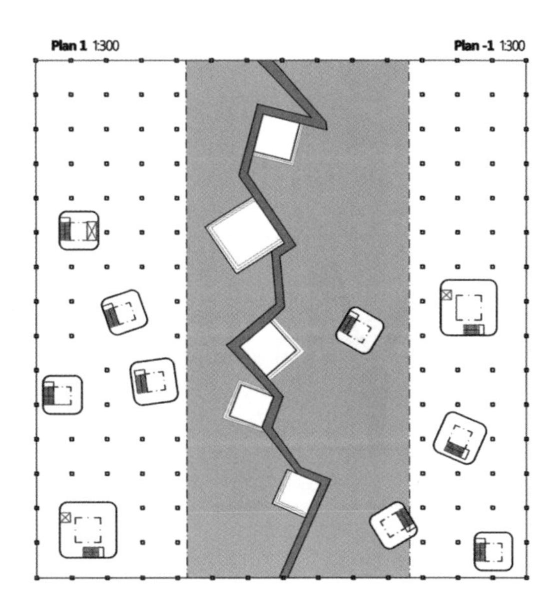

清华—轴线—补全

完整的清华南北轴线

清华东西景观轴线的“断尾”，场地作为景观轴线的补全

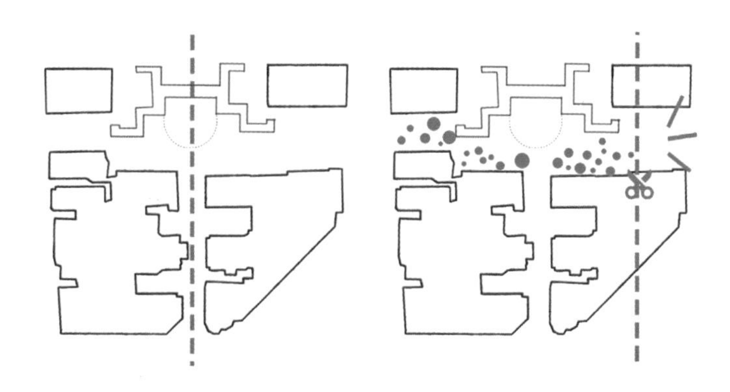

模件—深化—构造

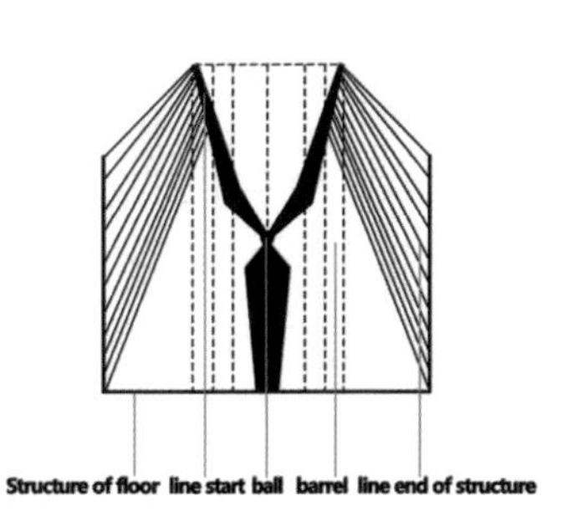

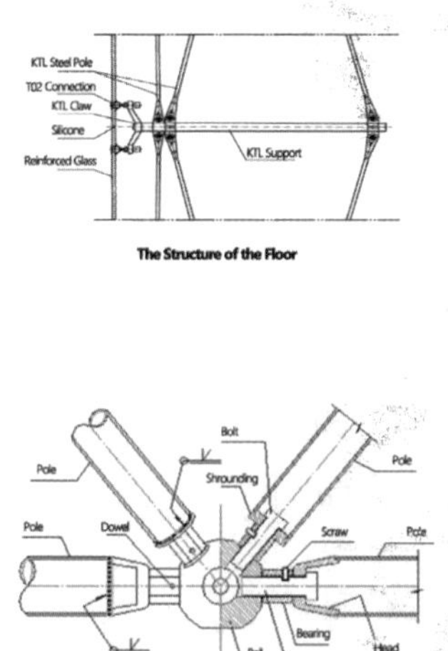

The Structure of the Floor

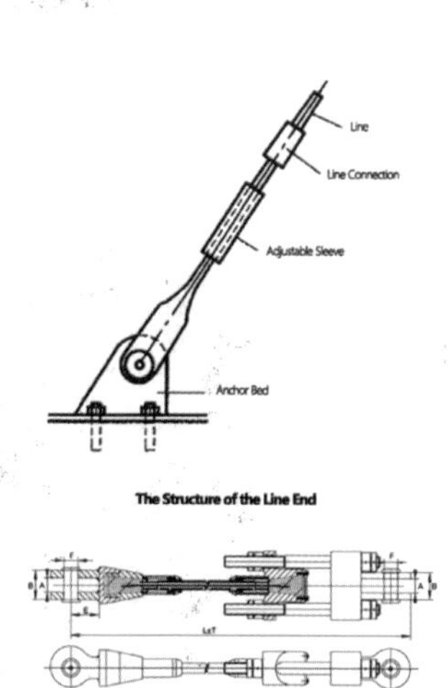

The Structure of the Line End

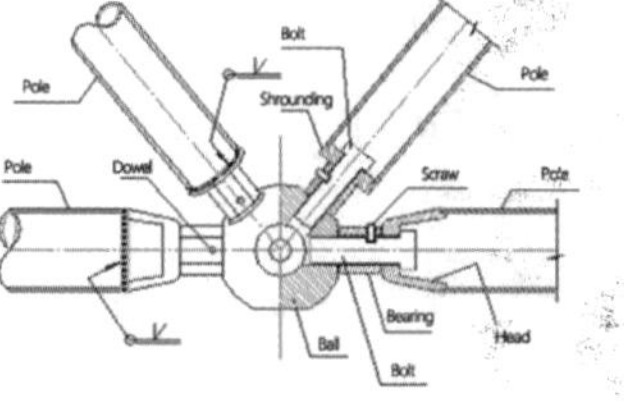

The Structure of the Ball

The Structure of the Line Start

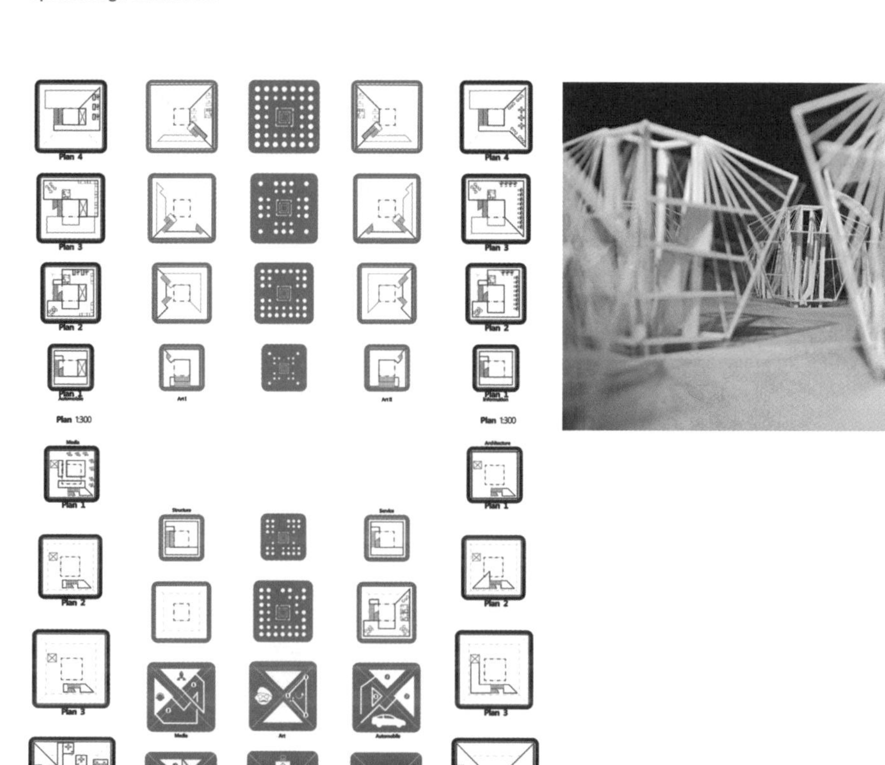

对页上图：轴线分析。**对页下图：**结构深化分析。**本页上图：**各单元平面图。**本页下图：**特写。

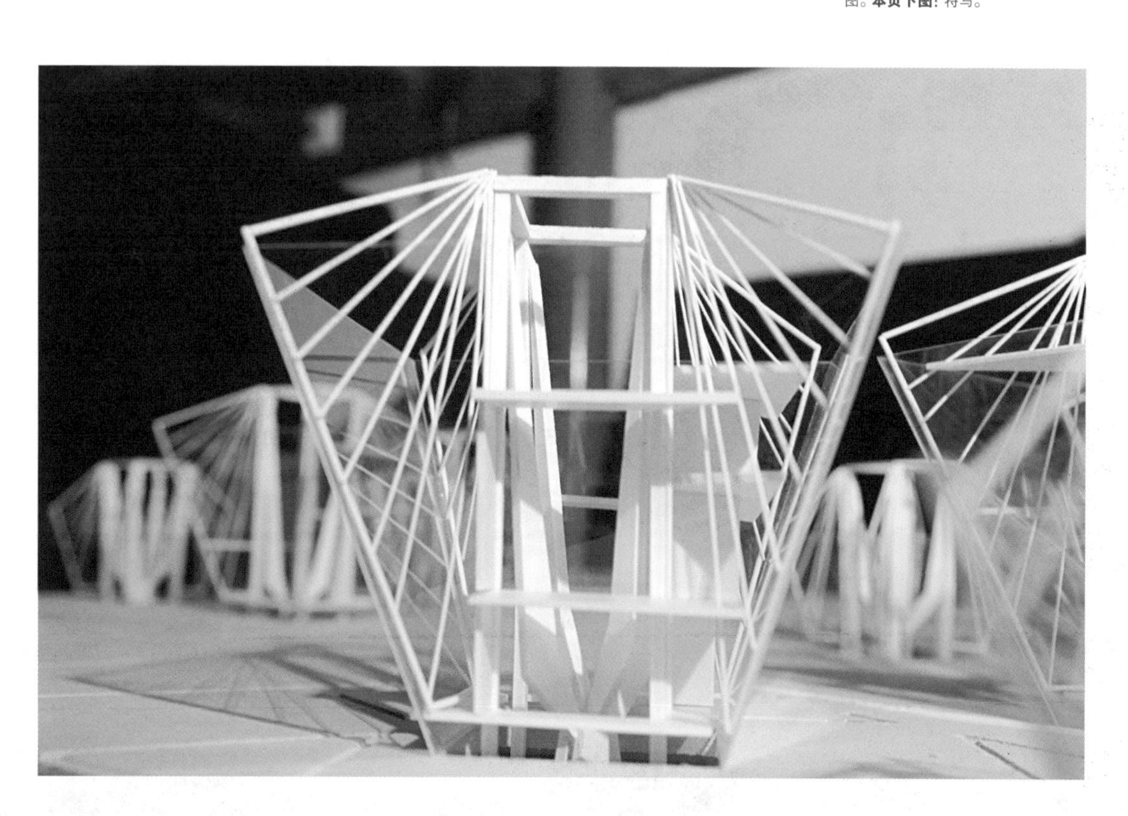

Accurately speaking, "Module" is not only a kind of design technique. Before this, it is a technique of manufacture, Ten Thousand Things by Lothar Ledderose places "modular" and "scale production" side by side, which displays the most magical side of module that is similar or even the same birth of ten thousand things. This is also the most attractive point to me in this course, at the same time, the training of "module" will consider the problem that is often ignored from the production to assembly, and it also makes me look forward to it.

The start point of the design is for the study of natural and artificial module, I choose a hive as the research object, for this natural module, I pay more attention to composition logic of the space, and the adaptation of the organization logic to flight and possibilities of transformation to adapt to the human; another is the artificial module of Coca Cola bottle, Coca Cola 2nd Lives "(the replacement of the cap makes the waste cola bottles have multiple purposes), the rich adaptability brought by the" variable "and" invariable" thing is an important idea that my design adapts to the multiple functions.

教师点评

本课程要求以“模件单元”作为设计的出发点和成果表达的重要部分， 试图传授给学生一种在面对大规模设计对象时的“规律性”设计原则。因此，形式和结构类似，尺度不同，功能互异的“模件单元”的研究和设计是对题目做出的符合逻辑的解读，也是大多数学生能迅速掌握的方法。王昭雨同学的“光的容器”便是通过上述方式做出的对“清华大学”设计地段的回应。

该设计通过地段的调查研究，提炼出“光”的主题，在对原有停车场进行功能置换，引入新建筑时，将建筑定位为具有“采光器”功能的模件单元，让被置换到地下的停车场在光中“重生”。

该设计选择了一种全能、自平衡的“模件单元”并经过推演变化实现其设计意图：通过地下停车场柱网尺寸与展览空间尺寸对“采光器”模件的尺寸进行确定；结合绿化、水池等手段对“采光器”模件的类型进行划分；分析各种结构对于光的表现对“采光器”模件的结构进行研究；考虑场地规划水池以及庞大体量建筑的情况，对“采光器”模件的布局进行控制。最终形成八种形式结构类似的模件单元，根据功能的不同采用了差异化的模件尺度和内部空间形式。外部空间则结合模件自身的转动形成宛如曲水流觞般的聚落形态。

跨页图：透视。

Teacher's comments

This course aims to impart a kind of "regularity" design principle in the face of large-scale design objects to the students by taking the "module unit" as the starting point of designs and an important part of the achievements expression. Therefore, the research & design of the "modular units" with similar forms and structures and different dimensions and functions is a logical interpretation of the subject and a method that most students can quickly master. The "light vessel" of Wang Zhaoyu is the response to the design section of "Tsinghua University" by the above method.

The design extracted the theme of "light" through the investigation and research of the section, and positioned the building as the "module unit" with the function of "light collecting device" when conducing function replacing to the existing parking lot in the introduction of new buildings, so that the parking lots replaced underground are "reborn" in the light.

The design chosen an all-round and self-balanced "module unit" to achieve the design intent through deductions and changes: the size of the "light collecting device" was determined according to the column network dimension of underground parking lot and exhibition space size; the type of "light collecting device" modules were divided combined with the greening, water pool and other means; the performance of the light of different structures on the structure of "light collecting device" modules was studied, considering the site planning pool and the situation of a large amount of construction, the layout of "light collecting device" modules was controlled. In the end, the eight modular units of similar forms and structures were formed, according to different functions, the differentiated module size and internal spatial form were adopted. The outer space is a settlement like a winding stream in combination of self-rotation of modules.

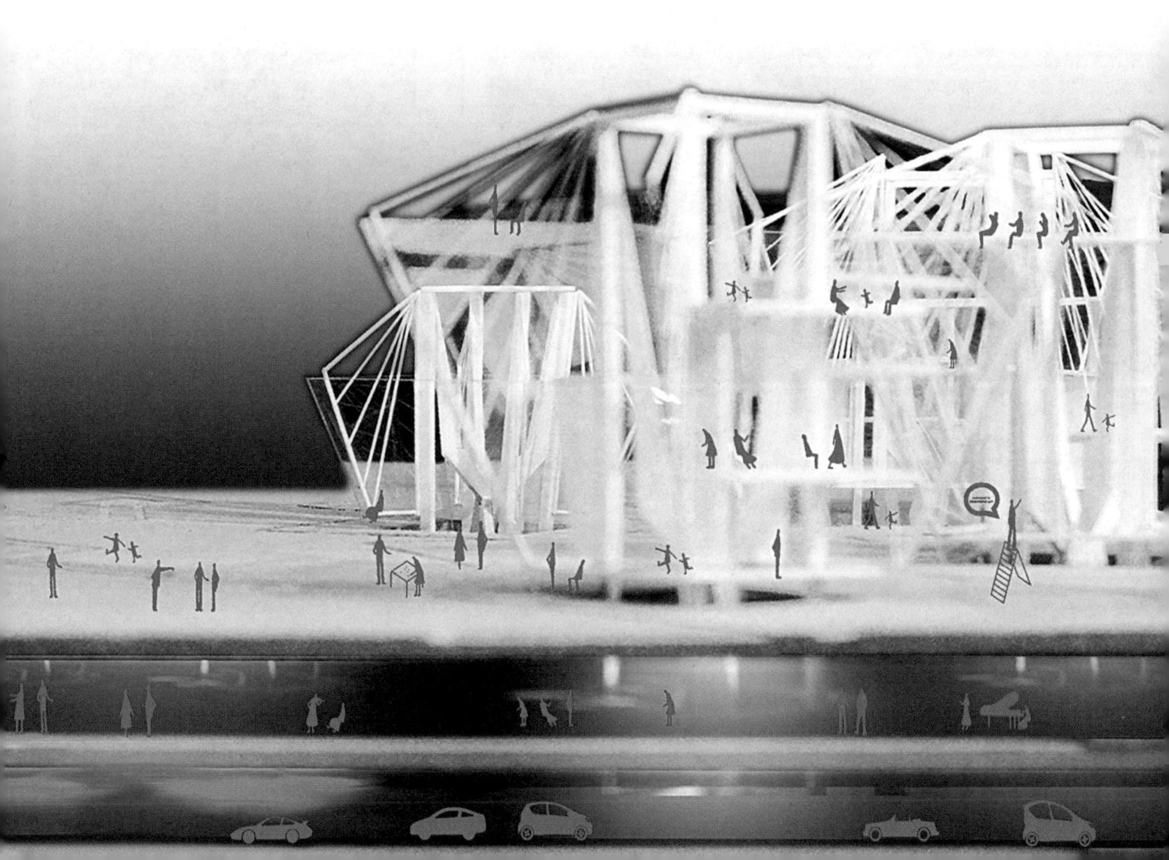

学研长廊/LEARNING RESEARCH GALLERY

项目选址：清华大学美术学院与博物馆之间的停车场

功能定位：学研产混合功能建筑

建筑面积：13000m^2 **用地面积：**4500m^2

方案设计：唐义琴
指导教师：崔彤
完成时间：2015

开篇：主要效果图。**本页上图：**单元簇分析图。**本页下图：**单元生成分析。**对页上图：**总平面图。**对页中图：**空间分析。**对页下图：**剖面图。

单元簇

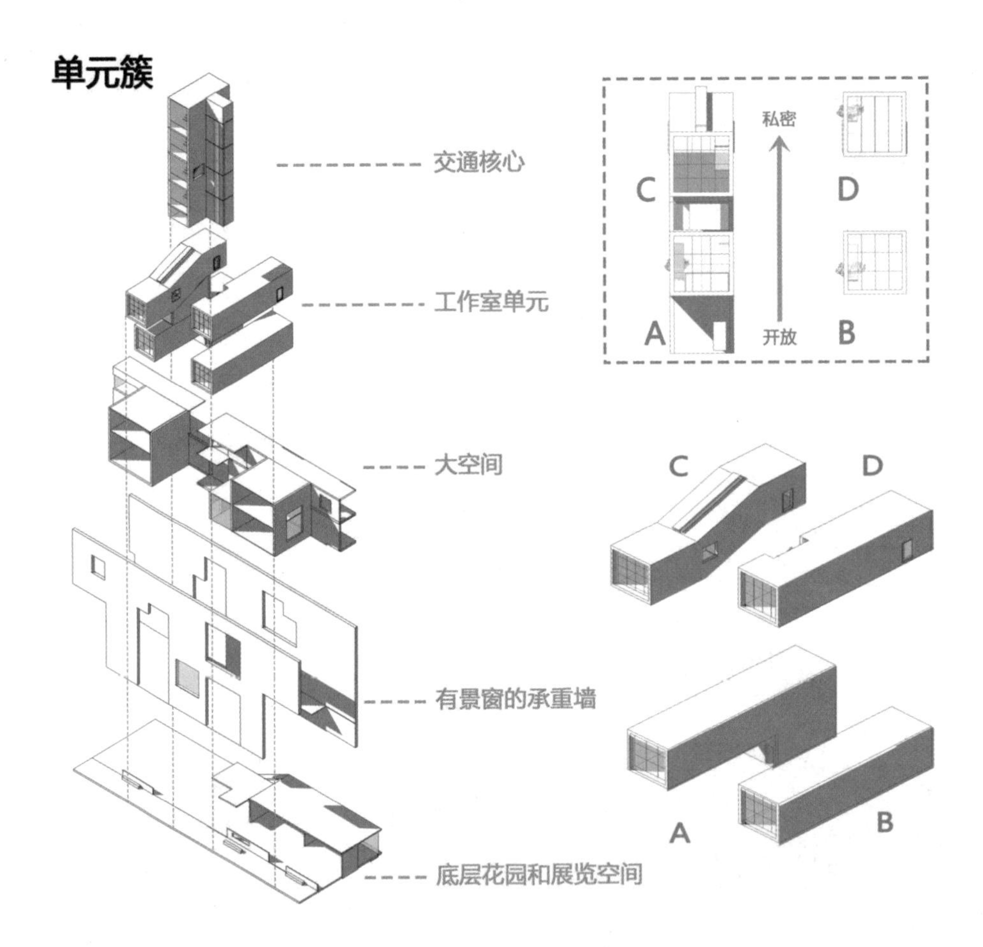

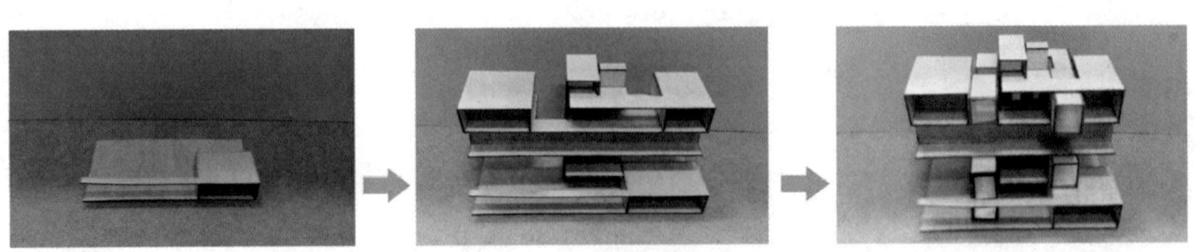

地下一层的展览空间和花园　　交通核心筒连接的空中街道和大空间　　在缝隙中漂浮着由外墙承重的单元盒子

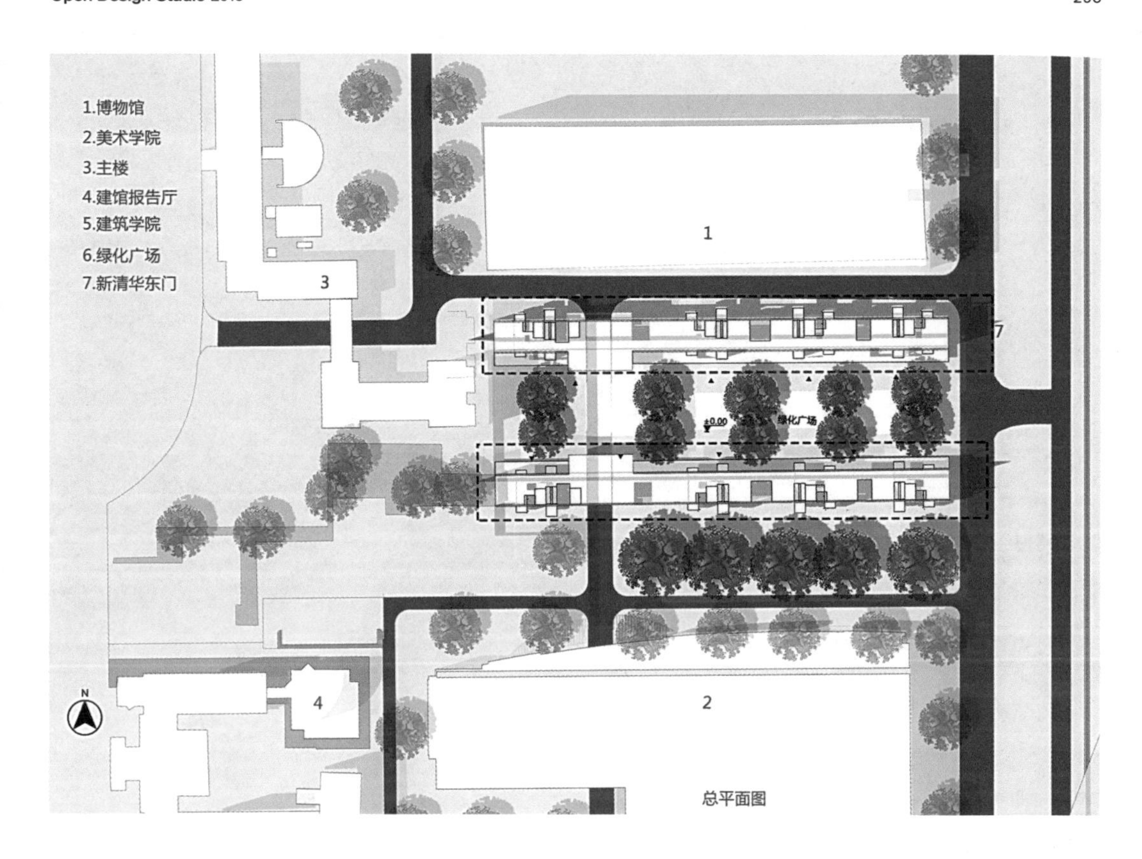

总平面图

模件体系是一种以标准化零件组装成物品的生产体系。这些标准化的零件可以被大量地预制，并能以不同的组合方式迅速装配在一起，从而以有限形式的构件创造出无限形式的单元。“设计聚落”，为建筑学院、美术学院、新闻传播学院、汽车系、机械系等院系相关专业师生提供“设计研发”、“工坊制作”、“产品销售”等一系列的科教融合，学研产一体化的模件单元。由ABCD四种简单单元组成的两条对称的建筑体里，形成商业街、天井、盒子等趣味横生的丰富空间，达到单元的模块建造化效果。

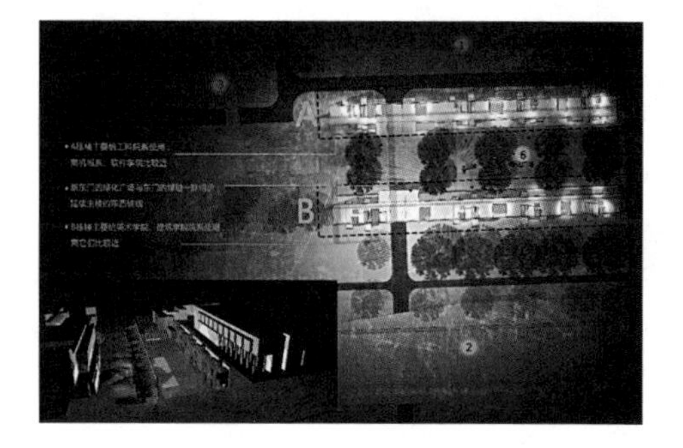

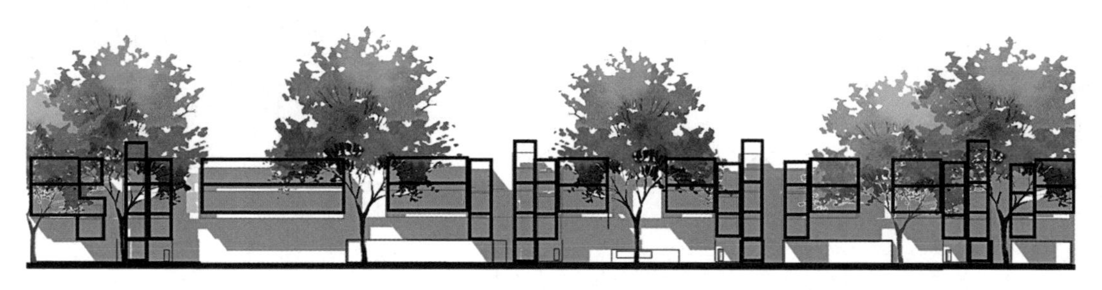

剖面图

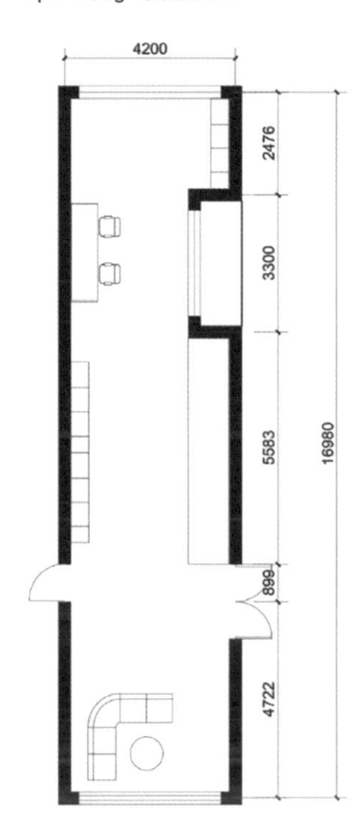

对页图：建筑外观图。**本页上图（从左至右）：**单元平面图。建筑入口图。**本页上图：**单元剖透视。

Modular system is a kind of production system that assembles the articles with standard parts. These standardized parts can be prefabricated in a large amount, and can be assembled together in different combination method, so as to create an infinite unit in the form of a finite element. The "design settlement" provides a series of module units by integrating industry-university-research cooperation, such as the "design research", "workshop production", "product sale", etc. for the school of construction, school of arts, school of journalism and communication, auto school and mechanical engineering school, etc. The interesting spaces of commercial street, patio, box, etc. are formed in the two symmetrical building structure composed by ABCD four simple units, achieving the module construction effect of the unit.

教师点评

基于校园中美院和美术馆之间的方形用地，以同构于场所的“平行墙”构想回应了场地，同时，在单元空间模件的介入中“平行墙”转化为“平行腔”。

在看似重复性的单元中，包含着内在的丰富性和空间趣味：几何化构成、水平向空间沟通、孔洞经营、立体园林的构想遵循着两个“墙式”大单元、八个方体空间模件及若干个小筒体一系列相对逻辑的空间体系。

在始终坚持着现象的透明性中，努力让校园中的“设计聚落”有一种新的空间秩序和自然属性，使空间单元既是“房子”也是“园子”。

Teacher's comments

Based on the square lands of the academy of fine arts and the museum of fine arts in the campus, the site is responded by the isomorphism of "parallel walls" concept. At the same time, in the intervention of modules in the unit space, the "parallel wall" is transformed into "parallel cavity".

The seemingly repetitive unit contains the inherent richness and spatial interest: the geometric composition, horizontal space communication, hole operation and three-dimensional garden concepts followed two "wall-type" large units, the eight cube space modules and several small barrels composed a series of relative logic space systems.

While always adhering to the transparency of the phenomenon, she strived to enable the "design settlements" in the campus to present a new space order and natural attributes, so that the space unit is both a "house" and a "garden".

对页上图及下图：建筑室内图。**本页上图**：建筑入口。**本页下图**：建筑室内。

众生相/
VISUALIZING
PEOPLE
项目选址：清华大学美术学院北侧
功能定位：多学科设计聚落
建筑面积：6700m²
用地面积：10000m²

方案设计：唐思齐
指导教师：崔彤
完成时间：2015

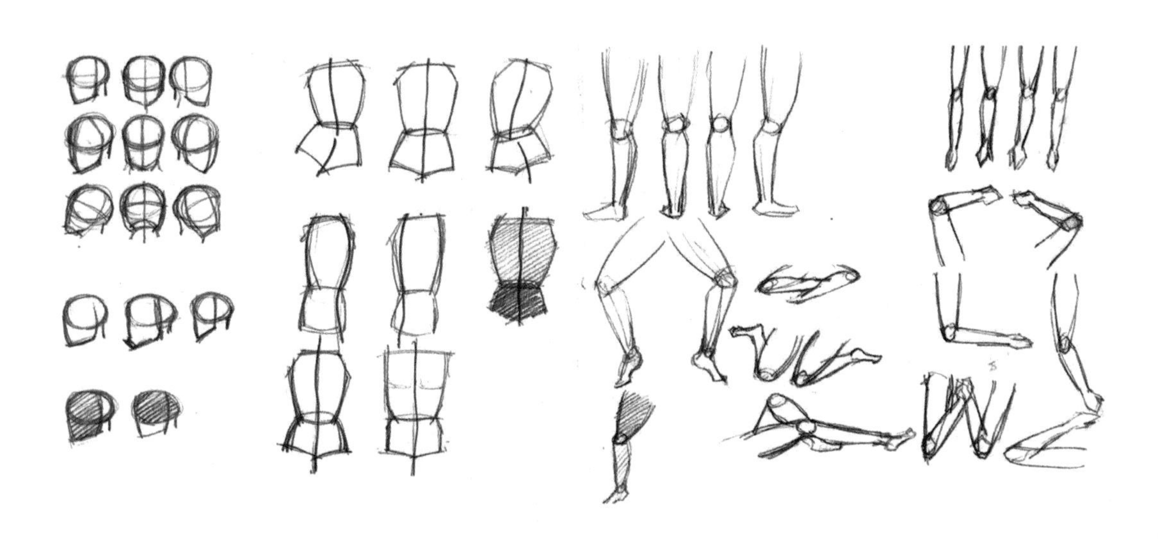

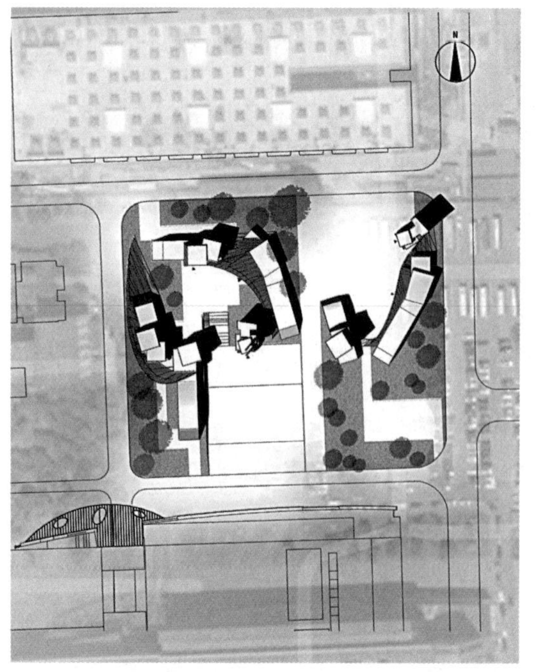

人本身作为最为自然的一种元素，其内在组成结构与建筑有着千丝万缕的联系。历史上，也不乏人体建筑化或建筑人体化的诸多尝试。本方案即选择了“人”作为一个完整的模件单元，试图通过研究人体的形态构成、组织逻辑和个体间的相互关系来找到一种其在建筑上的映射方法，在建筑和建筑群落的构成、组织上探索新的可能。每个建筑单元都被视作一个个人体形象，在组织逻辑上呼应的同时，造型上也与人体的趋势和特性相似。与此同时，结合各学科对于空间的不同要求，调整建筑内部的组织方式。

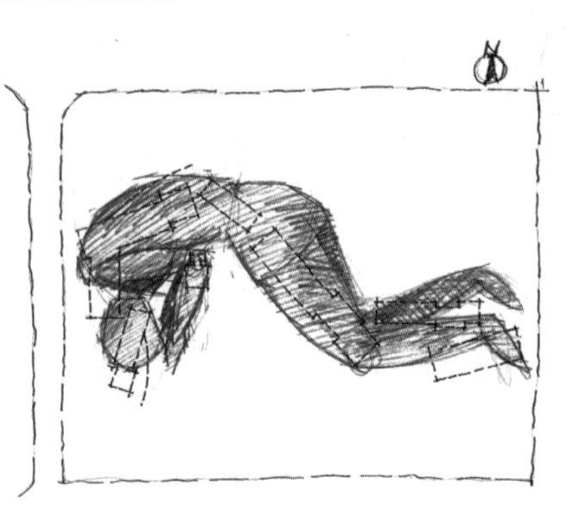

开篇：效果图。**对页上图及下图**：原概念草图。**对页左图**：总平面图。**本页图**：分析图。

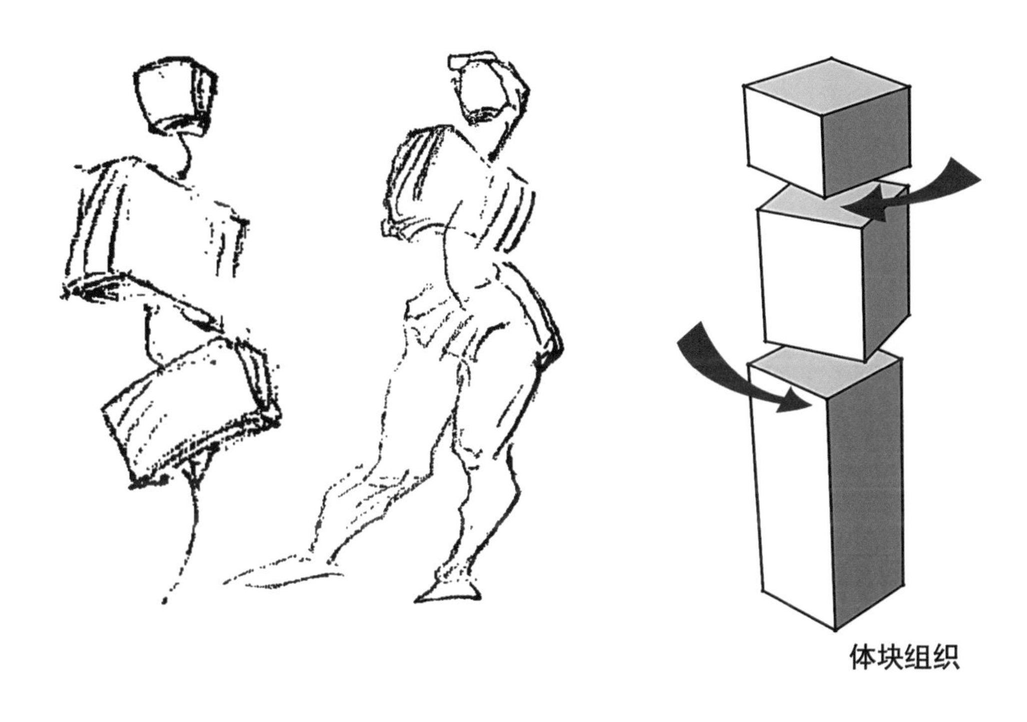

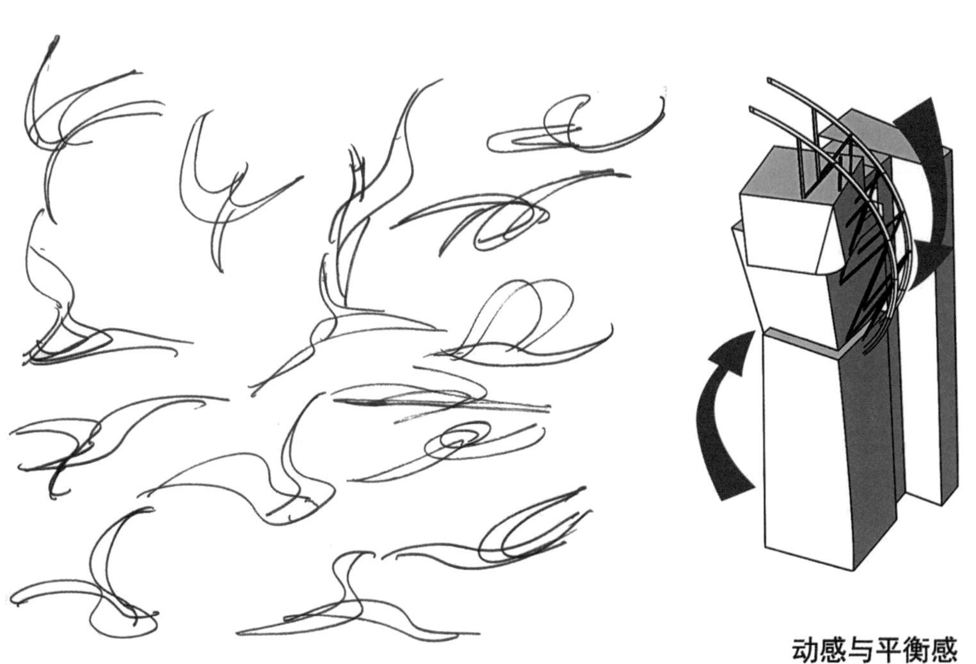

地段分析

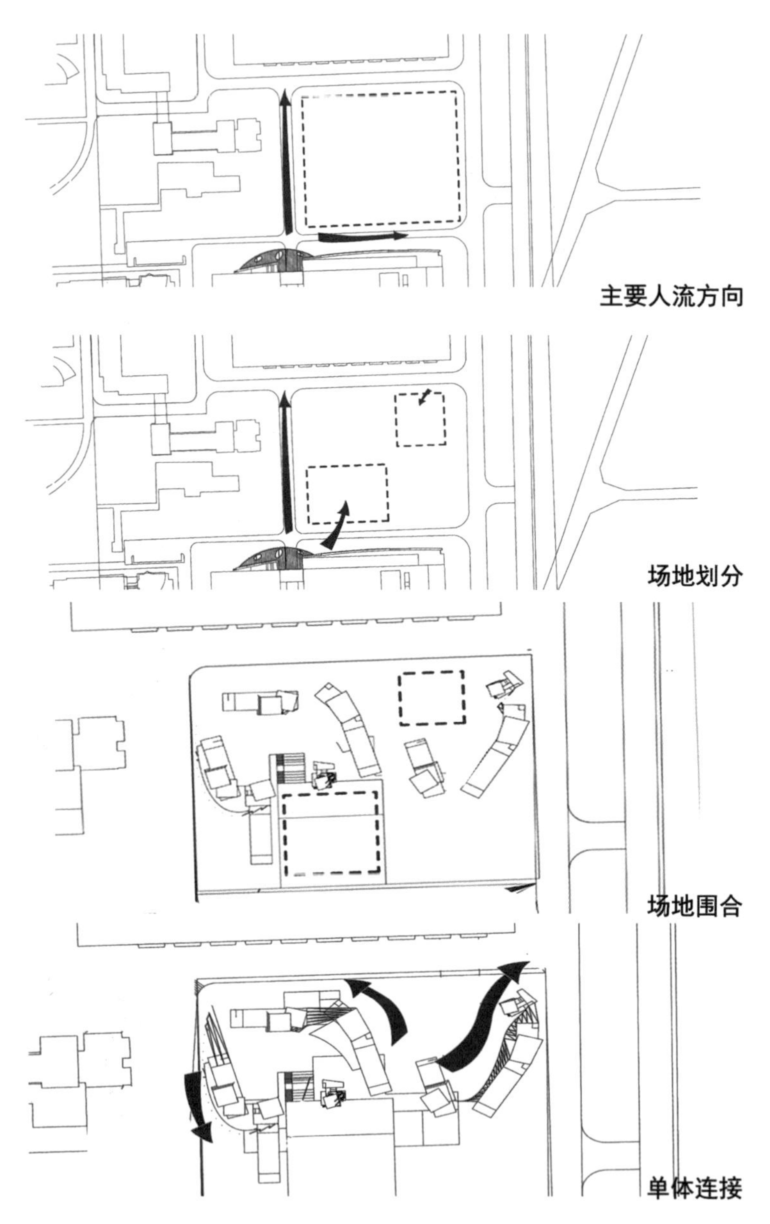

The internal structure of human being, the most natural element, is inextricably linked with the architecture. In history, there are many attempts of architectural human or humanized architecture. The project chooses the "human" as a complete module unit, trying to find a mapping method on architecture through the study of form structure, organization logic and the mutual relationship between individuals, and to explore new possibilities in the composition and structure of the building and building community. Each building unit is treated as an individual body image. While echoing in the organization logic, it's also similar to the trend and characteristics of the human body. At the same time, the internal structure of the building shall be adjusted according to the different requirements of different disciplines for space.

对页图：分析图。**本页图：**模型。

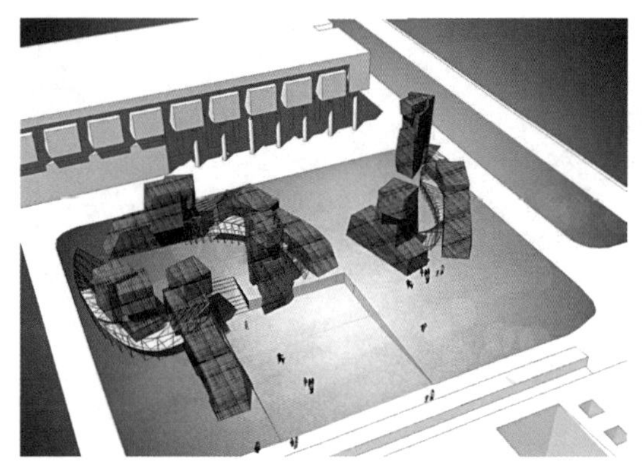

教师点评

除了常规意义上具有形态相似性的“模件单元”，一些设计将“模件”的概念演绎到更广阔和抽象的范畴。探讨模件内在形成机制和构成逻辑的类似性成为这些设计对“模件单元”的突破口,唐思齐同学的“人体模件”设计便是其中比较有意思的尝试。

该设计认为人体本身作为最自然的一种元素,其内在组成结构与建筑有着千丝万缕的联系。历史上,也不乏人体建筑化或建筑人体化的诸多尝试。该设计选择了“人”作为一个完整的模件单元,试图通过研究人体的形态构成、组织逻辑和个体间的相互关系来找到一种其在建筑上的映射方法,在建筑和建筑群落的构成、组织上探索新的可能。与此同时,该设计研究各学科功能对于空间的不同要求,以强化建筑内部空间的构成逻辑及外部空间的组合逻辑,最终形成一组“神态类似”但形态和功能均差异较大的建筑群。

Teacher's comments

In addition to the conventional sense of "modular unit" with similar form, some designs interpret the concept of "module" into a broader and more abstract category. The discussion on the intrinsic formation mechanism of the module and the similarity of logic composition becomes the breakthrough of these designs for "module unit". The "human body module" design of Tang Siqi is one of the interesting attempts.

In her opinion, the inherent composition structural of the human body, which is the most natural element, has innumerable links with the construction. In history, there were many attempts in combination of human body and architecture. This design chooses "person" as a complete module unit to find out its mapping method in the architecture and explores new possibilities in the composition of building and organization of building groups through the study of form composition and organizational logic of human body and the interrelationship between individuals. Meanwhile, this design studies different requirements of subject functions for space, so as to strengthen the composition logic of internal building space and combinational logic of external space and consequently construct a building group with "similar style" and different forms and functions.

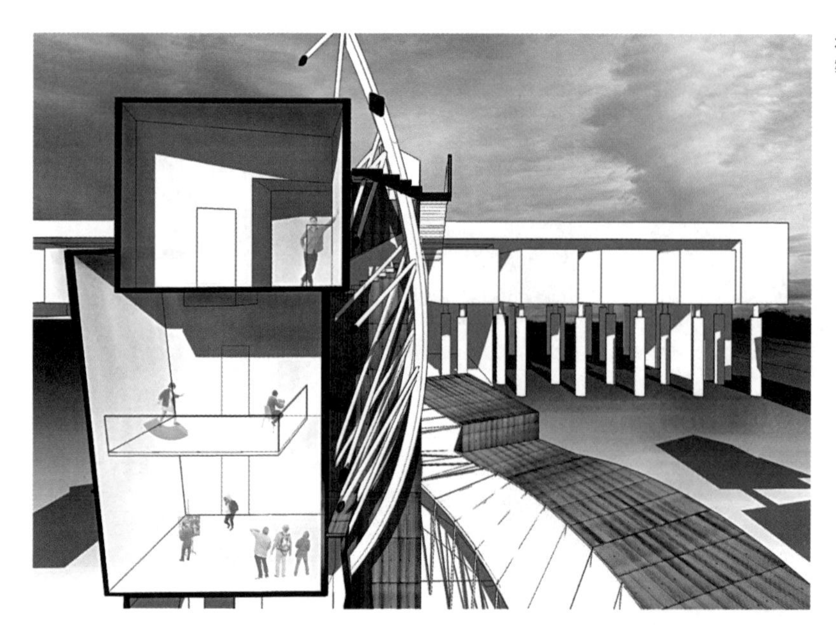

本页上图: 方案鸟瞰图; **本页下图:** 方案剖透视图; **对页图:** 广场人视效果

晶之馆/CRYSTAL HALL

项目选址：王府井大街西侧，金茂万丽酒店南侧

功能定位：商务会馆

建筑面积：30000m²

用地面积：15000m²

方案设计：于博柔

指导教师：崔彤

完成时间：2015

这是一座建设在王府井大街上的商务会馆。

王府井大街为北京市的繁华地带，商务会馆希望与地段北侧酒店和南侧大厦在体量上平衡。同时，地段的西侧为传统院落与胡同。会馆如何进行繁华界面和传统肌理之间的沟通，也是设计的重点之一。故在设计中考虑保持会馆整体体量的同时，将底部两层打开，与广场共同成为开放性的公共环境。广场引入水系，铺地分割，在空间尺度上拟合西侧合院与胡同。

设计的概念来自晶体。晶体的最主要特征之一是透明性，这既是会馆最后的整体形态呈现，也是繁华界面和传统肌理沟通的解读方式。另一方面，晶体在培养过程中会呈现出奇妙的结构形式，具有规律同时也富有变化。在向上生长的晶柱基础上，会出现晶错、孔洞等特殊结构。这是设计最终的形态依据。会馆各建筑单体具有相同的结构逻辑，东西朝向的墙面（城市界面）为传统框架，南北朝向的墙面以30°斜柱支撑，配合建筑内部核心筒共同构建。同时建筑内部设计了许多边庭、中庭，在沟通和开放视野的前提下，满足开放或封闭的不同需要。

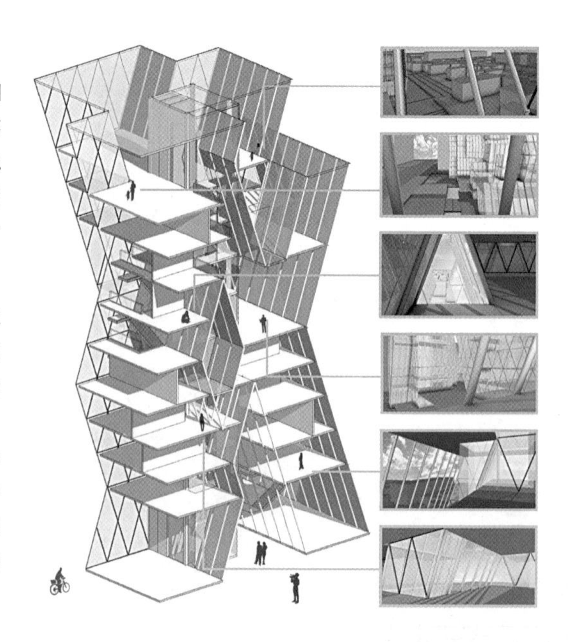

空间书写/SPACE WRITING

项目选址：清华大学新东门广场
项目类型：设计聚落
建筑面积：19200m² **用地面积：**16384m²
容积率：1.15 **绿化率：**30%
建筑高度：16.8

方案设计：陈爽云
指导教师：崔彤
完成时间：2015

场地变成宣纸，用建筑空间在三维上书写。从汉字的造字和书写中寻找规律，在繁杂的逻辑中挑拣、简化，与单元结合，最终成为本方案。草书太难，最终还是规规矩矩地、像小学生练字一样，写了块块字。

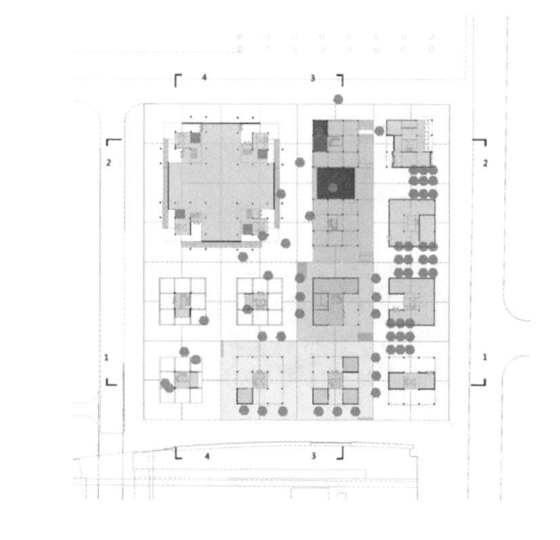

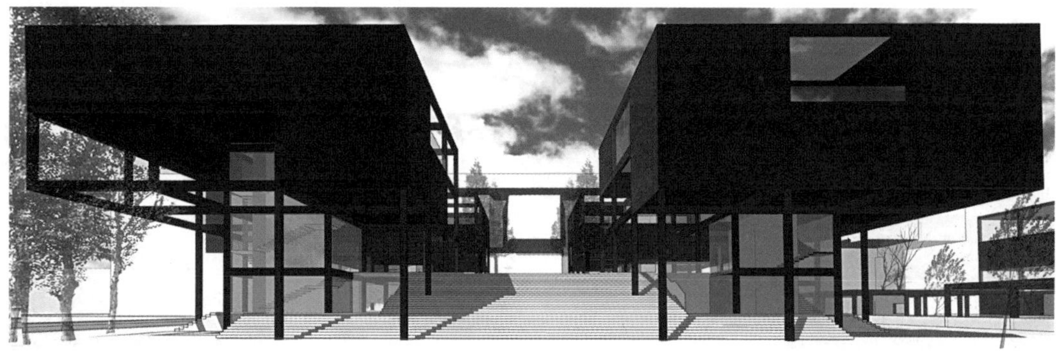

设计聚落/ DESIGNER' S SETTLEMENT

项目选址：清华大学美术学院北侧
建筑面积：11 242m²
用地面积：24 212m²

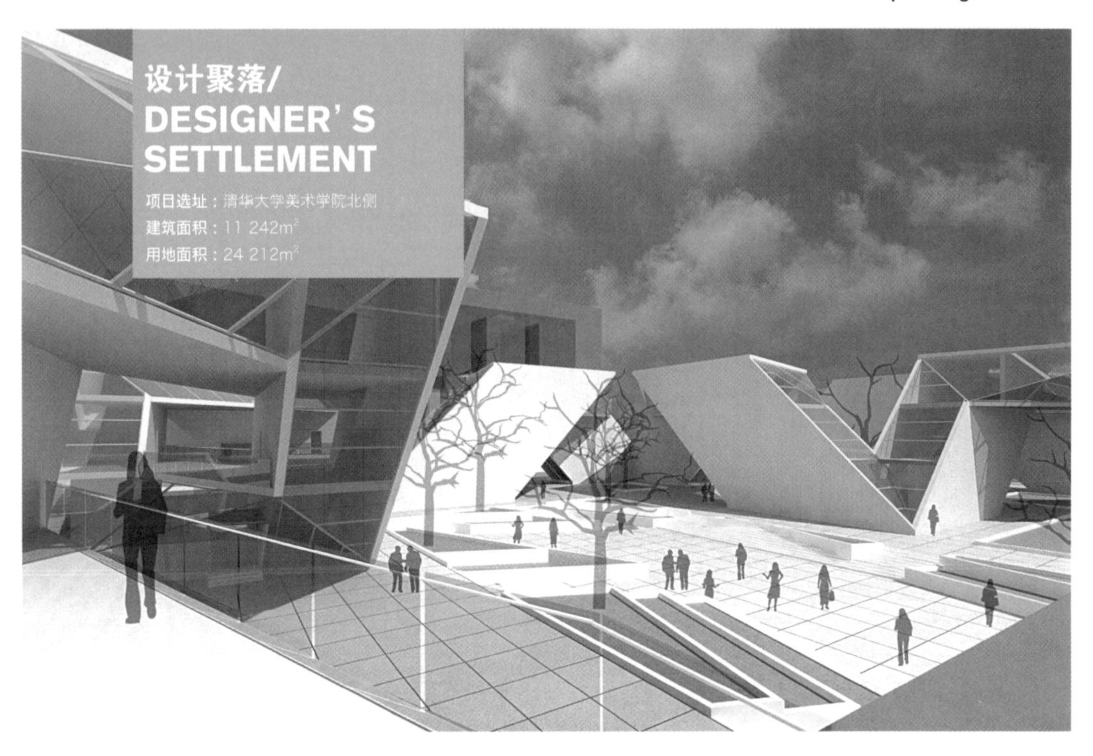

方案设计：高进赫
指导教师：崔彤
完成时间：2015

本项目是“空间单元”设计-通过单元形成一个聚落。

在作业项目设计中我是对从自然中成长的树木的成长方式进行了一个研究以后，从树枝的长相中得到灵感后形成一个不断变化方向的折叠形式。折叠中反复的各个单元最终形成了整个聚落。而且考虑了南北、东西的轴线关系，形成了十字形的景观道路。

方体/CUBES

项目选址： 王府井大街西侧
项目类型（功能）： 商务会所
建筑面积： 21300m²
用地面积： 7400m²

方案设计： 侯志荣
指导教师： 崔彤
完成时间： 2015

设计是以单元模块为设计手法出发。地段位于王府井，是一座以商务会所为主要功能的建筑。设计中将螺旋这一意向为一种单元，共有大小不同的6个单元落在起伏的基座上，提供了连续的建筑空间。在城市层面上，建筑表皮选用磨砂玻璃，试图营造一种虚幻的立面效果，为单调的城市界面提供变化。

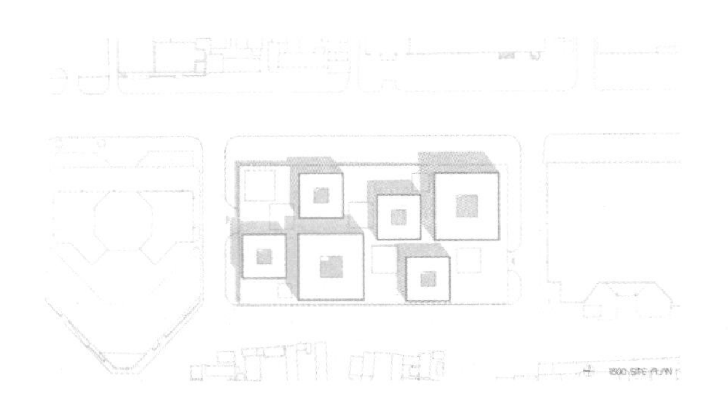

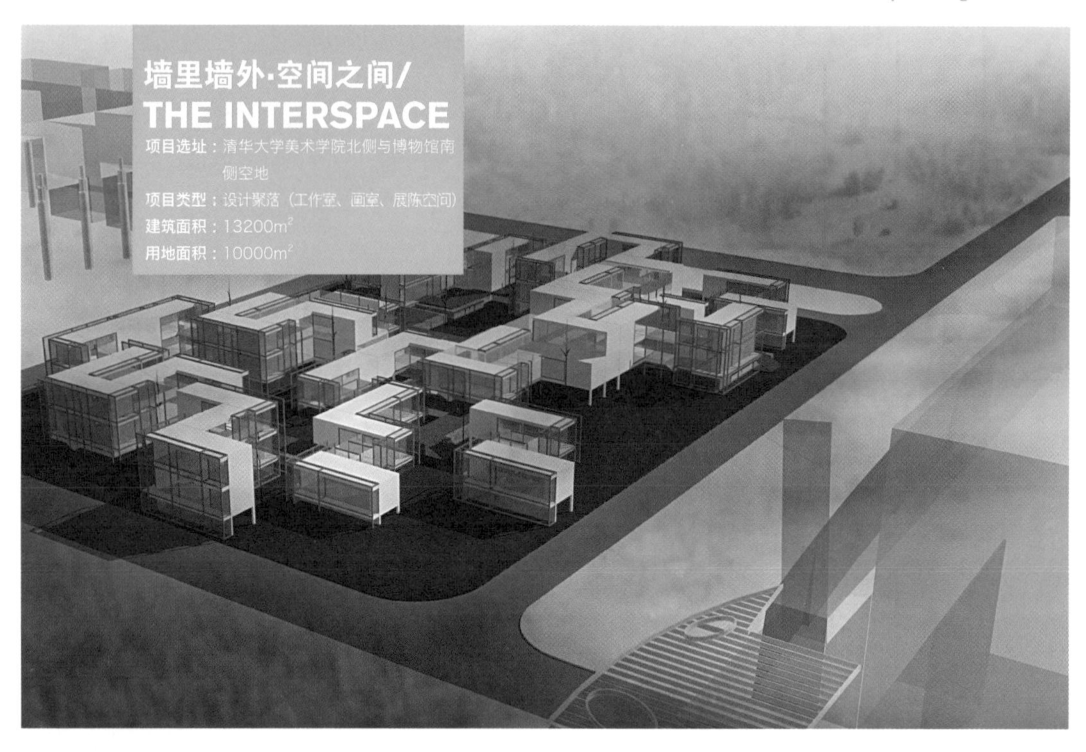

墙里墙外·空间之间/ THE INTERSPACE

项目选址： 清华大学美术学院北侧与博物馆南侧空地
项目类型： 设计聚落（工作室、画室、展陈空间）
建筑面积： 13200m²
用地面积： 10000m²

方案设计： 姜兴佳
指导教师： 崔彤
完成时间： 2015

本设计旨在于校园内外空间的交界节点处创造一片富有聚落感与生长感的设计聚落。
考虑到地段所处区位南北两向各有一栋体量较大的建筑，因此选择摒弃大体量或特别高的建筑形式，而是将建筑弱化、虚化、碎片化，并尽可能减少建筑对地面的扰动。
方案以多种形式的墙体为基本元素，以单元组合而成各种形式的院落与室内空间，试图探索空间之间的空间可能性。

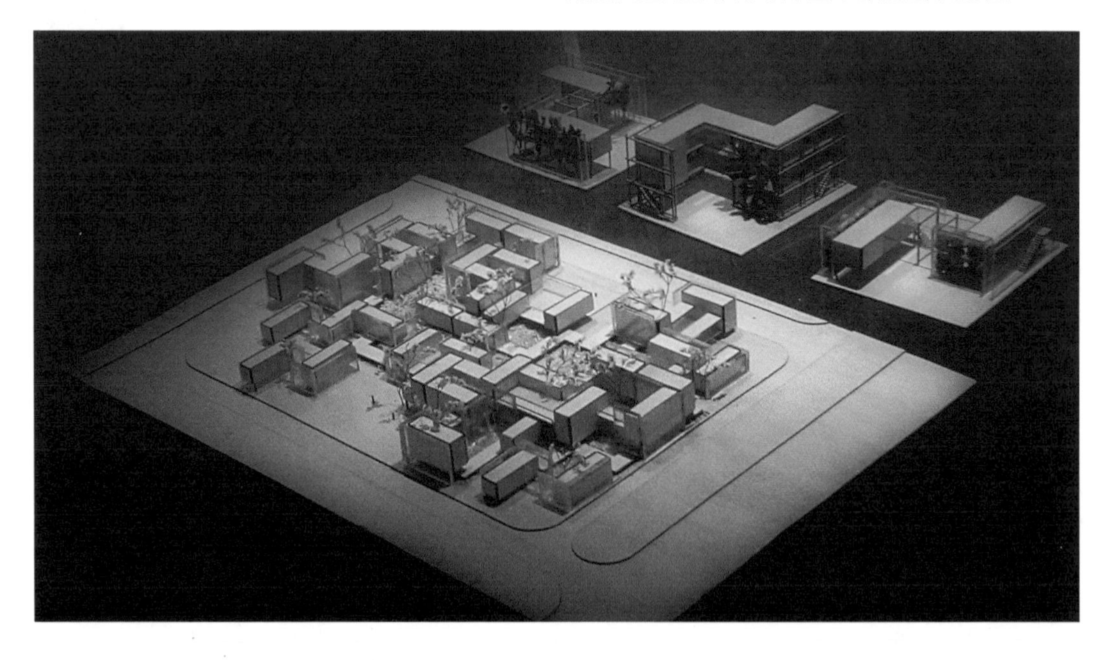

垂直院落/VERTICAL COURTYARD

项目选址：北京市东城区王府井大街北段西侧
项目类型：商务会所
建筑面积：$14800m^2$　**用地面积：**$8900m^2$
建筑高度：24m　**建筑层数：**4
容积率：1.66　**绿化率：**33%

方案设计：连璐
指导教师：崔彤
完成时间：2015

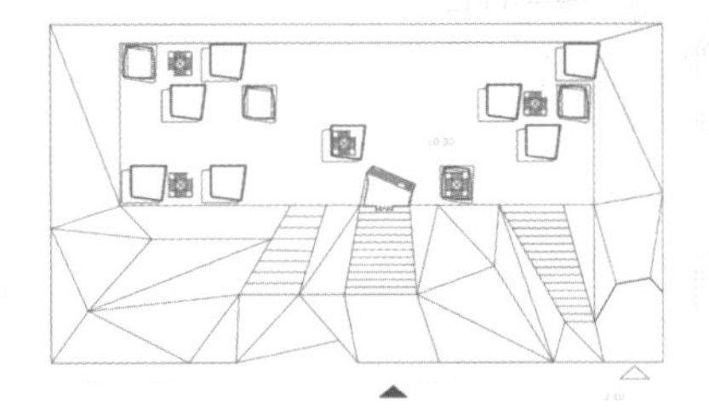

方案试图突破商务会所传统空间模式，以独立功能空间作为单元，“空儿”生成公共交往空间，形成垂直院落空间体系。模件单元及空间组合源于自然启示并基于建筑构成关系，希望以同质表达多样。建筑底层以开放的艺术商业业态面向城市，充分发掘地下空间可能性，整体形态完整谦逊。

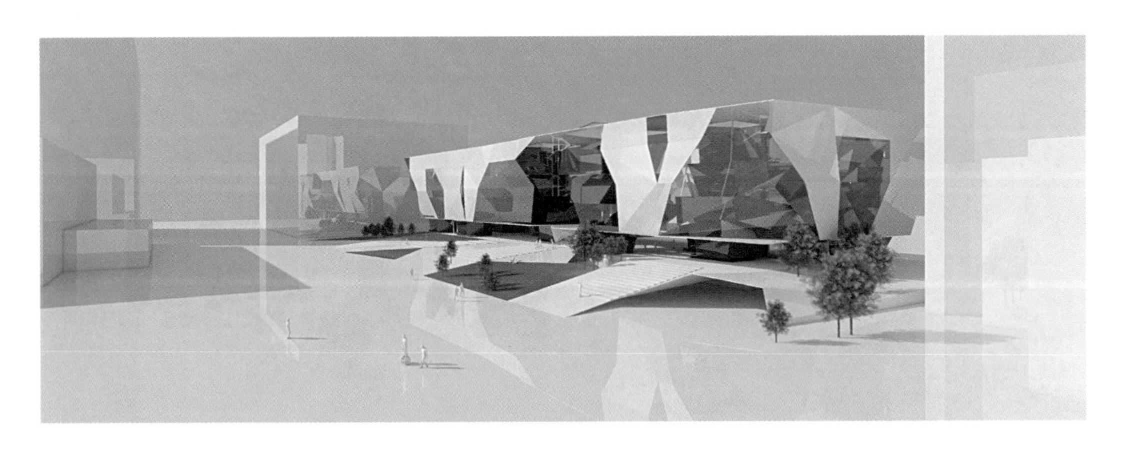

云端/CLOUD
项目选址：北京市王府井商务印书馆对面
功能定位：商务会馆
建筑面积：15000m²
用地面积：9000m²

方案设计：刘通
指导教师：崔彤
完成时间：2015

建筑地段位于北京市王府井商业区北侧，商务印书馆对面，功能为商务会馆。建筑形态和结构以“树”和“云”为概念，在王府井这一建筑密集区建造了一片人工自然景观。建筑主体由多个伞状结构架空，并合理开发地下空间，将城市地面还给市民，在拥挤的城市中心创造一片公共绿化空间。独特的造型也将成为京城一道亮丽的风景。

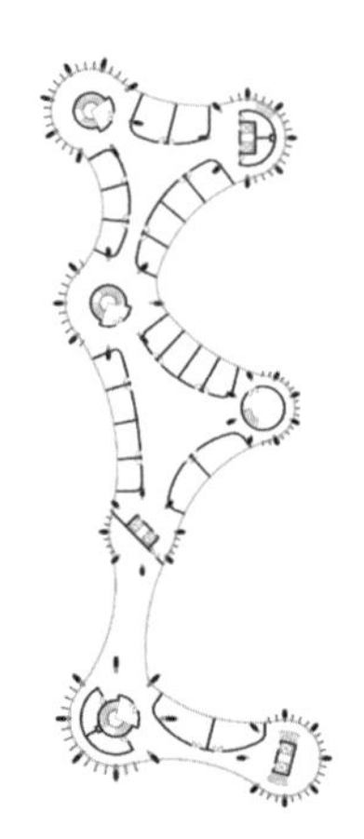

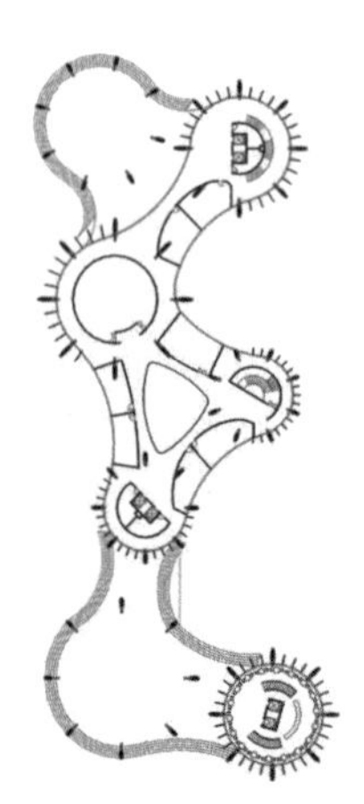

泥土和阳光/SOIL AND SUNSHINE

项目选址：清华大学美术学院北侧停车场
项目类型：美术学院、建筑学院、机械学院、汽车系、软件学院
建筑面积：24760m² **用地面积：**16800m²
容积率：1.47 **绿化率：**37%

方案设计：刘圆方
指导教师：崔彤
完成时间：2015

方案从单个单元到簇，到组，再到整体建筑的逻辑组织，选择了最为基本的立方块作为单元的原型，通过简单单元的组织交错和实体单元与虚的绿色单元的并用，满足各个簇的功能。同时簇与簇之间的连接逻辑则以一种地景式的形式有机地出现，使得匀质的单元组织之外又有了趣味性与公共活动场地。

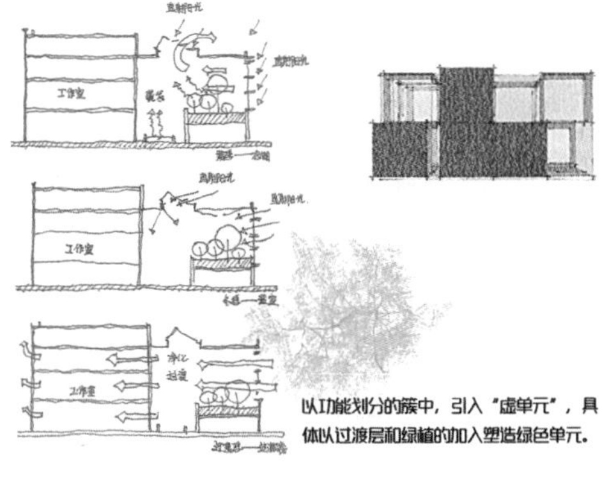

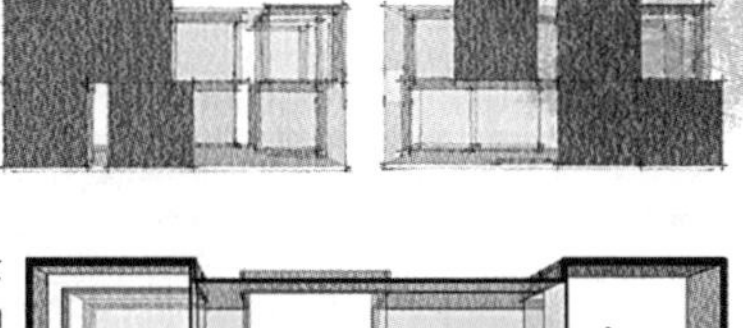

以功能划分的簇中，引入"虚单元"，具体以过渡层和绿植的加入塑造绿色单元。

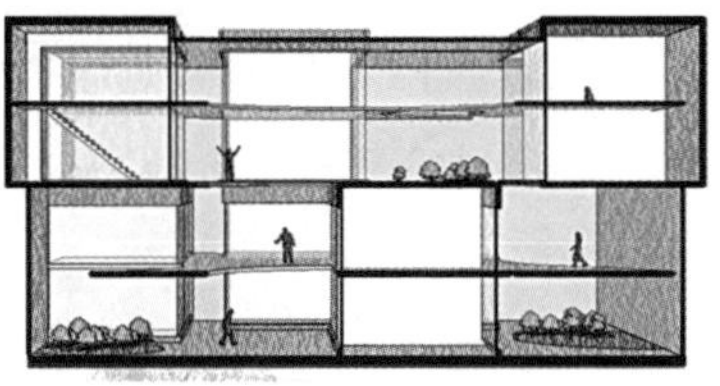

以立方体块为单元原型，进行上下以及同层的拆合以应对上住下工的工作室和同层较大空间的会议加工室。错动和虚单元为公共空间和廊道引入阳光与景观。

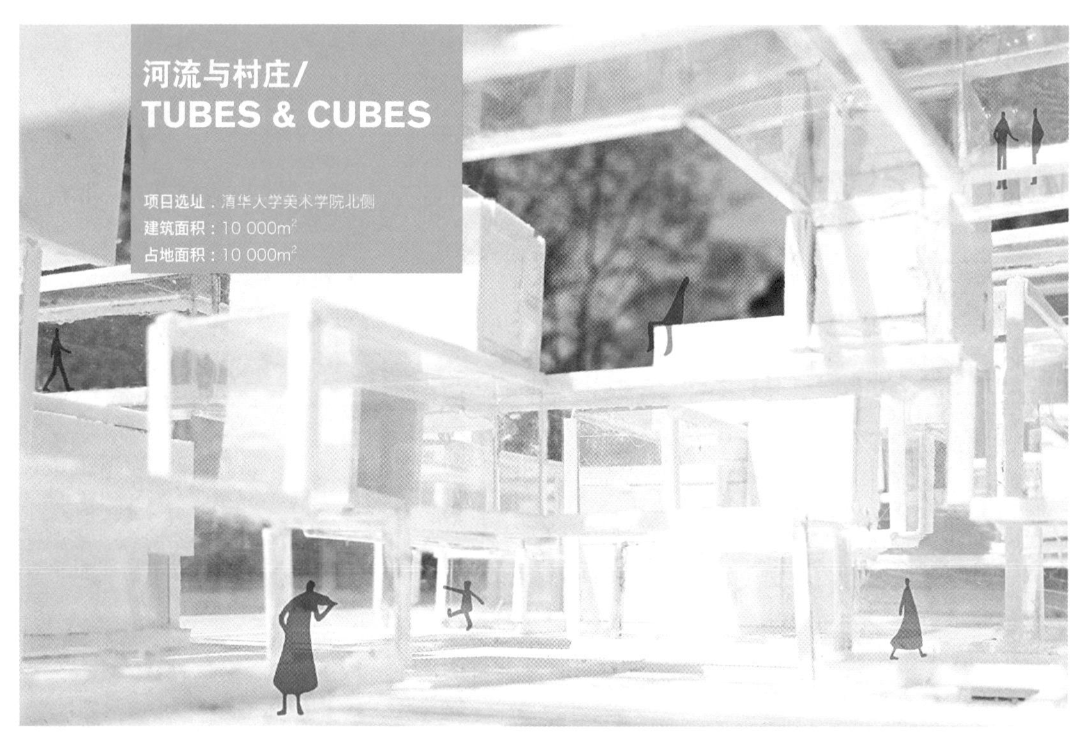

河流与村庄/ TUBES & CUBES

项目选址：清华大学美术学院北侧
建筑面积：10 000m²
占地面积：10 000m²

方案设计：祁盈
指导教师：崔彤
完成时间：2015

对于设计聚落这样一个对空间需求很灵活的场所，希望发挥“模件体系”可以灵活组装的优势，让建筑随着时间和人群演变，一直处在未完成的状态。起初设想让一些小盒子在场地上自由叠摞，位置和形式都由使用者选择，但结构管线交通上都面临问题，于是再引入管子体系，让盒子在此基础上自由生长。

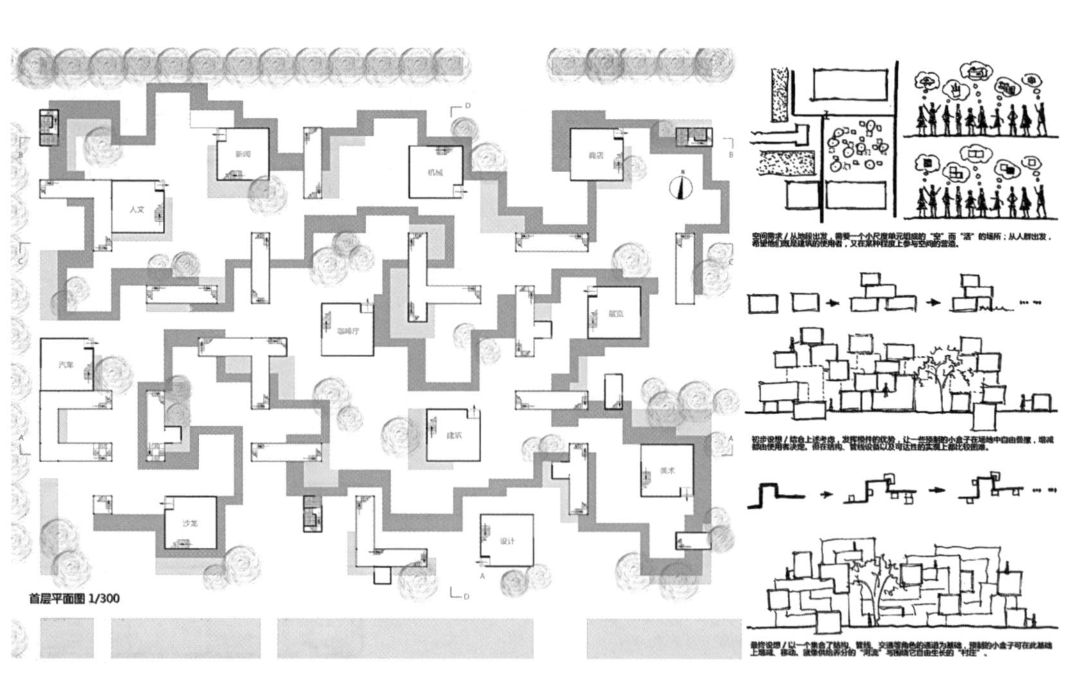

首层平面图 1/300

榫与卯/TENON AND MORTISE

项目选址：清华美术馆前广场

项目类型：集“学、产、研”一体的对外展示的设计聚落

建筑面积：8 000m²

占地面积：10 000m²

方案设计：熊芝锋

指导教师：崔彤

完成时间：2015

设计之初探索鲁班锁同构的组合形式，进而提出榫卯的概念，并简化得到建筑模件。之后，模件咬合组合的关系从二维转向三位空间。同时，基本单元也是阴阳互补，高度咬合的。最终通过场地分析，以线性流线串联起单元体生成设计聚落，并营造出聚落间的院落空间。

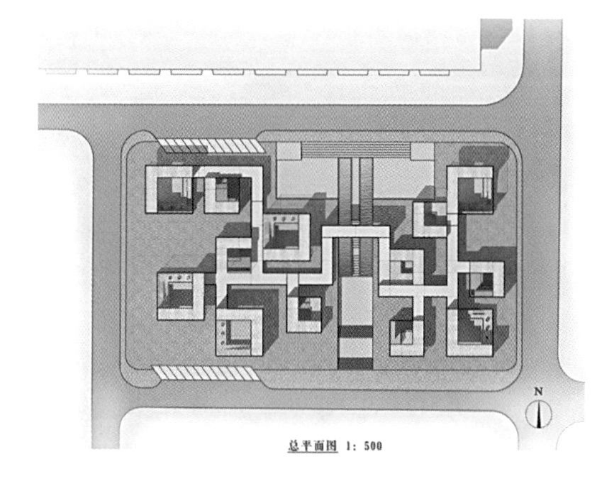

总平面图 1: 500

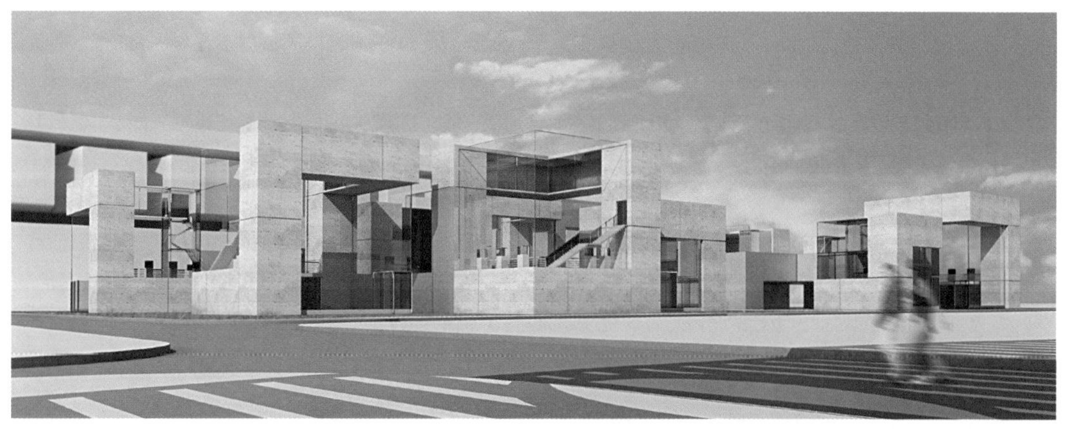

致谢

2014年春季学期，清华建筑学院在校长、人事处和教务处的支持下，一次性聘任15 位业界优秀建筑师作为“校聘设计导师”，进行了开放式建筑设计教学的尝试，同年春季学期，有8 位作为指导老师参与了三年级建筑设计课教学，其余7 位则参加了设计评图。2015年春季学期，同样有8 位设计导师参与了设计课程教学，他们是马岩松、王昀、王辉、朱锫、齐欣、李虎、徐全胜、崔彤，其余7 位设计导师董功、胡越、梁井宇、华黎、邵韦平、李兴钢、张轲参加了设计评图，在此对15位设计导师的辛勤工作深表敬意。

在教学过程中，清华建筑学院大部分设计教师参与了开放式教学的中期评图及最终评图，他们与设计导师们进行了多方面的讨论及交流，并对开放式教学提出许多宝贵的意见，在此深表谢意。

本作品集的编辑过程中，所有选修这门课的学生参与了作品的排版工作，感谢同学们的努力；这里还要特别感谢杜頔康、李晓岸、罗丹、卢倩、丁立南、赵一舟等研究生们所做的编辑、翻译、校核等工作；最后感谢中国建筑工业出版社的编辑们为本书的出版进行了不懈的努力。

2016年7月